Precedence-Type Tests
and Applications

Precedence-Type Tests and Applications

N. Balakrishnan

McMaster University
Department of Mathematics and Statistics
Hamilton, Ontario, Canada

H. K. Tony Ng

Southern Methodist University
Department of Statistical Science
Dallas, TX

A JOHN WILEY & SONS, INC., PUBLICATION

Library of Congress Cataloging-in-Publication Data:

Balakrishnan, N.
    Precedence-type tests and applications / N. Balakrishnan, H.K. Tony Ng.
      p. cm.
    Includes bibliographical references and index.
    ISBN-13 978-0-471-45720-6 (acid-free paper)
    ISBN-10 0-471-45720-5 (acid-free paper)
     1. Nonparametric statistics.   I. Ng, H. K. Tony, 1975– II. Title.

   QA278.8.B349 2006
   519.5—dc22   2005058222

Printed in the United States of America.

10 9 8 7 6 5 4 3 2 1

To

My Late Father and Mother,
R. Narayanaswamy Iyer and N. Lakshmi,
for their love, support, and encouragement!

(NB)

My Parents,
Cheong Leung Ng and Kit Ching Wong,
for their love and affection!

(HKTN)

Contents

List of Tables

List of Figures

Preface

Nonparametric statistics are intuitive and easily understood and inferential procedures based on ranks and runs are often heuristically simple to follow and implement. One such family of test procedures are the so-called *precedence-type tests*. These tests, which are quite useful in life-testing situations to make quick and reliable decisions early in the experiment, are also time- and cost-efficient as they are based on only a few early failures (instead of failures of all units in the life-test). However, the development of precedence-type tests requires care and usage of a wide range of statistical techniques. This volume provides a thorough and comprehensive overview of various theoretical as well as applied developments on a variety of problems in which precedence-type test procedures may be applied effectively.

This volume comprises 10 chapters, and may be broadly classified into four parts—Part A, comprising Chapter 3, deals with the original precedence test and some properties of precedence and related test procedures; Part B, comprising Chapters 4–6, deals with some alternatives to precedence test such as maximal precedence, weighted forms of precedence and maximal precedence, and Wilcoxon-type rank-sum precedence tests and their properties; Part C, comprising Chapter 7, deals with the extension of precedence, maximal precedence, and Wilcoxon-type rank-sum precedence tests to the situation when the sample arising from the life-testing experiment is progressively Type-II censored, and their properties; and Part D, comprising Chapters 8–10, deals with precedence-type tests in multisample situations and selection problems. Throughout the volume, several tables have been presented so as to facilitate the use of these tests in practical problems, and also some examples have been included in order to illustrate all the precedence-type procedures.

The length of this volume as well as the extensive bibliography at the end of the volume (with a good number of publications being in the last 25 years or so) provides ample testimony to the growth and continued interest in this topic of research. Even though we have discussed a number of variations of the precedence test and also different applications of these test procedures, there is clearly a lot more potential to develop new precedence-type tests as well as to apply them to diverse inferential problems. It is our sincere hope that this volume would enable and encourage this to happen.

In a volume of this nature and size, inevitably there will be omission of some results that should have been included. Any such omission is accidental and by no means due to personal nonscientific antipathy.

We encourage readers to comment on the contents of this volume and thank them in advance for informing us of any errors, omissions, or misrepresentations.

We acknowledge the support and encouragement of Mr. Steve Quigley, Editor, John Wiley & Sons, throughout the course of this project. The managerial, editorial, and production help provided by Ms. Susanne Steitz, Mr. Andrew Prince, and Ms. Shirley Thomas, respectively, of John Wiley & Sons, Hoboken, NJ, are gratefully acknowledged. We also acknowledge the kind permission provided by World Scientific Publishing Company (Singapore), Blackwell Publishing Company (UK), Elsevier Science B.V. (The Netherlands), and John Wiley & Sons (USA), for us to reproduce some of the previously published tables and figures. Thanks are also due to Mrs. Debbie Iscoe (Ontario, Canada) for her help and assistance during the typesetting of this volume. Finally, we express our sincere thanks to all our family members whose support, cooperation, understanding and immense patience we both enjoyed during the entire period we worked on this volume.

N. BALAKRISHNAN

H. K. TONY NG

Hamilton, Ontario, Canada

Dallas, Texas, USA

January 2006

Chapter 1

Introduction

1.1 PROBLEMS OF INTEREST

The comparison of the quality of products from different manufacturing processes or the effectiveness of different treatments for an illness is a commonly encountered problem in practice. For example, a manufacturer of a product may wish to compare a new manufacturing process with the existing process. If there is significant statistical evidence that the new process results in better product (meaning, more reliable or with longer lifetime), then the manufacturer may wish to abandon the existing process and implement the new process into production. Another example is when a medical researcher wishes to compare a new treatment with a control. In this case, patients may be assigned randomly to treatment and control groups, and their remission times (or lifetimes) are recorded. Based on these data, the researcher will be primarily interested in determining whether the treatment is effective.

The development of efficient statistical procedures for these problems is, therefore, of great interest and importance.

1.2 SPECIAL CONSIDERATIONS

In the examples described above, we may have some special considerations. For example, in the medical experiment, the treatment may be toxic and harmful to the patients; therefore, based on ethical grounds, the researcher may wish to terminate the experiment as soon as there is evidence to draw a reliable conclusion, one way or the other. In the other example involving quality or reliability of products, the manufacturer may want (1) to make quick and reliable decisions early in the life-testing experiment, and (2) to minimize the number of failures of units from the new process since their cost of production may be relatively high so that the units that had not failed could be used for some other testing purposes.

For these reasons, our main goal is to make decisions during the early stage of the experiment, not having observed many failures.

1.3   SPECIAL FORM OF TESTING

As we are concerned with the comparison of the lifetime distributions of units from the new process with those from the standard process, and because we would expect more failures to occur from the standard process than from the new process during the early stage of the experiment, we would naturally like to utilize this to collect data in this special form and then carry out a test suitably based on these data.

For this purpose, we assume throughout this book that sample units from the processes are placed simultaneously on a life-testing experiment and that failures are observed as they arise in a naturally time-increasing manner.

At this point, there are clearly two ways to proceed: one is to assume specific lifetime distributions for the samples and carry out the test under a parametric setup, and the other is to collect data in a nonparametric manner (for example, only the ranks of failure times rather than the failure times themselves) and carry out the test under a nonparametric setup. We have chosen the latter because we will have only very few failures (and so assumption of a family of lifetime distributions for data in hand may be difficult to justify or verify) and the decisions made will be somewhat robust (as compared to those from efficient tests based on some specific family of lifetime distributions).

1.4   PRECEDENCE TESTS

From the life-testing experiment described in the previous section, one form of (nonparametric) data that we could collect easily is the number of failures from the standard process that preceded the first failure from the new process, the number of failures from the standard process that occurred between the first and second failures from the new process, and so on.

If the experimenter had decided to allow only a certain number (say, r) of failures from the new process (for reasons stated earlier), then the life-test would be terminated as soon as this rth failure occurred from the new process. We would then have the data on the numbers of failures from the standard process only until this particular failure time.

This form of life-test is called a *precedence test* and any test statistic based on these "precedences" is called a *precedence-type statistic*; see, for example, Nelson (1963, 1986) and Ng and Balakrishnan (2006). Of course, the simplest precedence-type statistic is the number of failures from the standard process that preceded the rth failure from the new process; as a matter of fact, this is what Nelson (1963) has called as a *precedence statistic*. One may then change the functional form (instead of just the sum of the numbers of failures from the standard process) and come up with different precedence-type statistics, each with its own special features and properties. Furthermore, the idea of precedence-type statistics can also be extended to some other statistical problems. These, indeed, form the bases for all the developments in this book!

1.5 DEVELOPMENTS

For the problems described in Section 1.1, with the goals as stated in Section 1.2, many different precedence-type test procedures are developed in this book and their properties are evaluated.

First, in Chapter 2, we present briefly the basic concepts and results that are essential for the developments in the subsequent chapters. After describing the form and nature of data arising from a life-testing experiment, we introduce order statistics and present some important formulas and results concerning order statistics. We then explain the concept of censoring and different forms of censored data that could arise from a life-test. We pay special attention to progressive censoring and present some formulas and results concerning the progressively censored order statistics. Some useful lifetime distributions are described next, and these distributions are used throughout this book for evaluating the performance of all the test procedures. Since the test procedures are developed under a stochastically ordered alternative, it will be quite natural to compare their performance with the well-known Wilcoxon's rank-sum statistic for complete samples; so, we present a brief description of Wilcoxon's rank-sum test and also explain how a randomized test could be developed if one wishes to have a test that attains exactly a prespecified level of significance.

Next, in Chapter 3, we introduce the concept of precedence testing and present the precedence test statistic. We derive the exact null distribution of this test statistic by combinatorial method and also by means of an order statistics approach. We evaluate the power properties of this test through the exact power function under the Lehmann alternative as well as through the simulated power under the location-shift alternative. We then discuss various properties of the precedence test and some other related nonparametric tests, and present finally some examples to illustrate the use of precedence tests.

Since the precedence test developed in Chapter 3 suffers from a masking effect, we introduce in Chapter 4 the maximal precedence test statistic. We derive the exact null distribution of this test statistic by means of an order statistics approach. We evaluate the power properties of this test through the exact power function under the Lehmann alternative as well as through the simulated power under the location-shift alternative. We then make some comparisons of this test with the precedence test and show that the maximal precedence test, unlike the precedence test, does not suffer from the masking effect. Finally, we present some examples to illustrate the use of maximal precedence tests.

In Chapter 5, we introduce the concept of weighted precedence and weighted maximal precedence tests. We derive the exact null distributions of these test statistics. We evaluate their power properties through their exact power functions under the Lehmann alternative as well as through their simulated power values under the location-shift alternative. Finally, we present some examples to illustrate the use of weighted precedence and weighted maximal precedence tests.

In Chapter 6, we introduce three Wilcoxon-type rank-sum precedence test statistics—the minimal, maximal, and expected rank-sum statistics. We derive the exact null distributions of these three test statistics. Since the large-sample normal approximation for the null distributions is not satisfactory in the case of small or moderate sample sizes, we present an Edgeworth approximation to the significance probabilities. We evaluate the power properties of these three tests through their exact power functions under the Lehmann alternative as well as through their simulated power values under the location-shift alternative. We then make some comparisons of these three tests with the precedence test, the maximal precedence test, and the Wilcoxon's rank-sum test based on complete samples. Finally, we present some examples to illustrate the use of the three Wilcoxon-type rank-sum precedence tests.

In Chapter 7, we generalize the weighted precedence test, the weighted maximal precedence test, and the maximal Wilcoxon rank-sum precedence test to the situation when the available sample is progressively Type-II censored. This progressively censored life-test will facilitate saving of test units at the early stage of the life-testing experiment. We introduce the corresponding three test statistics and derive their exact null distributions. We evaluate the power properties of these three tests through their exact power functions under the Lehmann alternative as well as through their simulated power values under the location-shift alternative. Finally, we present some examples to illustrate the use of these three tests when the available sample is progressively Type-II censored.

In Chapter 8, we consider a generalization of the two-sample reliability testing problem discussed so far to a scenario when there are k manufacturing processes or treatments to be compared. We first outline some problems that will be of interest in this k-sample situation. We consider the problem of comparing $k - 1$ treatments with a control and propose precedence-type and maximal precedence-type test statistics based on a fixed sampling scheme. Next, we consider the problem of comparing k treatments and propose precedence-type and maximal precedence-type test statistics based on an inverse sampling scheme.

When we have several competing populations, it will naturally be of interest to select the "best population" if we reject the null hypothesis of homogeneity of populations. In Chapter 9, we formulate this problem as a multiple-decision problem and propose a precedence-type two-sample selection procedure. We derive the exact null distribution of the test statistic as well as the exact probability of correct selection under the Lehmann alternative. Next, we extend this precedence-type selection procedure to the k-sample case. We derive the exact null distribution of the test statistic and examine its properties under a location-shift through Monte Carlo simulations. We also present an example to illustrate the use of precedence-type selection procedure.

Finally, in Chapter 10, we introduce a minimal Wilcoxon rank-sum precedence-type two-sample selection procedure. We derive the exact null distribution of the test statistic as well as the exact probability of correct selection under the Lehmann alternative. We then extend this procedure to the k-sample case, and derive the exact null distribution of the test statistic.

Next, we examine the properties of this selection procedure and compare its performance to that of the precedence-type selection procedure. We conclude by presenting examples to illustrate the use of the minimal Wilcoxon rank-sum precedence-type selection procedure.

Chapter 2

Preliminaries

2.1 LIFE-TEST DATA

A manufacturer of products will naturally be interested in assessing the reliability of the product. The reliability may be evaluated, for example, by the median lifetime or some other suitable quantile of the lifetime distribution of the product. For this purpose, a random sample of n units may be selected, placed on a life-testing experiment, and the lifetimes (viz., times-to-failure) of these units may be observed. From such life-test data, our interest lies in developing inference about the reliability characteristics of the product.

Another situation that commonly arises in this context involves the comparison of two or more products in terms of their reliability. This problem can come up in different ways. For example, a manufacturer producing units using a standard process may be interested in determining whether a new manufacturing process, if implemented, could result in a better product (better meaning more reliable). Another scenario wherein this problem arises is when a dealer receives units from two (or more) firms and wishes to find out which firm supplies the best quality product. Yet another scenario where the problem arises is in evaluating the relative merits of two drugs and in determining whether Drug A results in increasing the lifetime of patients as compared to Drug B. For the reliability problem considered here, the experiment will involve independent samples being taken from the processes that are being compared, placed simultaneously on a life-test, and the times-to-failure of all units being observed. From such multisample life-test data, our interest lies in deciding whether one process is better than the other.

2.2   ORDER STATISTICS

In the life-testing experiments described in the preceding section, it is evident
that the lifetimes of units are observed naturally in an increasing order with
the time-to-failure of the least reliable unit being observed first, the time-to-
failure of the second least reliable unit being observed next, and so on, with
the time-to-failure of the most reliable unit being observed last. This aspect
of life-test data brings in *order statistics* in the analysis of lifetimes.

Let $X_1, X_2, \cdots, X_n$ denote the lifetimes of the n units under test, taken
from a continuous population with lifetime distribution function $F(x)$ and
probability density function $f(x)$. Let $X_{1:n} \leq X_{2:n} \leq \cdots \leq X_{n:n}$ denote the
order statistics obtained by arranging the n lifetimes in an increasing order
of magnitude.

2.2.1   Joint Densities and Markovian Property

The joint density function of all n order statistics is [see Arnold, Balakrishnan, and Nagaraja (1992) and David and Nagaraja (2003)]

$$f_{X_{1:n},\cdots,X_{n:n}}(x_{1:n},\cdots,x_{n:n}) = n! \prod_{l=1}^{n} f(x_{l:n}), \qquad x_{1:n} < \cdots < x_{n:n}. \quad (2.1)$$

From Eq. (2.1), upon integrating out appropriate variables, we obtain
the joint density functions of $(X_{1:n}, X_{2:n}, \cdots, X_{i:n})$ and $(X_{1:n}, \cdots, X_{i:n}, X_{j:n})$
for $1 \leq i < j \leq n$, as

$$\begin{aligned}
f_{X_{1:n},\cdots,X_{i:n}}&(x_{1:n},\cdots,x_{i:n}) \\
&= \frac{n!}{(n-i)!} \left\{ \prod_{l=1}^{i} f(x_{l:n}) \right\} \left\{ 1 - F(x_{i:n}) \right\}^{n-i}, \\
&\qquad\qquad\qquad\qquad x_{1:n} < \cdots < x_{i:n}, \qquad (2.2)
\end{aligned}$$

and

$$\begin{aligned}
f_{X_{1:n},\cdots,X_{i:n},X_{j:n}}&(x_{1:n},\cdots,x_{i:n},x_{j:n}) \\
&= \frac{n!}{(j-i-1)!(n-j)!} \left\{ \prod_{l=1}^{i} f(x_{l:n}) \right\} f(x_{j:n}) \\
&\quad \times \left\{ F(x_{j:n}) - F(x_{i:n}) \right\}^{j-i-1} \left\{ 1 - F(x_{j:n}) \right\}^{n-j}, \\
&\qquad\qquad\qquad x_{1:n} < \cdots < x_{i:n} < x_{j:n}, \qquad (2.3)
\end{aligned}$$

respectively. From Eqs. (2.2) and (2.3), we readily obtain the conditional density function of $X_{j:n}$, given $(X_{1:n} = x_{1:n}, \cdots, X_{i:n} = x_{i:n})$ for $1 \leq i < j \leq n$, as

$$f_{X_{j:n}|X_{1:n}=x_{1:n},\cdots,X_{i:n}=x_{i:n}}(x_{j:n}|x_{1:n},\cdots,x_{i:n})$$
$$= \frac{(n-i)!}{(j-i-1)!(n-j)!}$$
$$\times \frac{\{F(x_{j:n}) - F(x_{i:n})\}^{j-i-1} f(x_{j:n}) \{1 - F(x_{j:n})\}^{n-j}}{\{1 - F(x_{i:n})\}^{n-i}},$$
$$x_{i:n} < x_{j:n}. \qquad (2.4)$$

Theorem 2.1. Order statistics $\{X_{i:n}, 1 \leq i \leq n\}$ from a continuous distribution form a Markov chain.

Proof. From Eq. (2.4), we see that the conditional density function of $X_{j:n}$, given $(X_{1:n} = x_{1:n}, \cdots, X_{i:n} = x_{i:n})$ for $1 \leq i < j \leq n$, depends only on $x_{i:n}$ which implies that the order statistics form a Markov chain.

$\odot$

2.2.2 Marginal Densities

From the joint density function of all order statistics given in (2.1), upon integrating out $x_1, \cdots, x_{i-1}, x_{i+1}, \cdots, x_n$, we obtain the marginal density function of $X_{i:n}$ $(1 \leq i \leq n)$ as

$$f_{X_{i:n}}(x_{i:n}) = \frac{n!}{(i-1)!(n-i)!} \{F(x_{i:n})\}^{i-1} \{1 - F(x_{i:n})\}^{n-i} f(x_{i:n}),$$
$$-\infty < x_{i:n} < \infty. \qquad (2.5)$$

The cumulative distribution function of $X_{i:n}$ $(1 \leq i \leq n)$ is

$$F_{X_{i:n}}(x_{i:n}) = \sum_{l=i}^{n} \binom{n}{l} \{F(x_{i:n})\}^{l} \{1 - F(x_{i:n})\}^{n-l},$$
$$-\infty < x_{i:n} < \infty. \qquad (2.6)$$

Theorem 2.2. The conditional density function of $X_{j:n}$, given $X_{i:n} = x_{i:n}$, for $1 \le i < j \le n$, is exactly the same as the density function of the $(j-i)$th order statistic in a sample of size $n - i$ from the distribution $F(\cdot)$ truncated on the left at $x_{i:n}$, i.e., from the distribution with density function $f(x)/\{1 - F(x_{i:n})\}$, $x > x_{i:n}$.

Proof. Upon writing the conditional density function in (2.4) as

$$\frac{(n-i)!}{(j-i-1)!(n-j)!} \left\{ \frac{F(x_{j:n}) - F(x_{i:n})}{1 - F(x_{i:n})} \right\}^{j-i-1} \left\{ \frac{1 - F(x_{j:n})}{1 - F(x_{i:n})} \right\}^{n-j}$$
$$\times \frac{f(x_{j:n})}{1 - F(x_{i:n})}, \quad x_{i:n} < x_{j:n},$$

the result follows immediately by comparing this expression to (2.5).

$$\odot$$

Proceeding similarly, we can establish the following result.

Theorem 2.3. The conditional density function of $X_{i:n}$, given $X_{j:n} = x_{j:n}$, for $1 \le i < j \le n$, is exactly the same as the density function of the ith order statistic in a sample of size $j - i$ from the distribution $F(\cdot)$ truncated on the right at $x_{j:n}$, i.e., from the distribution with density function $f(x)/\{F(x_{j:n})\}$, $x < x_{j:n}$.

Proof. Upon multiplying the conditional density function of $X_{j:n}$, given $X_{i:n} = x_{i:n}$, in (2.4) with the marginal density function of $X_{i:n}$ in (2.5), we readily obtain the joint density function of $X_{i:n}$ and $X_{j:n}$ (for $1 \le i < j \le n$) as

$$f_{X_{i:n}, X_{j:n}}(x_{i:n}, x_{j:n})$$
$$= \frac{n!}{(i-1)!(j-i-1)!(n-j)!} \{F(x_{i:n})\}^{i-1} \{F(x_{j:n}) - F(x_{i:n})\}^{j-i-1}$$
$$\times \{1 - F(x_{j:n})\}^{n-j} f(x_{i:n}) f(x_{j:n}), \quad x_{i:n} < x_{j:n}. \tag{2.7}$$

$$\odot$$

2.2.3 Moments

From Eqs. (2.5) and (2.7), we simply have the single and product moments of order statistics as

$$
\begin{aligned}
E(X_{i:n}^l) &= \frac{n!}{(i-1)!(n-i)!} \\
&\times \int_{-\infty}^{\infty} x_{i:n}^l \left\{F(x_{i:n})\right\}^{i-1} \left\{1 - F(x_{i:n})\right\}^{n-i} f(x_{i:n})dx_{i:n}, \quad (2.8)
\end{aligned}
$$

and

$$
\begin{aligned}
E(X_{i:n}^{l_1} X_{j:n}^{l_2}) &= \frac{n!}{(i-1)!(j-i-1)!(n-j)!} \\
&\times \int_{-\infty}^{\infty} \int_{x_{i:n}}^{\infty} x_{i:n}^{l_1} x_{j:n}^{l_2} \left\{F(x_{i:n})\right\}^{i-1} \\
&\times \left\{F(x_{j:n}) - F(x_{i:n})\right\}^{j-i-1} \left\{1 - F(x_{j:n})\right\}^{n-j} \\
&\times f(x_{i:n}) f(x_{j:n}) dx_{j:n} dx_{i:n}, \quad (2.9)
\end{aligned}
$$

respectively.

Upon making the probability integral transformation $u_i = F(x_{i:n})$, for $1 \le i \le n$, and noting that it is order-preserving, we can rewrite the single and product moments in (2.8) and (2.9) as

$$
\begin{aligned}
E(X_{i:n}^l) &= \frac{n!}{(i-1)!(n-i)!} \\
&\times \int_0^1 \left\{F^{-1}(u_i)\right\}^l u_i^{i-1} (1-u_i)^{n-i} du_i, \quad (2.10)
\end{aligned}
$$

and

$$
\begin{aligned}
E(X_{i:n}^{l_1} X_{j:n}^{l_2}) &= \frac{n!}{(i-1)!(j-i-1)!(n-j)!} \\
&\times \int_0^1 \int_{u_i}^1 \left\{F^{-1}(u_i)\right\}^{l_1} \left\{F^{-1}(u_j)\right\}^{l_2} u_i^{i-1} \\
&\times (u_j - u_i)^{j-i-1} (1-u_j)^{n-j} du_j du_i. \quad (2.11)
\end{aligned}
$$

2.2.4 Results for the Uniform Distribution

For the *Uniform(0,1)* distribution with density function

$$
f(x) = 1 \quad \text{for} \ \ 0 < x < 1, \quad (2.12)
$$

expressions for distributions and moments of order statistics simplify considerably. For example, the joint density of all n order statistics in (2.1) simply becomes

$$f_{X_{1:n},\cdots,X_{n:n}}(x_{1:n},\cdots,x_{n:n}) = n!, \quad 0 < x_{1:n} < \cdots < x_{n:n} < 1. \qquad (2.13)$$

Similarly, the joint density of $(X_{1:n},\cdots,X_{i:n})$ in (2.2) becomes

$$f_{X_{1:n},\cdots,X_{i:n}}(x_{1:n},\cdots,x_{i:n}) = \frac{n!}{(n-i)!}(1-x_{i:n})^{n-i},$$
$$0 < x_{1:n} < \cdots < x_{i:n} < 1. \qquad (2.14)$$

The marginal density function of $X_{i:n}$ (for $1 \le i \le n$) in (2.5) becomes

$$f_{X_{i:n}}(x_{i:n}) = \frac{n!}{(i-1)!(n-i)!}\, x_{i:n}^{i-1}(1-x_{i:n})^{n-i},$$
$$0 < x_{i:n} < 1, \qquad (2.15)$$

which simply reveals that $X_{i:n}$ in this case is distributed as *Beta(i,n-i+1)*; see Chapter 25 of Johnson, Kotz, and Balakrishnan (1995) for a detailed discussion on beta distributions. From (2.15), we immediately obtain

$$E(X_{i:n}^k) = \frac{B(i+k, n-i+1)}{B(i, n-i+1)} = \frac{(i+k-1)!}{(n+k)!} \times \frac{n!}{(i-1)!},$$

from which we get

$$E(X_{i:n}) = \frac{i}{n+1} \quad \text{and} \quad Var(X_{i:n}) = \frac{i(n-i+1)}{(n+1)^2(n+2)}.$$

The joint density function of $X_{i:n}$ and $X_{j:n}$ (for $1 \le i < j \le n$) in (2.7) becomes

$$f_{X_{i:n},X_{j:n}}(x_{i:n},x_{j:n}) = \frac{n!}{(i-1)!(j-i-1)!(n-j)!} x_{i:n}^{i-1}(x_{j:n}-x_{i:n})^{j-i-1}$$
$$\times (1-x_{j:n})^{n-j}, \quad 0 < x_{i:n} < x_{j:n} < 1. \qquad (2.16)$$

From (2.16), we can show that

$$E(X_{i:n}X_{j:n}) = \frac{i(j+1)}{(n+1)(n+2)} \quad \text{and} \quad Cov(X_{i:n},X_{j:n}) = \frac{i(n-j+1)}{(n+1)^2(n+2)}.$$

From the joint density of all n uniform order statistics given in (2.13), we can establish the following result originally due to Malmquist (1950).

Theorem 2.4. The random variables

$$V_1 = \frac{X_{1:n}}{X_{2:n}}, \quad \cdots \quad , V_{n-1} = \frac{X_{n-1:n}}{X_{n:n}}, \; V_n = X_{n:n} \tag{2.17}$$

are statistically independent.

Proof. Noting that

$$X_{i:n} = V_i \, V_{i+1} \cdots V_n, \quad 1 \le i \le n$$

is the inverse transformation and that the Jacobian of the transformation is $\prod_{i=1}^{n} V_i^{i-1}$ we obtain from Eq. (2.13) the joint density function of $V_1, \cdots, V_n$ as

$$f_{V_1, \cdots, V_n}(v_1, \cdots, v_n) = n! \prod_{i=1}^{n} v_i^{i-1}, \quad 0 < v_1, \cdots, v_n < 1. \tag{2.18}$$

The independence of $V_1, \cdots, V_n$ follows readily from (2.18) with the use of the factorization theorem.

$$\odot$$

Theorem 2.5. The probability integral transformation $F(X_{i:n})$, $1 \le i \le n$, transforms the order statistics $\{X_{i:n}\}_{i=1}^{n}$ from any continuous distribution $F(\cdot)$ into order statistics from *Uniform(0,1)* distribution with joint density function as in (2.13).

2.2.5 Results for the Exponential Distribution

For the standard exponential distribution with density function

$$f(x) = e^{-x}, \quad x > 0, \tag{2.19}$$

order statistics possess some very nice properties. For example, the joint density of all n order statistics in (2.1) becomes

$$f_{X_{1:n}, \cdots, X_{n:n}}(x_{1:n}, \cdots, x_{n:n}) = n! \, \exp\left\{ -\sum_{l=1}^{n} x_{l:n} \right\}, \quad 0 < x_{1:n} < \cdots < x_{n:n}. \tag{2.20}$$

From Eq. (2.20), we can establish the following result originally due to Sukhatme (1937).

Theorem 2.6. The random variables

$$S_1 = nX_{1:n}, \ S_2 = (n-1)(X_{2:n} - X_{1:n}), \ \cdots, \ S_n = X_{n:n} - X_{n-1:n} \quad (2.21)$$

are independent standard exponential random variables.

Proof. Noting that

$$X_{i:n} = \frac{S_1}{n} + \cdots + \frac{S_i}{n-i+1}, \quad 1 \le i \le n$$

is the inverse transformation and that the Jacobian of the transformation is $1/n!$, we obtain from Eq. (2.20) the joint density function of $S_1, \cdots, S_n$ as

$$f_{S_1,\cdots,S_n}(s_1,\cdots,s_n) = \exp\left\{-\sum_{l=1}^{n} s_l\right\}, \quad 0 < s_1,\cdots,s_n < \infty. \quad (2.22)$$

The result then follows readily from (2.22) with the use of the factorization theorem.

$$\odot$$

Theorem 2.6, in addition to showing that the order statistics from an exponential distribution form an additive Markov chain, gives a convenient representation for $X_{i:n}$ as

$$X_{i:n} \overset{d}{=} \sum_{l=1}^{i} \frac{S_l}{n-l+1}, \quad 1 \le i \le n, \quad (2.23)$$

where S_l's are independent standard exponential random variables. From (2.23), it follows immediately that

$$E(X_{i:n}) = \sum_{l=1}^{i} \frac{1}{n-l+1}, \quad Var(X_{i:n}) = \sum_{l=1}^{i} \frac{1}{(n-l+1)^2}$$

and

$$Cov(X_{i:n}, X_{j:n}) = Var(X_{i:n}) = \sum_{l=1}^{i} \frac{1}{(n-l+1)^2} \quad \text{for } 1 \le i \le j \le n.$$

Theorem 2.7. For $r = 1, 2, \cdots, n$, the random variable

$$T_r = X_{1:n} + X_{2:n} + \cdots + X_{r:n} + (n-r)X_{r:n} \qquad (2.24)$$

has a standard *Gamma(r)* distribution with density function

$$f_{T_r}(t) = \frac{1}{\Gamma(r)}\, e^{-t}\, t^{r-1}, \quad 0 < t < \infty; \qquad (2.25)$$

consequently, $2T_r$ is distributed as chi-square with $2r$ degrees of freedom.

Proof. Noting that, for $1 \le r \le n$,

$$\begin{aligned}
T_r &= X_{1:n} + X_{2:n} + \cdots + X_{r:n} + (n-r)X_{r:n} \\
&= nX_{1:n} + (n-1)(X_{2:n} - X_{1:n}) + \cdots + (n-r+1)(X_{r:n} - X_{r-1:n}) \\
&\stackrel{d}{=} S_1 + S_2 + \cdots + S_r, \qquad (2.26)
\end{aligned}$$

and that S_l's are independent standard exponential random variables, we immediately see that T_r has a standard *Gamma(r)* distribution with density as in (2.25). The fact that $2T_r$ is distributed as chi-square with $2r$ degrees of freedom follows from the well-known relationship between gamma and chi-square distributions; see, for example, Chapters 17 and 18 of Johnson, Kotz, and Balakrishnan (1994).

$$\odot$$

2.3 CENSORED DATA

Suppose, as described in Section 2.1, n units are placed on a life-testing experiment. Further, suppose $X_1, \cdots, X_n$ denote the lifetimes of these n units taken from a population with lifetime distribution function $F(x; \theta)$ and density function $f(x; \theta)$, where θ is an unknown parameter(s) of interest. Let $X_{1:n} \le \cdots \le X_{n:n}$ denote the corresponding ordered lifetimes observed from the life-test.

2.3.1 Type-I Censoring

Suppose it is planned that the life-testing experiment will be terminated at a pre-fixed time T. Then, only the failures until time T will be observed. The data obtained from such a restrained life-test will be referred to as a *Type-I censored sample.* Note that the number of failures observed here is random and, in fact, has a *Binomial(n, F(T))* distribution. Inferential procedures based on Type-I censored samples have been discussed quite extensively in the literature; see, for example, Cohen and Whitten (1988), Bain and Engelhardt (1991), Cohen (1991), and Balakrishnan and Cohen (1991).

For example, if the lifetimes are assumed to come from an *Exponential(θ)* distribution with density function

$$f(x; \theta) = \frac{1}{\theta} e^{-x/\theta}, \quad x > 0, \theta > 0, \tag{2.27}$$

then the likelihood function based on the Type-I censored sample $\{X_{1:n} < \cdots < X_{r:n} \leq T\}$ is given by

$$L(\theta) = \begin{cases} \exp\left\{-\frac{nT}{\theta}\right\} & \text{if } r = 0 \\ \frac{n!}{(n-r)!\theta^r} \exp\left\{-\frac{1}{\theta}\left(\sum\limits_{i=1}^{r} x_{i:n} + (n-r)T\right)\right\} & \text{if } r \geq 1. \end{cases} \tag{2.28}$$

From Eq. (2.28), it is evident that the maximum likelihood estimator (MLE) of θ exists only if $r \geq 1$ and is given by

$$\hat{\theta}_c = \frac{1}{r} \left\{\sum\limits_{i=1}^{r} x_{i:n} + (n-r)T\right\}; \tag{2.29}$$

here, $\hat{\theta}_c$ is used in order to emphasise that it is a constrained MLE. Because of this, the exact distribution of $\hat{\theta}_c$ in (2.29) becomes quite complicated; see Bartholomew (1963).

2.3.2 Type-II Censoring

Suppose it is planned that the life-testing experiment will be terminated as soon as the r-th (where r is pre-fixed) failure is observed. Then, only the first r failures out of n units under test will be observed. The data obtained from such a restrained life-test will be referred to as a *Type-II censored sample.* Note that in this case, in contrast to Type-I censoring, the number of failures observed is fixed (viz., r) while the duration of the experiment is random (viz., $X_{r:n}$). Inferential procedures based on Type-II censored samples have been discussed extensively in the literature; see, for example, Nelson (1982), Bain and Engelhardt (1991), Cohen (1991), and Balakrishnan and Cohen (1991).

If the lifetimes are assumed to come from an *Exponential(θ)* distribution with density function as in (2.27) , we have the likelihood function based on the Type-II censored sample $\{X_{1:n} < \cdots < X_{r:n}\}$ as [from Eq. (2.2)]

$$L(\theta) = \frac{n!}{(n-r)!\theta^r} \ \exp\left[-\frac{1}{\theta}\left\{\sum_{i=1}^{r} x_{i:n} + (n-r)x_{r:n}\right\}\right]. \tag{2.30}$$

From (2.30), it is evident that the MLE of θ is

$$\hat{\theta} = \frac{1}{r}\left\{\sum_{i=1}^{r} x_{i:n} + (n-r)x_{r:n}\right\}. \tag{2.31}$$

Now, upon rewriting $\hat{\theta}$ in (2.31) as

$$\hat{\theta} = \frac{1}{r}\left[nx_{1:n} + (n-1)(x_{2:n} - x_{1:n}) + \cdots + (n-r+1)(x_{r:n} - x_{r-1:n})\right],$$

we have

$$
\begin{aligned}
\frac{r\hat{\theta}}{\theta} &= \frac{1}{\theta}\left[nx_{1:n} + (n-1)(x_{2:n} - x_{1:n}) + \cdots + (n-r+1)(x_{r:n} - x_{r-1:n})\right] \\
&\stackrel{d}{=} S_1 + S_2 + \cdots + S_r, \tag{2.32}
\end{aligned}
$$

as in (2.26), where S_l's are independent standard exponential random variables. From (2.32), we readily find

$$E(\hat{\theta}) = \theta \quad \text{and} \quad Var(\hat{\theta}) = \frac{\theta^2}{r}.$$

Furthermore, due to Theorem 2.7, we also have $2r\hat{\theta}/\theta$ to be distributed exactly as chi-square with $2r$ degrees of freedom. This fact may be used to carry out an exact chi-square test and construct exact confidence intervals for the parameter θ; see, for example, Epstein and Sobel (1953, 1954).

2.4 PROGRESSIVELY CENSORED DATA

2.4.1 General Properties

The censored life-testing experiments described in the previous section can be extended to situations wherein censoring occurs in multiple stages. Data arising from such life-tests are referred to as *progressively censored data.* Naturally, progressive censoring can be introduced in both Type-I and Type-II forms.

For example, a progressive Type-II censored life-testing experiment will be carried out in the following manner. As in the preceding section, n units will be placed on a life-testing experiment and it is assumed that these n units come from a population with lifetime distribution function $F(x; \theta)$ and density function $f(x; \theta)$, where θ is an unknown parameter(s) of interest. It is planned that only r complete failures will be observed and the remaining $n-r$ lifetimes will be censored progressively. More specifically, at the time of the first failure (denoted by $X_{1:r:n}$), R_1 of the $n-1$ surviving units will be removed randomly from the life-test; next, at the time of the second failure (denoted by $X_{2:r:n}$), R_2 of the $n - 2 - R_1$ surviving units will be removed randomly from the life-test, and so on; finally, at the time of the rth failure (denoted by $X_{r:r:n}$), all the remaining $R_r = n - r - R_1 - \cdots - R_{r-1}$ surviving units will be removed from the life-test. The data so obtained, viz., $(X_{1:r:n} < X_{2:r:n} < \cdots < X_{r:r:n})$, form a *progressively Type-II censored sample* of size r from a sample of size n with the progressive censoring scheme $(R_1, \cdots, R_r)$. Clearly, $n = r + R_1 + \cdots + R_r$ and when $r = n$ (i.e., $R_1 = \cdots = R_r = 0$) we obtain the set of usual order statistics. When $R_1 = \cdots = R_{r-1} = 0$ and $R_r = n - r$, we obtain the usual Type-II censored sample of size r described in Section 2.3.2. Inferential procedures based on progressively Type-II censored samples have been discussed quite extensively in the literature, and a booklength account of these developments is due to Balakrishnan and Aggarwala (2000).

The joint density of $(X_{1:r:n}, \cdots, X_{r:r:n})$ is given by [see Balakrishnan and Aggarwala (2000, p. 8) and Kamps (1995a,b)]

$$
\begin{aligned}
f_{X_{1:r:n}, \cdots, X_{r:r:n}} &(x_{1:r:n}, \cdots, x_{r:r:n}) \\
&= n(n - R_1 - 1) \cdots (n - R_1 - \cdots - R_{r-1} - r + 1) \\
&\quad \times \prod_{l=1}^{r} f(x_{l:r:n}) \left\{ 1 - F(x_{l:r:n}) \right\}^{R_l}, \quad x_{1:r:n} < \cdots < x_{r:r:n}.
\end{aligned}
$$

$$(2.33)$$

From the joint density in (2.33), the following results can be easily established and these are analogous to the corresponding results for the usual order statistics presented earlier in Section 2.2.

Theorem 2.8.

(i) The marginal distribution of $X_{i:r:n}$ (for $1 \leq i \leq r$) is free of $R_i, R_{i+1}, \cdots, R_r$;

(ii) $(X_{1:r:n}, \cdots, X_{i:r:n})$, for $1 \leq i \leq r$, form a progressively Type-II censored sample of size i from a sample of size n with the progressive censoring scheme $(R_1, \cdots, R_{i-1}, n - R_1 - \cdots - R_{i-1} - i)$.

Theorem 2.9. The progressively Type-II censored order statistics $\{X_{i:r:n}\}_{i=1}^r$ from a continuous distribution form a Markov chain; that is, given $X_{i:r:n} = x_{i:r:n}$, $X_{j:r:n}$ (for $i < j \leq r$) is independent of $X_{1:r:n}, \cdots, X_{i-1:r:n}$.

Theorem 2.10. Given $X_{i:r:n} = x_{i:r:n}$, $(X_{i+1:r:n}, \cdots, X_{r:r:n})$ are distributed exactly as a progressively Type-II censored sample of size $r - i$ from a sample of size $n - R_1 - \cdots - R_i - i$, with the progressive censoring scheme $(R_{i+1}, \cdots, R_r)$, from the distribution $F(\cdot)$ truncated on the left at $x_{i:r:n}$, i.e., from the distribution with density function $f(x)/\{1 - F(x_{i:r:n})\}$, $x > x_{i:r:n}$.

2.4.2 Results for the Uniform Distribution

For the *Uniform(0,1)* distribution with density as in (2.12), the joint density of $(X_{1:r:n}, \cdots, X_{r:r:n})$ in (2.33) simply becomes

$$f_{X_{1:r:n},\cdots,X_{r:r:n}}(x_{1:r:n}, \cdots, x_{r:r:n})$$
$$= n(n - R_1 - 1) \cdots (n - R_1 - \cdots - R_{r-1} - r + 1)$$
$$\times \prod_{l=1}^{r}(1 - x_{l:r:n})^{R_l}, \quad 0 < x_{1:r:n} < \cdots < x_{r:r:n} < 1. \quad (2.34)$$

Then, Balakrishnan and Sandhu (1995) and Aggarwala and Balakrishnan (1998) have generalized Theorem 2.4 to the case of progressive Type-II censoring as follows.

Theorem 2.11.

(i) The random variables

$$V_1 = \frac{1 - X_{r:r:n}}{1 - X_{r-1:r:n}}, \quad \cdots \quad, V_{r-1} = \frac{1 - X_{2:r:n}}{1 - X_{1:r:n}}, \quad V_r = 1 - X_{1:r:n} \quad (2.35)$$

are statistically independent;

(ii)

$$V_i \stackrel{d}{=} Beta\left(i + \sum_{j=r-i+1}^{r} R_j, 1\right), \quad i = 1, 2, \cdots, r; \qquad (2.36)$$

(iii)

$$W_i = V_i^{i+R_{r-i+1}+\cdots+R_r}$$
$$\stackrel{d}{=} Uniform(0,1), \quad i = 1, 2, \cdots, r \qquad (2.37)$$

independently.

Proof. The proof is along the same lines as that of Theorem 2.4 using the Jacobian method.

$$\odot$$

The stochastic representation in Theorem 2.11 will facilitate easy calculation of moments of $X_{i:r:n}$. For example, from (2.35) and (2.36), we have

$$X_{i:r:n} \stackrel{d}{=} 1 - \prod_{j=r-i+1}^{r} V_j, \quad i = 1, 2, \cdots, r,$$

and consequently

$$E(X_{i:r:n}) = 1 - \prod_{j=r-i+1}^{r} E(V_j)$$
$$= 1 - \prod_{j=r-i+1}^{r} \left\{ \frac{j + R_{r-j+1} + \cdots + R_r}{j + 1 + R_{r-j+1} + \cdots + R_r} \right\}.$$

2.4.3 Results for the Exponential Distribution

For the standard exponential distribution with density as in (2.19), the joint density of $(X_{1:r:n}, \cdots, X_{r:r:n})$ in (2.33) simply becomes

$$f_{X_{1:r:n},\cdots,X_{r:r:n}}(x_{1:r:n}, \cdots, x_{r:r:n})$$
$$= n(n - R_1 - 1) \cdots (n - R_1 - \cdots - R_{r-1} - r + 1)$$
$$\times \exp\left\{ -\sum_{l=1}^{r} (R_i + 1)x_{i:r:n} \right\}, \quad 0 < x_{1:r:n} < \cdots < x_{r:r:n} < \infty.$$

$$(2.38)$$

Then, Thomas and Wilson (1972) and Viveros and Balakrishnan (1994) have generalized Theorem 2.6 to the case of progressive Type-II censoring as follows.

Theorem 2.12. The random variables

$$S_1 = nX_{1:r:n}, \ S_2 = (n - R_1 - 1)(X_{2:r:n} - X_{1:r:n}),$$
$$\cdots, S_r = (n - R_1 - \cdots - R_{r-1} - r + 1)(X_{r:r:n} - X_{r-1:r:n})$$
$$(2.39)$$

are independent standard exponential random variables.

Proof. The proof is along the same lines as that of Theorem 2.6 using the Jacobian method.

$$\odot$$

Theorem 2.12, in addition to showing that the progressively Type-II censored order statistics from an exponential distribution form an additive Markov chain, gives a convenient representation for $X_{i:r:n}$ as

$$X_{i:r:n} \stackrel{d}{=} \sum_{l=1}^{i} \frac{S_l}{n - R_1 - \cdots - R_{l-1} - l + 1}, \quad 1 \le i \le r, \qquad (2.40)$$

where S_l's are independent standard exponential random variables. From (2.40), it follows immediately that

$$E(X_{i:r:n}) = \sum_{l=1}^{i} \frac{1}{n - R_1 - \cdots - R_{l-1} - l + 1}$$

and

$$Var(X_{i:r:n}) = Cov(X_{i:r:n}, X_{j:r:n}) = \sum_{l=1}^{i} \frac{1}{(n - R_1 - \cdots - R_{l-1} - l + 1)^2}$$

for $1 \le i \le j \le r$.

Theorem 2.13. For $r = 1, 2, \cdots, n$, the random variable

$$T_r = \sum_{l=1}^{r} (R_l + 1)X_{l:r:n} \qquad (2.41)$$

has a standard *Gamma(r)* distribution with density as in (2.25), and so $2T_r$ has a chi-square distribution with $2r$ degrees of freedom.

Proof. The result follows along the same lines as that of Theorem 2.7 upon noting that

$$T_r = \sum_{l=1}^{r}(R_l + 1)X_{l:r:n} = \sum_{l=1}^{r} S_l.$$

$\odot$

Now, let us assume that the progressively Type-II censored sample $(X_{1:r:n}, \cdots, X_{r:r:n})$ comes from an *Exponential(θ)* distribution with density function as in (2.27). Then, from Eq. (2.33), we have the likelihood function as

$$\begin{aligned}
L &= n(n - R_1 - 1)\cdots(n - R_1 - \cdots - R_{r-1} - r + 1) \\
&\quad \times \frac{1}{\theta^r}\exp\left\{-\frac{1}{\theta}\sum_{l=1}^{r}(R_l + 1)x_{l:r:n}\right\}.
\end{aligned} \tag{2.42}$$

From (2.42), it is evident that the MLE of θ is

$$\hat{\theta} = \frac{1}{r}\sum_{l=1}^{r}(R_l + 1)x_{l:r:n}. \tag{2.43}$$

Now, upon rewriting $\hat{\theta}$ in (2.43) as

$$\begin{aligned}
\hat{\theta} &= \frac{1}{r}[nx_{1:r:n} + (n - R_1 - 1)(x_{2:r:n} - x_{1:r:n}) \\
&\quad + \cdots + (n - R_1 - \cdots - R_{r-1} - r + 1)(x_{r:r:n} - x_{r-1:r:n})],
\end{aligned}$$

we have

$$\frac{r\hat{\theta}}{\theta} \overset{d}{=} S_1 + S_2 + \cdots + S_r, \tag{2.44}$$

where S_l's are independent standard exponential random variables. From (2.44), we readily find

$$E(\hat{\theta}) = \theta \quad \text{and} \quad Var(\hat{\theta}) = \frac{\theta^2}{r}.$$

Furthermore, due to Theorem 2.13, we also have $2r\hat{\theta}/\theta$ to be distributed exactly as chi-square with $2r$ degrees of freedom. This fact may be used to carry out an exact chi-square test and construct exact confidence intervals for the parameter θ.

2.5 SOME USEFUL LIFETIME DISTRIBUTIONS

In this section, we describe briefly some useful lifetime distributions and present some of their special features. These distributions will be used in all subsequent chapters while evaluating the power performance of all the reliability test procedures discussed therein.

2.5.1 Exponential Distribution

The *Exponential(θ)* random variable, X, has its density function as in (2.27) and cumulative distribution function as

$$F_X(x; \theta) = 1 - e^{-x/\theta}, \quad x > 0, \theta > 0.$$

The hazard function of X is

$$h_X(x; \theta) = \frac{f_X(x)}{1 - F_X(x)} = \frac{1}{\theta}, \tag{2.45}$$

which is a constant. In fact, the constant hazard function is a characterization of the exponential distribution. Moreover, since

$$\begin{aligned}
\Pr(X \le x + y | X \ge x) &= \frac{\Pr(x \le X \le x + y)}{\Pr(X \ge x)} = \frac{e^{-x/\theta} - e^{-(x+y)/\theta}}{e^{-x/\theta}} \\
&= 1 - e^{-y/\theta} = \Pr(X \le y),
\end{aligned}$$

the exponential distribution possesses *lack of memory* property.

From (2.27), we get the nth moment of X as

$$E(X^n) = \int_0^\infty x^n f_X(x)dx = \frac{1}{\theta} \int_0^\infty e^{-x/\theta} x^n dx = \Gamma(n+1)\theta^n. \tag{2.46}$$

In particular, we have

$$E(X) = \theta \quad \text{and} \quad Var(X) = \theta^2,$$

and Pearson's coefficients of skewness and kurtosis are

$$\sqrt{\beta_1(X)} = 2 \quad \text{and} \quad \beta_2(X) = 9,$$

respectively.

For an elaborate discussion on various properties, inferential procedures, and applications of the exponential distribution, interested readers may refer to the books by Balakrishnan and Basu (1995) and Johnson, Kotz, and Balakrishnan (1994, Chapter 19).

2.5.2   Gamma Distribution

The *Gamma(α, θ)* random variable, X, has its density function as

$$f_X(x) = \frac{1}{\Gamma(\alpha)\theta^\alpha} \, e^{-x/\theta} x^{\alpha-1}, \quad x > 0, \alpha > 0, \theta > 0, \qquad (2.47)$$

and cumulative distribution function as

$$\begin{aligned}
F_X(x) &= \frac{1}{\Gamma(\alpha)\theta^\alpha} \int_0^x e^{-t/\theta} t^{\alpha-1} dt \\
&= \frac{1}{\Gamma(\alpha)} \int_0^{x/\theta} e^{-u} u^{\alpha-1} du \\
&= \frac{\Gamma_{x/\theta}(\alpha)}{\Gamma(\alpha)}, \qquad (2.48)
\end{aligned}$$

where

$$\Gamma_y(\alpha) = \int_0^y e^{-u} u^{\alpha-1} du, \quad y > 0,$$

is the incomplete gamma function.

From (2.47), we get the nth moment of X as

$$\begin{aligned}
E(X^n) &= \frac{1}{\Gamma(\alpha)\theta^\alpha} \int_0^\infty e^{-t/\theta} t^{\alpha+n-1} dt \\
&= \frac{\Gamma(\alpha+n)}{\Gamma(\alpha)} \, \theta^n \\
&= (\alpha+n-1)(\alpha+n-2)\cdots\alpha \, \theta^n. \qquad (2.49)
\end{aligned}$$

In particular, we have

$$E(X) = \alpha\theta \quad \text{and} \quad Var(X) = \alpha\theta^2,$$

and Pearson's coefficients of skewness and kurtosis are

$$\sqrt{\beta_1(X)} = \frac{2}{\sqrt{\alpha}} \quad \text{and} \quad \beta_2(X) = 3 + \frac{6}{\alpha},$$

respectively. Note that $\beta_1(X)$ and $\beta_2(X)$ have a linear relationship of the form $\beta_2(X) = 3 + 1.5\,\beta_1(X)$.

The limiting distribution of $(X - \alpha\theta)/\sqrt{\alpha\theta^2}$, as $\alpha \to \infty$, can be shown to be standard normal. Thus, even though gamma distributions are positively skewed and have heavy tails, when α becomes large gamma distributions become nearly normal.

For a detailed discussion on various properties, inferential procedures, and applications of the gamma distribution, interested readers may refer to the books by Bowman and Shenton (1988) and Johnson, Kotz, and Balakrishnan (1994, Chapter 17).

2.5.3 Weibull Distribution

The *Weibull*(c, θ) random variable, X, has its density function as

$$f_X(x) = \frac{c}{\theta^c} x^{c-1} e^{-(x/\theta)^c}, \quad x > 0, c > 0, \theta > 0, \tag{2.50}$$

and cumulative distribution function as

$$F_X(x) = 1 - e^{-(x/\theta)^c}, \quad x > 0, c > 0, \theta > 0. \tag{2.51}$$

The hazard function of X is

$$h_X(x) = \frac{f_X(x)}{1 - F_X(x)} = \frac{c}{\theta^c} x^{c-1}, \quad x > 0, c > 0, \theta > 0. \tag{2.52}$$

It is clear that the hazard function in (2.52) is decreasing for $c < 1$, constant for $c = 1$ (since the distribution is exponential in this case), and increasing for $c > 1$.

From (2.50), we get the nth moment of X as

$$\begin{aligned} E(X^n) &= \frac{c}{\theta^c} \int_0^\infty x^n e^{-(x/\theta)^c} x^{c-1} dx \\ &= \int_0^\infty (\theta u^{1/c})^n e^{-u} du \\ &= \theta^n \, \Gamma\left(\frac{n}{c} + 1\right). \end{aligned} \tag{2.53}$$

In particular, we have

$$E(X) = \theta \, \Gamma\left(\frac{1}{c} + 1\right) \quad \text{and} \quad Var(X) = \theta^2 \left\{ \Gamma\left(\frac{2}{c} + 1\right) - \left[\Gamma\left(\frac{1}{c} + 1\right)\right]^2 \right\}.$$

With $c_0 \approx 3.602$, the coefficient of skewness ($\sqrt{\beta_1}$) is a decreasing function of c in the interval $c < c_0$, an increasing function of c in the interval $c > c_0$, and is nearly 0 when $c = c_0$. Similarly, with $c_1 = 3.35$, the coefficient of kurtosis (β_2) is a decreasing function of c in the interval $c < c_1$, an increasing function of c in the interval $c > c_1$, and attains its minimal value of about 2.71 when $c = c_1$. As a matter of fact, when c is in the neighborhood of 3.602, the Weibull distribution is similar in shape to a normal distribution.

For a detailed discussion on various properties, inferential procedures, and applications of the Weibull distribution, interested readers may refer to the books by Murthy, Xie, and Jiang (2003) and Johnson, Kotz, and Balakrishnan (1994, Chapter 21).

2.5.4   Extreme Value Distribution

The $EV(\mu, \sigma)$ random variable, X, has its density function as

$$f_X(x) = \frac{1}{\sigma} e^{(x-\mu)/\sigma} e^{-e^{(x-\mu)/\sigma}}, \quad -\infty < x < \infty, -\infty < \mu < \infty, \sigma > 0, \quad (2.54)$$

and cumulative distribution function as

$$F_X(x) = 1 - e^{-e^{(x-\mu)/\sigma}}, \quad -\infty < x < \infty, -\infty < \mu < \infty, \sigma > 0. \quad (2.55)$$

The hazard function of X is

$$h_X(x) = \frac{f_X(x)}{1 - F_X(x)} = \frac{1}{\sigma} e^{(x-\mu)/\sigma}, \quad -\infty < x < \infty, -\infty < \mu < \infty, \sigma > 0.$$

$$(2.56)$$

It is clear that the hazard function in (2.57) is an increasing function. From (2.54), we can show that

$$E(X) \;=\; \mu + \sigma\psi(1) = \mu - \gamma\sigma \approx \mu - 0.57722\sigma,$$

where $\psi(\cdot)$ denotes the psi function and γ is Euler's constant, and

$$Var(X) \;=\; \frac{\pi^2 \sigma}{6} \approx 1.64493\sigma^2.$$

If Y is distributed as $Weibull(c, \theta)$ with density function as in (2.50), then it can be readily seen that $X = \log Y$ has a density function

$$f_X(x) \;=\; \frac{c}{\theta^c} \, e^{cx} \, e^{-e^{cx}/\theta^c}$$

$$=\; c \, e^{c(x-\log\theta)} \, e^{-e^{c(x-\log\theta)}}, \quad -\infty < x < \infty, c > 0, \theta > 0;$$

this, when compared with (2.54), immediately reveals that $X = \log Y$ has an $EV(\mu = \log \theta, \sigma = 1/c)$ distribution.

For a detailed account of various developments on the extreme value distribution, interested readers may refer to the book by Johnson, Kotz, and Balakrishnan (1995, Chapter 22).

2.5.5 Lognormal Distribution

The $LN(\mu, \sigma)$ random variable, X, has its density function as

$$f_X(x) = \frac{1}{\sqrt{2\pi}\sigma x} \exp\left\{ -\frac{1}{2\sigma^2}(\log x - \mu)^2 \right\}, \ x > 0, -\infty < \mu < \infty, \sigma > 0.$$

(2.57)

It is evident from (2.57) that $Z = (\log X - \mu)/\sigma$ has a standard normal distribution. From this relationship, we readily find the nth moment of X as

$$E(X^n) = E(e^{n\mu + n\sigma Z}) = e^{n\mu + n^2\sigma^2/2}.$$

(2.58)

In particular, with $\omega = e^{\sigma^2}$, we have

$$E(X) = e^\mu \sqrt{\omega} \quad \text{and} \quad Var(X) = e^{2\mu}\omega(\omega - 1),$$

and Pearson's coefficients of skewness and kurtosis are

$$\sqrt{\beta_1(X)} = \sqrt{\omega - 1}(\omega + 2) \quad \text{and} \quad \beta_2(X) = \omega^4 + 2\omega^3 + 3\omega^2 - 3,$$

respectively. Note that, therefore, the lognormal distributions are positively skewed and have heavy tails. When ω gets close to 1, i.e., when σ is very small, the lognormal distributions become nearly normal.

If Y is distributed as $LN(\mu, \sigma)$ with density function as in (2.57), then $X = \log Y$ will have a density function

$$f_X(x) = \frac{1}{\sqrt{2\pi}\sigma} \exp\left\{ -\frac{1}{2\sigma^2}(x - \mu)^2 \right\}, \quad -\infty < x, \ \mu < \infty, \ \sigma > 0,$$

(2.59)

which is the density function of a normal distribution with mean μ and variance σ^2.

For a detailed discussion on various properties, inferential procedures, and applications of the lognormal distribution, interested readers may refer to the books by Crow and Shimizu (1988) and Johnson, Kotz, and Balakrishnan (1994, Chapter 14).

2.6   WILCOXON'S RANK-SUM STATISTIC

Suppose $X_1, \cdots, X_{n_1}$ is a random sample from a continuous distribution with cdf $F_X(x)$ and pdf $f_X(x)$, and $Y_1, \cdots, Y_{n_2}$ is a random sample from another continuous distribution with cdf $F_Y(x)$ and pdf $f_Y(x)$. A natural problem that is of interest is to test the null hypothesis $H_0 : F_X(x) = F_Y(x)$ for all x; i.e., to test whether two processes have the same lifetime distributions. In the reliability context, it will often be of interest to consider the alternative hypothesis as one that specifies Y is stochastically larger than X, viz., $F_X(x) \geq F_Y(x)$ for all x with strict inequality for at least one x. Note that this alternative implies that the Y-process results in more reliable products (i.e., having a longer lifetime) than the X-process. Evidently, under a location-shift model $Y \stackrel{d}{=} X + \delta$, we have

$$F_Y(x) = \Pr(Y \leq x) = \Pr(X \leq x - \delta) = F_X(x - \delta) \leq F_X(x)$$

for all x if $\delta > 0$, and so Y is stochastically larger than X in this case.

For this hypothesis testing problem, *Wilcoxon's rank-sum statistic* is constructed as follows [Wilcoxon (1945)]. This test is also often called the *Mann-Whitney-Wilcoxon statistic* [Mann and Whitney (1947)]; for more details, see Lehmann (1975), Hettmansperger and McKean (1998), Hollander and Wolfe (1999), and Gibbons and Chakraborti (2003). We should first combine the two samples and then rank the observations in the combined sample from 1 to $n_1 + n_2$. Then, the Wilcoxon's rank-sum statistic is defined to be

$$W_R = \sum_{i=1}^{n_1} \mathrm{Rank}(X_i),$$

which is simply the sum of ranks of all X-observations in the combined sample. Since under the alternative hypothesis that Y is stochastically larger than X we would expect the X-sample to take on most of the smaller ranks, we would tend to reject H_0 in favor of H_1 if $W_R \leq s$, where s is a suitable critical value.

Under $H_0 : F_X(x) = F_Y(x)$ for all x, we simply have

$$\Pr\left[\mathrm{Rank}(X_1) = i_1, \cdots, \mathrm{Rank}(X_{n_1}) = i_{n_1} \,\middle|\, H_0 : F_X = F_Y\right] = \dfrac{1}{\dbinom{n_1 + n_2}{n_1}}$$

for any subset $\{i_1, \cdots, i_{n_1}\}$ of $\{1, 2, \cdots, n_1 + n_2\}$. Hence, we can express

$$\Pr\left(W_R = w | H_0 : F_X = F_Y\right) = \frac{N_w}{\binom{n_1 + n_2}{n_1}}, \qquad (2.60)$$

where N_w is the number of subsets $\{i_1, \cdots, i_{n_1}\}$ of $\{1, 2, \cdots, n_1 + n_2\}$ with $i_1 + \cdots + i_{n_1} = w$. From (2.60), we can determine the critical value s for a chosen level of significance α (as close to α as possible since W_R is a discrete random variable). As a matter of fact, these critical values and the exact levels of significance have been tabulated extensively by Wilcoxon, Katti, and Wilcox (1970), and they have been used in all the simulations in the subsequent chapters.

For large sample sizes, the null distribution of W_R can be approximated by a normal distribution with $n_1(n_1 + n_2 + 1)/2$ and $n_1 n_2(n_1 + n_2 + 1)/12$ as mean and variance, respectively.

As Wilcoxon's rank-sum statistic W_R is known to provide a good nonparametric test for the hypothesis testing problem described above, it will be used for comparison with all the reliability test procedures developed in the subsequent chapters.

2.7 RANDOMIZED TEST

Since the Wilcoxon's rank-sum statistic W presented in the last section is a discrete random variable, it will often be difficult to attain a prespecified level of significance. In this case, we will have

$$\Pr\left(W_R \leq s | H_0 : F_X = F_Y\right) = \alpha_L, \ \Pr\left(W_R \leq s + 1 | H_0 : F_X = F_Y\right) = \alpha_U,$$
$$(2.61)$$

where $\alpha_L < \alpha < \alpha_U$. If we now wish to have an exact α level test, it could be achieved by the following randomization procedure. Independent of the given data, produce a Bernoulli random variable Z that takes on the value 1 with probability

$$\pi = \frac{\alpha - \alpha_L}{\alpha_U - \alpha_L}$$

and the value 0 with probability $1 - \pi$. Then, a test procedure with critical region

$$\{W_R \leq s\} \ \cup \ \{W_R = s + 1 \cap (Z = 1)\}$$

will have its level of significance to be exactly α.

Though such a randomized test will have its level of significance to be exactly α, it has the undesirable feature that two individuals could arrive at different decisions even though they make the same assumptions, have the same data, and apply the same Wilcoxon's rank-sum test.

For this reason, we have not used the randomization process in the implementation of any of the tests proposed in the subsequent chapters. If one wishes to use it, however, one can easily do so from the exact null distributions presented therein.

Chapter 3

Precedence Test

3.1 INTRODUCTION

The precedence test is a distribution-free two-sample life-test based on the order of early failures, and was first proposed by Nelson (1963); see also Nelson (1986, 1993) and Ng and Balakrishnan (2006). The precedence test allows the experimenter to make decisions in the early stage of the life-test or when the data contain some right censored observations. The precedence test will be useful (1) when a life-test involves expensive units as the units that had not failed could be used for some other testing purposes, and (2) to make quick and reliable decisions early in the life-testing experiment. Here are two examples where the precedence test is desirable to use in a two-sample setting.

Example 3.1. A manufacturer of electronic components wishes to compare a new design with the design in use in terms of life. Specifically, he or she wants to abandon the design in use if there is significant evidence that it produces components with shorter life. Since new designs of electronic components are under development, the cost of production may be relatively high. Therefore, the manufacturer may wish to fail only r out of n_1 components produced under the new design and save the remaining $n_1 - r$ components for future use. In this situation, a precedence life-test can be employed.

Example 3.2. In a medical study, an experimenter wants to compare a new treatment with a control. Cancer patients with similar characteristics are assigned to treatment and control groups randomly. The remission times of the patients are recorded. Since the treatment can be toxic and harmful to the patients, based on ethical considerations, the experiment should be terminated as soon as there is enough evidence to draw a reliable conclusion, one way or the other. Precedence life-test is useful in this situation since it would enable the experimenter to terminate the experiment and make a decision in the early stage of the experiment.

We will introduce in Section 3.2 the concept of precedence testing and present the precedence test statistic. In Section 3.3, the null distribution is derived via two different methods. The exact power function under the Lehmann alternative is derived next in Section 3.4. Monte Carlo method is then used in Section 3.5 to study the power properties of the precedence test under the location-shift alternative and the corresponding results are discussed in Section 3.6. In Section 3.7, the properties of the precedence test and some related nonparametric tests are discussed. Finally, some illustrative examples are presented in Section 3.8.

3.2 CONCEPT OF PRECEDENCE TEST

The precedence test allows a simple and robust comparison of two distribution functions. Assume that a random sample of size n_1 is from F_X, another independent sample of size n_2 is from F_Y, and that all these sample units are placed simultaneously on a life-testing experiment. We use $X_1, X_2, \ldots, X_{n_1}$ to denote the sample from distribution F_X, and $Y_1, Y_2, \ldots, Y_{n_2}$ to denote the sample from distribution F_Y. A natural null hypothesis is that the two distributions are equal and we are generally concerned with the alternative models wherein one distribution is stochastically larger than the other, for example, the alternative that F_Y is stochastically larger than F_X, which supports the hypothesis that the Y-sample is more reliable than the X-sample. For instance, in Example 3.1, the X-sample may correspond to the components from the design in use while the Y-sample may correspond to components from the new design.

Suppose there are two failure time distributions F_X and F_Y and that we are interested in testing the hypotheses

$$H_0 : F_X = F_Y \quad \text{against} \quad H_1 : F_X > F_Y. \qquad (3.1)$$

Note that some specific alternatives such as the location-shift alternative and the Lehmann alternative are subclasses of the general alternative considered here.

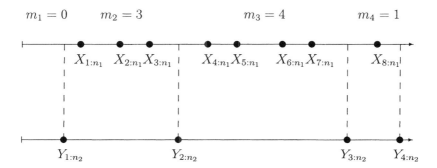

Figure 3.1: Schematic representation of a precedence life-test

We denote the order statistics from the X-sample and the Y-sample by $X_{1:n_1} \leq X_{2:n_1} \leq \cdots \leq X_{n_1:n_1}$ and $Y_{1:n_2} \leq Y_{2:n_2} \leq \cdots \leq Y_{n_2:n_2}$, respectively. Without loss of generality, we assume that $n_1 \leq n_2$. Moreover, we let M_1 be the number of X-failures before $Y_{1:n_2}$ and M_i be the number of X-failures between $Y_{i-1:n_2}$ and $Y_{i:n_2}$, $i = 2, 3, \ldots, r$. We denote the observed value of M_i by m_i, for $i = 1, 2, \ldots, r$.

It is of interest to mention here that these M_i's are related to the so-called "exceedance statistics" whose distributional properties have been discussed, for example, by Fligner and Wolfe (1976) and Randles and Wolfe (1979).

The precedence test statistic $P_{(r)}$ is then defined as the number of failures from the X-sample that precede the rth failure from the Y-sample. That is,

$$P_{(r)} = \sum_{i=1}^{r} M_i. \tag{3.2}$$

For example, from Figure 3.1, with $r = 4$, the precedence test statistic takes on the value $P_{(4)} = 0 + 3 + 4 + 1 = 8$.

It is quite clear that large values of $P_{(r)}$ lead to the rejection of H_0 and in favor of H_1 in (3.1).

For a fixed level of significance α, the critical region will be $\{s, s + 1, \ldots, n_1\}$, where

$$\alpha = \Pr(P_r \geq s | H_0 : F_X = F_Y). \tag{3.3}$$

3.3   EXACT NULL DISTRIBUTION

In this section, we shall derive the null distribution of the precedence test statistic $P_{(r)}$ by two different approaches—the combinatorial approach and the order statistics approach.

Combinatorial Approach

Under $H_0 : F_X = F_Y$, the two distributions are identical and so the number of ways of choosing a sample of size n_1 for the X-sample from a total of $n_1 + n_2$ units is $\binom{n_1 + n_2}{n_1}$. Next, we note that the event $P_{(r)} = j$ may be viewed as the number of ways of choosing j from the first $j + r - 1$ failures for the X-sample and then $n_1 - j$ from the last $n_1 + n_2 - j - r$ failures for the remaining X-sample. Hence, we can write

$$\Pr(P_{(r)} = j | H_0 : F_X = F_Y) = \frac{\binom{j + r - 1}{j} \binom{n_1 + n_2 - j - r}{n_1 - j}}{\binom{n_1 + n_2}{n_2}}.$$

Order Statistics Approach

Conditional on $Y_{r:n_2} = y$, we have the probability of j failures from the X-sample preceding $Y_{r:n_2}$ as

$$\Pr(\text{no. of } X\text{-failures before } Y_{r:n_2} = j | Y_{r:n_2} = y)$$
$$= \binom{n_1}{j} [F_X(y)]^j [1 - F_X(y)]^{n_1 - j}.$$

Then, unconditioning with respect to $Y_{r:n_2}$ by using the density function of the rth order statistic from a sample of size n_2 [see David and Nagaraja (2003) and Arnold, Balakrishnan, and Nagaraja (1992)], we obtain

$$\Pr(P_{(r)} = j)$$

$$= \int_{-\infty}^{\infty} \Pr(\text{no. of } X\text{-failures before } Y_{r:n_2} = j | Y_{r:n_2} = y) f_{r:n_2}(y) dy$$

$$= \binom{n_1}{j} \int_{-\infty}^{\infty} [F_X(y)]^j [1 - F_X(y)]^{n_1-j}$$

$$\times \frac{n_2!}{(r-1)!(n_2-r)!} [F_Y(y)]^{r-1} [1 - F_Y(y)]^{n_2-r} f_Y(y) \ dy. \quad (3.4)$$

Under $H_0 : F_X = F_Y = F$, we have

$$\Pr(P_{(r)} = j | H_0 : F_X = F_Y)$$

$$= \binom{n_1}{j} \frac{n_2!}{(r-1)!(n_2-r)!}$$

$$\times \int_{-\infty}^{\infty} [F(y)]^{r+j-1} [1 - F(y)]^{n_1+n_2-r-j} f(y) \ dy$$

$$= \binom{n_1}{j} \frac{n_2!}{(r-1)!(n_2-r)!} \frac{(r+j-1)!(n_1+n_2-r-j)!}{(n_1+n_2)!}$$

$$= \frac{\binom{j+r-1}{j} \binom{n_1+n_2-j-r}{n_1-j}}{\binom{n_1+n_2}{n_2}}.$$

Therefore, for specified values of n_1, n_2, s and r, an expression for α in (3.3) is given by

$$\alpha = \frac{\sum_{j=s}^{n_1} \binom{j+r-1}{j} \binom{n_1+n_2-j-r}{n_1-j}}{\binom{n_1+n_2}{n_2}}$$

with the summation terminating as soon as any of the factorials involve negative arguments.

The critical value s and the exact level of significance α as close as possible to 5% and 10% for different choices of the sample sizes n_1 and n_2 and $r = 1(1)10$ are presented in Tables 3.1 and 3.2, respectively.

3.4 EXACT POWER FUNCTION UNDER LEHMANN ALTERNATIVE

In this section, we derive an explicit expression for the power function of the precedence test under the Lehmann alternative $H_1 : [F_X]^\gamma = F_Y$ for some

Table 3.1: Near 5% critical values (s) and exact levels of significance (α) for the precedence test statistic $P_{(r)}$

		\multicolumn{2}{c}{$r = 1$}		\multicolumn{2}{c}{$r = 2$}		\multicolumn{2}{c}{$r = 3$}		\multicolumn{2}{c}{$r = 4$}		\multicolumn{2}{c}{$r = 5$}	
n_1	n_2	s	α	s	α	s	α	s	α	s	α
10	10	4	0.0433	6	0.0286	7	0.0349	8	0.0349	9	0.0286
10	15	3	0.0522	4	0.0640	5	0.0618	6	0.0533	7	0.0416
15	15	4	0.0498	6	0.0400	7	0.0543	8	0.0641	10	0.0328
15	20	3	0.0695	5	0.0401	6	0.0461	7	0.0472	8	0.0449
20	20	4	0.0530	6	0.0457	7	0.0637	9	0.0412	10	0.0479
20	25	4	0.0325	5	0.0523	6	0.0635	8	0.0340	9	0.0357
25	25	4	0.0549	6	0.0491	8	0.0369	9	0.0477	10	0.0568
25	30	4	0.0371	5	0.0611	7	0.0383	8	0.0450	9	0.0492
30	30	4	0.0562	6	0.0514	8	0.0399	9	0.0521	10	0.0626
30	50	3	0.0494	4	0.0629	5	0.0649	6	0.0623	7	0.0575
		\multicolumn{2}{c}{$r = 6$}		\multicolumn{2}{c}{$r = 7$}		\multicolumn{2}{c}{$r = 8$}		\multicolumn{2}{c}{$r = 9$}		\multicolumn{2}{c}{$r = 10$}	
n_1	n_2	s	α	s	α	s	α	s	α	s	α
10	10	9	0.0704	10	0.0433	10	0.1053	10	0.2368	10	0.5000
10	15	8	0.0287	8	0.0576	9	0.0339	9	0.0654	10	0.0283
15	15	11	0.0328	11	0.0697	12	0.0641	13	0.0543	14	0.0400
15	20	9	0.0404	10	0.0345	10	0.0646	11	0.0521	12	0.0392
20	20	11	0.0527	12	0.0555	13	0.0564	14	0.0555	15	0.0527
20	25	10	0.0355	11	0.0340	11	0.0628	12	0.0572	13	0.0507
25	25	12	0.0359	13	0.0396	14	0.0423	15	0.0438	16	0.0444
25	30	10	0.0514	11	0.0520	12	0.0514	13	0.0496	14	0.0470
30	30	12	0.0420	13	0.0473	14	0.0517	15	0.0551	16	0.0577
30	50	8	0.0519	9	0.0460	10	0.0402	11	0.0347	11	0.0560

Table 3.2: Near 10% critical values (s) and exact levels of significance (α) for the precedence test statistic $P_{(r)}$

		$r = 1$		$r = 2$		$r = 3$		$r = 4$		$r = 5$	
n_1	n_2	s	α	s	α	s	α	s	α	s	α
10	10	3	0.1053	5	0.0704	6	0.0849	7	0.0894	8	0.0849
10	15	3	0.0522	4	0.0640	5	0.0618	5	0.1281	6	0.1059
15	15	3	0.1121	5	0.0843	6	0.1074	7	0.1225	9	0.0697
15	20	3	0.0695	4	0.0933	5	0.1007	6	0.1000	7	0.0947
20	20	3	0.1154	5	0.0909	6	0.1176	8	0.0776	9	0.0880
20	25	3	0.0803	4	0.1118	5	0.1256	7	0.0691	8	0.0709
25	25	3	0.1173	5	0.0947	6	0.1234	8	0.0853	9	0.0982
25	30	3	0.0877	4	0.1244	6	0.0760	7	0.0850	8	0.0900
30	30	3	0.1186	5	0.0973	6	0.1271	8	0.0903	9	0.1046
30	50	2	0.1377	4	0.0629	5	0.0649	5	0.1252	6	0.1121
		$r = 6$		$r = 7$		$r = 8$		$r = 9$		$r = 10$	
n_1	n_2	s	α	s	α	s	α	s	α	s	α
10	10	9	0.0704	9	0.1517	10	0.1053	10	0.2368	10	0.5000
10	15	7	0.0820	8	0.0576	8	0.1048	9	0.0654	9	0.1175
15	15	10	0.0716	11	0.0697	11	0.1318	12	0.1225	13	0.1074
15	20	8	0.0865	9	0.0763	9	0.1300	10	0.1109	11	0.0909
20	20	10	0.0954	11	0.1001	12	0.1025	13	0.1025	14	0.1001
20	25	8	0.1273	9	0.1218	10	0.1145	11	0.1059	12	0.0962
25	25	10	0.1083	11	0.1160	12	0.1218	14	0.0768	15	0.0782
25	30	9	0.0922	10	0.0922	11	0.0905	12	0.0875	13	0.0834
30	30	10	0.1163	12	0.0790	13	0.0851	14	0.0899	15	0.0936
30	50	7	0.0993	8	0.0871	9	0.0758	9	0.1159	10	0.0996

γ, which was first proposed by Lehmann (1953). For more details on the Lehmann alternative, see Davies (1971), Lehmann (1975), Hettmansperger and McKean (1998), and Gibbons and Chakraborti (2003).

We note that $H_1 : [F_X]^\gamma = F_Y$ is a subclass of the alternative $H_1 : F_X > F_Y$ when $\gamma > 1$. The power of a test is the probability of rejecting the null hypothesis when the alternative hypothesis is indeed true.

From Eq. (3.4), under $H_1 : [F_X]^\gamma = F_Y$, we have

$$
\Pr\left(P_{(r)} = j | H_1 : [F_X]^\gamma = F_Y\right)
$$

$$
= \binom{n_1}{j} \int_{-\infty}^{\infty} [F_X(y)]^j \, [1 - F_X(y)]^{n_1-j} \, \frac{n_2!}{(r-1)!(n_2-r)!}
$$

$$
\times [F_X(y)]^{\gamma(r-1)} \left\{ \sum_{k=0}^{n_2-r} (-1)^k \binom{n_2-r}{k} [F_X(y)]^{\gamma k} \right\}
$$

$$
\times \left\{ \gamma [F_X(y)]^{\gamma-1} f_X(y) \right\} \, dy.
$$

$$
= \frac{n_1! n_2! \gamma}{j!(n_1-j)!(r-1)!(n_2-r)!} \sum_{k=0}^{n_2-r} \left[(-1)^k \binom{n_2-r}{k} \right.
$$

$$
\left. \times \int_{-\infty}^{\infty} [F_X(y)]^{j+\gamma(r+k)-1} \, [1 - F_X(y)]^{n_1-j} \, f_X(y) \, dy \right].
$$

The integral in the formula can be transformed into a Beta integral by letting $u = F_X(y)$ and $du = f_X(y)dy$:

$$
\Pr\left(P_{(r)} = j | H_1 : [F_X]^\gamma = F_Y\right)
$$

$$
= \frac{n_1! n_2! \gamma}{j!(n_1-j)!(r-1)!(n_2-r)!}
$$

$$
\times \sum_{k=0}^{n_2-r} (-1)^k \binom{n_2-r}{k} \int_0^1 u^{j+\gamma(r+k)-1}(1-u)^{n_1-j} \, du
$$

$$
= \frac{n_1! n_2! \gamma}{j!(n_1-j)!(r-1)!(n_2-r)!}
$$

$$
\times \sum_{k=0}^{n_2-r} \left[(-1)^k \binom{n_2-r}{k} \frac{\Gamma(j+\gamma(r+k)) \, \Gamma(n_1-j+1)}{\Gamma(n_1+\gamma(r+k)+1)} \right.
$$

$$
\left. \frac{n_1! n_2! \gamma}{j!(r-1)!(n_2-r)!} \sum_{k=0}^{n_2-r} (-1)^k \binom{n_2-r}{k} \frac{\Gamma(j+\gamma(r+k))}{\Gamma(n_1+\gamma(r+k)+1)} \right].
$$

So, under the Lehmann alternative $H_1 : [F_X]^\gamma = F_Y$, the power function

Table 3.3: Power values of the precedence test $P_{(r)}$ under the Lehmann alternative for $n_1 = n_2 = 10, r = 2(1)5$, and $\gamma = 2(1)6$

	Power computed from Eq. (3.5)			
γ	$r = 2$	$r = 3$	$r = 4$	$r = 5$
1	0.02864	0.03489	0.03489	0.02864
2	0.23970	0.24581	0.21679	0.16106
3	0.49048	0.48087	0.41996	0.31528
4	0.67299	0.65212	0.57708	0.44601
5	0.78963	0.76460	0.68834	0.54833
6	0.86216	0.83731	0.76610	0.62715

of the precedence test is given by

$$\Pr\left(P_{(r)} \geq s \mid H_1 : [F_X]^\gamma = F_Y\right)$$
$$= \sum_{j=s}^{n_1} \left[\frac{n_1! n_2! \gamma}{j!(r-1)!(n_2-r)!} \right.$$
$$\left. \times \sum_{k=0}^{n_2-r} (-1)^k \binom{n_2-r}{k} \frac{\Gamma(j + \gamma(r+k))}{\Gamma(n_1 + \gamma(r+k) + 1)} \right]. \qquad (3.5)$$

For $n_1 = n_2 = 10, r = 2(1)5$ and $\gamma = 2(1)6$, the power values of the precedence test under the Lehmann alternative computed from Eq.(3.5) are presented in Table 3.3.

3.5 MONTE CARLO SIMULATION UNDER LOCATION-SHIFT ALTERNATIVE

In order to assess the power properties of the precedence test, we consider the precedence test under the location-shift alternative $H_1 : F_X(x) = F_Y(x + \theta)$ for some $\theta > 0$, where θ is a shift in location, along the lines of Nelson (1993). The power values of the precedence test with $r = 1(1)6$ were estimated through Monte Carlo simulations when $\theta = 0.5$ and $\theta = 1.0$. The following lifetime distributions were considered in order to demonstrate the power performance of the precedence test under this location-shift alternative:

1. Standard normal distribution

2. Standard exponential distribution

3. Gamma distribution with shape parameter a and standardized by mean a and standard deviation $\sqrt{a}$

4. Lognormal distribution with shape parameter σ and standardized by mean $e^{\sigma^2/2}$ and standard deviation $\sqrt{e^{\sigma^2}(e^{\sigma^2}-1)}$

A brief description of these distributions and their properties has been provided in Section 2.5. For different choices of sample sizes, we generated 100,000 sets of data, utilizing the IMSL subroutines RNNOR, RNEXP, RNGAM, and RNLNL, in order to obtain the estimated rejection rates.

In Tables 3.4–3.8, we have presented the power values of the precedence tests with $r = 1(1)10$ for different choices of sample sizes for the underlying standard normal, standard exponential, standardized gamma, and standardized lognormal distributions, with location-shift being equal to 0.5 and 1.0. For comparison proposes, the corresponding critical values and the exact levels of significance are also presented. The corresponding power values of the Wilcoxon's rank-sum test W_R (based on complete samples) are also presented for the sake of comparison. A brief description of the Wilcoxon's rank-sum statistic has been provided in Section 2.6.

3.6   EVALUATION AND COMPARATIVE REMARKS

From Tables 3.4–3.8, we see that the power of the precedence test increases with increasing sample sizes as well as with increasing location-shift. We also observe that, in most cases, more information from the data leads to better decisions. However, since the precedence test statistic $P_{(r)}$ is the number of failures from the X-sample before the rth failure from the Y-sample, the amount of information relative to the sample size is important. For small sample sizes such as 10, 15, and 20, the precedence test with $r = 1$ gives higher power than the precedence tests with larger values of r under some right-skewed distributions such as the exponential distribution, gamma distribution with small values of shape parameter a, and lognormal distribution with large values of shape parameter σ. Roughly speaking, this means that

Table 3.4: Power of precedence tests $(P_{(r)})$ and Wilcoxon's rank-sum test (W_R) when $n_1 = n_2 = 10$

Test	r	Critical value	Exact l.o.s.	Location shift	Distribution N(0,1)	Exp(1)	LN(0.1)	LN(0.5)
$P_{(r)}$	1	4	0.0433	0.5	0.1736	0.7197	0.1936	0.3944
				1.0	0.4208	0.9779	0.4817	0.8709
	2	6	0.0286	0.5	0.1496	0.3916	0.1603	0.2693
				1.0	0.4243	0.8376	0.4571	0.7253
	3	7	0.0349	0.5	0.1779	0.3073	0.1828	0.2571
				1.0	0.4813	0.7239	0.4964	0.6676
	4	8	0.0349	0.5	0.1771	0.2251	0.1777	0.2149
				1.0	0.4799	0.5767	0.4783	0.5656
	5	9	0.0286	0.5	0.1473	0.1453	0.1444	0.1527
				1.0	0.4245	0.3960	0.4038	0.4144
	6	9	0.0704	0.5	0.2659	0.2385	0.2567	0.2531
				1.0	0.5832	0.4991	0.5545	0.5314
	7	10	0.0433	0.5	0.1721	0.1280	0.1614	0.1405
				1.0	0.4240	0.2757	0.3791	0.3074
	8	10	0.1053	0.5	0.3032	0.2253	0.2841	0.2427
				1.0	0.5777	0.3902	0.5265	0.4243
	9	10	0.2368	0.5	0.4880	0.3828	0.4598	0.4002
				1.0	0.7394	0.5407	0.6910	0.5720
	10	10	0.5	0.5	0.7290	0.6267	0.7020	0.6395
				1.0	0.8887	0.7415	0.8546	0.7598
W_R		82	0.0446	0.5	0.2536	0.4332	0.2605	0.3508
				1.0	0.6480	0.8163	0.6554	0.7819

Table 3.4: (Continued)

Test	r	Critical value	Exact l.o.s.	Location shift	Distribution Gamma(2)	Gamma(5)	Gamma(10)
$P_{(r)}$	1	4	0.0433	0.5	0.4139	0.2644	0.2249
				1.0	0.8806	0.6731	0.5744
	2	6	0.0286	0.5	0.2519	0.1936	0.1731
				1.0	0.6877	0.5590	0.5046
	3	7	0.0349	0.5	0.2330	0.2013	0.1893
				1.0	0.6157	0.5462	0.5181
	4	8	0.0349	0.5	0.1935	0.1796	0.1755
				1.0	0.5103	0.4847	0.4764
	5	9	0.0286	0.5	0.1378	0.1371	0.1388
				1.0	0.3730	0.3760	0.3853
	6	9	0.0704	0.5	0.2349	0.2400	0.2436
				1.0	0.4962	0.5131	0.5280
	7	10	0.0433	0.5	0.1325	0.1411	0.1475
				1.0	0.2912	0.3205	0.3390
	8	10	0.1053	0.5	0.2345	0.2529	0.2616
				1.0	0.4146	0.4536	0.4794
	9	10	0.2368	0.5	0.3959	0.4218	0.4342
				1.0	0.5719	0.6148	0.6425
	10	10	0.5000	0.5	0.6422	0.6653	0.6777
				1.0	0.7657	0.8016	0.8228
W_R		82	0.0446	0.5	0.3337	0.2822	0.2663
				1.0	0.7513	0.6964	0.6719

Table 3.5: Power of precedence tests $(P_{(r)})$ and Wilcoxon's rank-sum test (W_R) when $n_1 = 10, n_2 = 15$

Test	r	Critical value	Exact l.o.s.	Location shift	Distribution			
					N(0,1)	Exp(1)	LN(0.1)	LN(0.5)
$P_{(r)}$	1	3	0.0522	0.5	0.1971	0.8733	0.2290	0.5146
				1.0	0.4648	0.9952	0.5431	0.9395
	2	4	0.0640	0.5	0.2607	0.7638	0.2886	0.5243
				1.0	0.5942	0.9821	0.6500	0.9309
	3	5	0.0618	0.5	0.2741	0.6340	0.2956	0.4775
				1.0	0.6335	0.9480	0.6704	0.8933
	4	6	0.0533	0.5	0.2574	0.4933	0.2745	0.4037
				1.0	0.6260	0.8779	0.6488	0.8308
	5	7	0.0416	0.5	0.2246	0.3567	0.2324	0.3137
				1.0	0.5864	0.7587	0.5958	0.7311
	6	8	0.0287	0.5	0.1750	0.2323	0.1756	0.2189
				1.0	0.5096	0.5854	0.5041	0.5844
	7	8	0.0576	0.5	0.2704	0.3091	0.2687	0.3052
				1.0	0.6357	0.6524	0.6217	0.6658
	8	9	0.0339	0.5	0.1839	0.1789	0.1783	0.1847
				1.0	0.5056	0.4375	0.4769	0.4679
	9	9	0.0654	0.5	0.2759	0.2499	0.2646	0.2603
				1.0	0.6195	0.5137	0.5858	0.5509
	10	10	0.0283	0.5	0.1430	0.1084	0.1313	0.1160
				1.0	0.3927	0.2526	0.3510	0.2840
W_R		99	0.0455	0.5	0.2934	0.5020	0.2998	0.4103
				1.0	0.7306	0.8637	0.7331	0.8399

Table 3.5: (Continued)

Test	r	Critical value	Exact l.o.s.	Location shift	Distribution Gamma(2)	Gamma(5)	Gamma(10)
$P_{(r)}$	1	3	0.0522	0.5	0.2783	0.3384	0.2783
				1.0	0.6656	0.7798	0.6656
	2	4	0.0640	0.5	0.3335	0.3780	0.3335
				1.0	0.7327	0.8022	0.7327
	3	5	0.0618	0.5	0.3266	0.3575	0.3266
				1.0	0.7262	0.7736	0.7262
	4	6	0.0533	0.5	0.2926	0.3102	0.2926
				1.0	0.6840	0.7179	0.6840
	5	7	0.0416	0.5	0.2414	0.2500	0.2414
				1.0	0.6122	0.6307	0.6122
	6	8	0.0287	0.5	0.1781	0.1811	0.1781
				1.0	0.5048	0.5066	0.5048
	7	8	0.0576	0.5	0.2679	0.2690	0.2679
				1.0	0.6138	0.6115	0.6138
	8	9	0.0339	0.5	0.1720	0.1703	0.1720
				1.0	0.4533	0.4443	0.4533
	9	9	0.0654	0.5	0.2549	0.2505	0.2549
				1.0	0.5568	0.5410	0.5568
	10	10	0.0283	0.5	0.1229	0.1167	0.1229
				1.0	0.3146	0.2953	0.3146
W_R		99	0.0455	0.5	0.3142	0.3323	0.3142
				1.0	0.7495	0.7685	0.7495

Table 3.6: Power of precedence tests $(P_{(r)})$ and Wilcoxon's rank-sum test (W_R) when $n_1 = n_2 = 15$

Test	r	Critical value	Exact l.o.s.	Location shift	Distribution N(0,1)	Exp(1)	LN(0.1)	LN(0.5)
$P_{(r)}$	1	4	0.0498	0.5	0.2117	0.9360	0.2443	0.5676
				1.0	0.5025	0.9996	0.5867	0.9729
	2	6	0.0400	0.5	0.2206	0.7682	0.2464	0.4908
				1.0	0.5719	0.9927	0.6379	0.9468
	3	7	0.0543	0.5	0.2847	0.6963	0.3074	0.5158
				1.0	0.6719	0.9819	0.7212	0.9421
	4	8	0.0641	0.5	0.3254	0.6230	0.3434	0.5095
				1.0	0.7268	0.9621	0.7606	0.9276
	5	10	0.0328	0.5	0.2264	0.3687	0.2321	0.3226
				1.0	0.6251	0.8318	0.6441	0.7975
	6	11	0.0328	0.5	0.2272	0.3014	0.2262	0.2834
				1.0	0.6248	0.7393	0.6293	0.7312
	7	11	0.0697	0.5	0.3514	0.4024	0.3451	0.3944
				1.0	0.7525	0.8016	0.7488	0.8073
	8	12	0.0641	0.5	0.3292	0.3258	0.3174	0.3356
				1.0	0.7265	0.6990	0.7103	0.7258
	9	13	0.0543	0.5	0.2858	0.2494	0.2694	0.2625
				1.0	0.6726	0.5657	0.6406	0.6075
	10	14	0.0400	0.5	0.2226	0.1668	0.2024	0.1792
				1.0	0.5721	0.3968	0.5256	0.4437
W_R		192	0.0488	0.5	0.3600	0.5938	0.3637	0.4927
				1.0	0.8240	0.9406	0.8317	0.9216

Table 3.6: (Continued)

Test	r	Critical value	Exact l.o.s.	Location shift	Distribution Gamma(2)	Gamma(5)	Gamma(10)
$P_{(r)}$	1	4	0.0498	0.5	0.6390	0.3713	0.2991
				1.0	0.9844	0.8362	0.7214
	2	6	0.0400	0.5	0.4893	0.3311	0.2852
				1.0	0.9425	0.8081	0.7286
	3	7	0.0543	0.5	0.4900	0.3768	0.3414
				1.0	0.9275	0.8336	0.7794
	4	8	0.0641	0.5	0.4755	0.3938	0.3683
				1.0	0.9036	0.8326	0.7976
	5	10	0.0328	0.5	0.2899	0.2518	0.2412
				1.0	0.7473	0.6840	0.6622
	6	11	0.0328	0.5	0.2533	0.2331	0.2285
				1.0	0.6756	0.6389	0.6317
	7	11	0.0697	0.5	0.3630	0.3463	0.3433
				1.0	0.7636	0.7469	0.7463
	8	12	0.0641	0.5	0.3079	0.3046	0.3089
				1.0	0.6790	0.6820	0.6919
	9	13	0.0543	0.5	0.2426	0.2501	0.2588
				1.0	0.5646	0.5875	0.6079
	10	14	0.0400	0.5	0.1706	0.1805	0.1881
				1.0	0.4138	0.4533	0.4788
W_R		192	0.0488	0.5	0.4707	0.3987	0.3797
				1.0	0.9042	0.8640	0.8481

Table 3.7: Power of precedence tests $(P_{(r)})$ and Wilcoxon's rank-sum test (W_R) when $n_1 = 15, n_2 = 20$

Test	r	Critical value	Exact l.o.s.	Location shift	Distribution N(0,1)	Exp(1)	LN(0.1)	LN(0.5)
$P_{(r)}$	1	3	0.0695	0.5	0.2610	0.9803	0.3004	0.6898
				1.0	0.5661	0.9999	0.6559	0.9902
	2	5	0.0401	0.5	0.2280	0.8712	0.2587	0.5644
				1.0	0.5893	0.9978	0.6678	0.9710
	3	6	0.0461	0.5	0.2698	0.7941	0.2992	0.5654
				1.0	0.6691	0.9937	0.7284	0.9648
	4	7	0.0472	0.5	0.2868	0.7041	0.3130	0.5347
				1.0	0.7056	0.9829	0.7494	0.9499
	5	8	0.0449	0.5	0.2889	0.6064	0.3109	0.4867
				1.0	0.7182	0.9603	0.7495	0.9250
	6	9	0.0404	0.5	0.2796	0.5049	0.2932	0.4289
				1.0	0.7120	0.9183	0.7343	0.8864
	7	10	0.0345	0.5	0.2572	0.4012	0.2651	0.3622
				1.0	0.6912	0.8495	0.6987	0.8285
	8	10	0.0646	0.5	0.3644	0.4854	0.3708	0.4608
				1.0	0.7913	0.8822	0.7932	0.8754
	9	11	0.0521	0.5	0.3220	0.3813	0.3212	0.3772
				1.0	0.7528	0.7938	0.7438	0.8020
	10	12	0.0392	0.5	0.2683	0.2580	0.2626	0.2890
				1.0	0.6943	0.6463	0.6717	0.6950
W_R		220	0.0497	0.5	0.3990	0.6438	0.4078	0.5445
				1.0	0.8710	0.9560	0.8735	0.9426

Table 3.7: (Continued)

Test	r	Critical value	Exact l.o.s.	Location shift	Distribution Gamma(2)	Gamma(5)	Gamma(10)
$P_{(r)}$	1	3	0.0695	0.5	0.7777	0.4727	0.3804
				1.0	0.9951	0.9071	0.7998
	2	5	0.0401	0.5	0.5814	0.3737	0.3143
				1.0	0.9723	0.8599	0.7747
	3	6	0.0461	0.5	0.5495	0.3945	0.3459
				1.0	0.9579	0.8662	0.8071
	4	7	0.0472	0.5	0.5033	0.3873	0.3494
				1.0	0.9351	0.8554	0.8081
	5	8	0.0449	0.5	0.4474	0.3637	0.3359
				1.0	0.9012	0.8274	0.7913
	6	9	0.0404	0.5	0.3878	0.3295	0.3092
				1.0	0.8519	0.7881	0.7625
	7	10	0.0345	0.5	0.3233	0.2858	0.2745
				1.0	0.7833	0.7317	0.7166
	8	10	0.0646	0.5	0.4161	0.3869	0.3768
				1.0	0.8407	0.8085	0.8004
	9	11	0.0521	0.5	0.3382	0.3238	0.3208
				1.0	0.7599	0.7397	0.7409
	10	12	0.0392	0.5	0.2798	0.2552	0.2564
				1.0	0.6682	0.6456	0.6543
W_R		220	0.0497	0.5	0.5180	0.4451	0.4229
				1.0	0.9298	0.8994	0.8845

Table 3.8: Power of precedence tests $(P_{(r)})$ and Wilcoxon's rank-sum test (W_R) when $n_1 = n_2 = 20$

Test	r	Critical value	Exact l.o.s.	Location shift	Distribution			
					N(0,1)	Exp(1)	LN(0.1)	LN(0.5)
$P_{(r)}$	1	4	0.0530	0.5	0.2386	0.9887	0.2804	0.6855
				1.0	0.5553	1.0000	0.6543	0.9943
	2	6	0.0457	0.5	0.2662	0.9355	0.3038	0.6453
				1.0	0.6529	0.9998	0.7338	0.9901
	3	7	0.0637	0.5	0.3480	0.9008	0.3851	0.6841
				1.0	0.7625	0.9994	0.8199	0.9910
	4	9	0.0412	0.5	0.2947	0.7499	0.3219	0.5614
				1.0	0.7352	0.9932	0.7877	0.9716
	5	10	0.0479	0.5	0.3317	0.6904	0.3525	0.5555
				1.0	0.7787	0.9864	0.8163	0.9648
	6	11	0.0527	0.5	0.3565	0.6291	0.3706	0.5369
				1.0	0.8055	0.9733	0.8295	0.9540
	7	12	0.0555	0.5	0.3686	0.5670	0.3782	0.5107
				1.0	0.8186	0.9532	0.8337	0.9367
	8	13	0.0564	0.5	0.3736	0.5045	0.3780	0.4777
				1.0	0.8241	0.9214	0.8297	0.9135
	9	14	0.0555	0.5	0.3708	0.4410	0.3665	0.4347
				1.0	0.8190	0.8732	0.8173	0.8792
	10	15	0.0527	0.5	0.3558	0.3784	0.3455	0.3846
				1.0	0.8066	0.8064	0.7939	0.8283
W_R		348	0.0482	0.5	0.4403	0.7011	0.4462	0.5941
				1.0	0.9136	0.9802	0.9179	0.9716

Table 3.8: (Continued)

Test	r	Critical value	Exact l.o.s.	Location shift	Distribution Gamma(2)	Gamma(5)	Gamma(10)
$P_{(r)}$	1	4	0.0530	0.5	0.7882	0.4525	0.3561
				1.0	0.9982	0.9145	0.8027
	2	6	0.0457	0.5	0.6709	0.4367	0.3646
				1.0	0.9914	0.9126	0.8373
	3	7	0.0637	0.5	0.6795	0.4998	0.4390
				1.0	0.9888	0.9341	0.8877
	4	9	0.0412	0.5	0.5308	0.4014	0.3575
				1.0	0.9616	0.8876	0.8422
	5	10	0.0479	0.5	0.5165	0.4189	0.3813
				1.0	0.9488	0.8876	0.8539
	6	11	0.0527	0.5	0.4967	0.4208	0.3920
				1.0	0.9319	0.8794	0.8570
	7	12	0.0555	0.5	0.4663	0.4129	0.3930
				1.0	0.9083	0.8649	0.8498
	8	13	0.0564	0.5	0.4324	0.3968	0.3850
				1.0	0.8779	0.8447	0.8353
	9	14	0.0555	0.5	0.3933	0.3736	0.3675
				1.0	0.8373	0.8154	0.8119
	10	15	0.0527	0.5	0.3489	0.3422	0.3418
				1.0	0.7822	0.7753	0.7771
W_R		348	0.0482	0.5	0.5701	0.4905	0.4646
				1.0	0.9623	0.9385	0.9275

the number of failures from the X-sample before the first failure in the Y-sample of size 10 provides enough information to make a good decision about the difference in locations of the two distributions.

We would ideally like the precedence test with $r = 3$ to be at least as powerful as the test with $r = 1$. Clearly, this is not true (as seen in Tables 3.4–3.8) and we therefore suspect that the information given by $r = 3$ is getting masked. For some distributions such as the gamma distribution with shape parameter $a = 2.0$, the power is negatively associated with r, meaning that the power decreases as r increases for fixed sample sizes. This may be due to the fact that since the precedence test statistic $P_{(r)}$ takes into account only the total number of failures from the X-sample before the rth failure from the Y-sample, there is no masking when $r = 1$, but there is masking effect when $r = 3$.

When we compare the power values of the precedence tests with those of Wilcoxon's rank-sum test, we find that the Wilcoxon's rank-sum test performs better than the precedence tests if the underlying distributions are close to symmetry, such as the normal distribution, gamma distribution with large values of shape parameter a, and lognormal distribution with small values of shape parameter σ. However, under some right-skewed distributions such as the exponential distribution, gamma distribution with shape parameter $a = 2.0$, and lognormal distribution with shape parameter $\sigma = 0.5$, the precedence tests have higher power values than the Wilcoxon's rank-sum test for small values of r. It is evident that the more right-skewed the underlying distribution is, the more powerful the precedence test is. For example, in Table 3.8, when $n_1 = n_2 = 20$, the location-shift equals 0.5 and the underlying distribution is exponential distribution, the simulated power of the precedence test with $r = 4$ is 0.7499 (exact level of significance is 0.0412), while the power of Wilcoxon's rank-sum test is 0.7011 (exact level of significance is 0.0482). Note that the precedence test gives better power performance in this case even though it is based on only the first 4 Y-failures.

Through this simulation study, we have observed that the precedence test gives good power properties when the underlying distributions are right-skewed as many lifetime distributions usually are. However, it suffers from the masking effect and for this reason some variations of the precedence test will be considered in subsequent chapters.

3.7   PROPERTIES OF PRECEDENCE AND RELATED TESTS

As can be seen from the extensive bibliography, due to the simplistic nature of precedence tests, several authors have discussed various properties as well as precedence-type tests for different inferential problems. In this section, we provide a description of these developments.

3.7.1   Powerful Precedence Tests

It appears that one of the earliest investigations of the use of precedence testing as a quick and efficient nonparametric procedure in life-test situations was due to Epstein (1955). He specifically considered the case when the two populations are normal and conducted an empirical investigation of the properties of a test based on the number of exceedances (i.e., the number of X-values that exceed the rth Y-value) for detecting a difference in the location of the two normal populations. It was, therefore, quite natural for Nelson (1963) to propose the precedence test statistic as a simple and efficient nonparametric life-test procedure that would allow for decisions to be taken (with regard to the quality or reliability) after relatively few lifetimes have been observed.

Eilbott and Nadler (1965) discussed the power properties of the precedence test by considering the case when both populations are exponential. In this situation, under the assumption that Type-II right censored samples are available from the two samples, an exact uniformly most powerful (UMP) test is available based on a central F-distribution for testing the equality of the two mean lifetimes (see Section 3.8). Eilbott and Nadler (1965), after deriving the exact power function of the precedence test, compared it with that of the UMP test. They then considered the asymptotic power function of the precedence test when the sample sizes get indefinitely large with the proportion of failures from each sample remaining fixed, and showed that it is precisely the same as the power function of the UMP test; this reveals that the precedence test makes highly efficient use of the information in the samples at least in the exponential case.

Shorack (1967), in continuing on the lines of Eilbott and Nadler (1965), pointed out that the expression of their power function of the precedence test derived under exponentiality will continue to hold under the Lehmann alternative $F_X = 1 - (1 - F_Y)^\gamma$, with $\gamma > 1$. This, of course, implies that the precedence test will be robust within this family of distributions (that includes exponential). This is in direct contrast to the parametric test procedures based on exponentiality, like the UMP F-test mentioned above, which are known to perform quite poorly when the underlying distributions are in fact Weibull with shape parameter different from 1; see, for example, Zelen and Dannemiller (1961).

Lin and Sukhatme (1992) considered the exact power function of the precedence test under the Lehmann alternative $F_X = 1 - (1 - F_Y)^\gamma$ $(\gamma > 1)$ with k and r failures from the X- and Y-samples, respectively, and discussed the optimal choice of k and r for a specified value of γ. Thus, for given values of n_1, n_2, α (the level of significance), and γ (the shape parameter under the Lehmann alternative), the optimal values of k and r that maximize the power give rise to the best precedence test. van der Laan and Chakraborti (1999, 2001) subsequently showed that the best precedence test of Lin and Sukhatme (1992) can also be used to determine the best precedence tests under three other Lehmann-type alternatives, viz., $F_Y = F_X^\gamma$ $(\gamma > 1)$, $F_Y = F_X^\gamma$ $(\gamma < 1)$, and $F_Y = 1 - (1 - F_X)^\gamma$ $(\gamma > 1)$. Continuing on these lines, Slud (1992) considered an extension of precedence test based on a comparison of rth Kaplan-Meier quantile from one sample to the sth Kaplan-Meier quantile from the other sample and discussed the optimal power against the local proportional-hazards alternatives.

3.7.2 Median Tests

In the course of screening chemical compounds for possible effects against ionizing radiation in mice, Kimball, Burnett, and Doherty (1957) considered the following sampling procedure. Two groups of mice were exposed to radiation, one having been previously injected with the chemical compound. The mice were observed until the death of the median mouse in the control group, at which time the number of dead mice in the experimental group were counted, which was then used in order to decide to accept or reject the compound from further consideration. If the times-to-death of the mice in the experimental group and the control group are treated as X- and Y-samples, respectively, then the test statistic used by Kimball, Burnett, and Doherty (1957) is nothing but the precedence test statistic $P_{(r)}$ with $r = (n_2 + 1)/2$. This *control median test* was proposed earlier by Mathisen (1943). Though Bowker (1944) pointed out that this test may not be consistent in certain situations (for example, when F_X and F_Y are identical in the neighborhood of their medians), it will indeed be consistent for location-shift alternatives.

Gart (1963) discussed some properties of this control median test and showed that it is asymptotically as efficient as the joint median test of Mood (1954). After recognizing that the control median test is indeed a precedence test with $r = (n_2 + 1)/2$ and that it only uses the sum of the frequencies M_1, $M_2, \cdots, M_r$, one can also exploit some other functions of these M_i's, as done in subsequent chapters. Some attempts had also been made by dividing the samples at fractiles and quantiles rather than at the median; see, for example, Massey (1951) and Chakravarti, Leone, and Alanen (1961).

As Gastwirth (1968) aptly pointed out, the control median test described above not only reduces the number of observations needed in a life-test but also has the advantage that it reduces the expected duration of the experiment if $\theta > 0$, where $F_Y(x) = F_X(x - \theta)$. Though such a control median test (or a precedence test) could be used as a two-sided test to test for the alternative $\theta \neq 0$, the expected duration of the experiment would unfortunately increase when $\theta < 0$. For this reason, Gastwirth (1968) proposed and discussed the *first median test*, which is based on the smaller median and the number of deaths from the other sample that precede this first-median value. He has further shown that in large sample sizes, when curtailed sampling is applied, a decision will be reached earlier by the control median test or the first-median test (as compared to the standard median test).

3.7.3 Precedence-type Tests for Complete and Censored Data

As mentioned in Section 3.7.2, the control median test of Mathisen (1943) is a special case of the precedence test when $r = (n_2+1)/2$. Similarly, Rosenbaum (1954) discussed the special case of the precedence test when $r = 1$. Subsequently, Gumbel and von Schelling (1950), Harris (1952), Epstein (1954), and Sarkadi (1957) all discussed various properties of these tests. Katzenbeisser (1989) and Liu (1992) discussed the power of precedence tests under location alternatives for exponential, logistic, and uniform distributions. Sukhatme (1992) studied the power of precedence tests under the Lehmann alternative and, in particular, obtained the exact power of Mathisen's and Rosenbaum's tests. van der Laan (1970) also derived the exact power of precedence tests under the uniform shift alternative, the exponential shift alternative, and the Lehmann alternative. While Young (1973) discussed a normal approximation to the power function of a precedence test, Chakraborti and Mukerjee

(1989) discussed the asymptotic distribution of the precedence statistic and showed it to be normal under some conditions.

Some other forms of statistics (as functions of precedences) have also been discussed in the literature. For example, with M_1 and M_{n_2+1} denoting the number of X-observations less than $Y_{1:n_2}$ and greater than $Y_{n_2:n_2}$, and similarly M_1^* and $M_{n_1+1}^*$ denoting the number of Y-observations less than $X_{1:n_1}$ and greater than $X_{n_1:n_1}$, respectively, Haga (1959) and Hájèk and Sidák (1967) discussed test statistics of the form $M_{n_2+1} + M_1^* - (M_{n_1+1}^* + M_1)$ and $\min(M_{n_2+1}, M_1^*) - \min(M_{n_1+1}^*, M_1)$.

A sequential version of the precedence test, based on the instantaneous difference between the number of failures from the new design and the design in use, has been discussed by Little (1974) for making an early termination of the test. Hackl and Katzenbeisser (1984), Chakraborti and Mukerjee (1989), and Schlittgen (1979) have also discussed precedence-type procedures for some other inferential problems. Two excellent survey articles by Chakraborti and van der Laan (1996, 1997) provide a detailed overview of various precedence tests and precedence-type tests for complete and censored data and analyze their features and properties.

Chakraborti, van der Laan, and van de Wiel (2004) have utilized a precedence statistic to construct a nonparametric control chart based on a test (future) sample quantile such as the median in comparison to two selected quantiles from a reference sample available from an in-control process.

3.7.4 Exceedance Statistics and Placement Statistics

A statistic that is closely related to the precedence statistic is the so-called *exceedance statistic*. To fix the idea, as before, let M_1 denote the number of X-failures before $Y_{1:n_2}$, M_i denote the number of X-failures between $Y_{i-1:n_2}$ and $Y_{i:n_2}$, and M_{n_2+1} denote the number of X-failures after $Y_{n_2:n_2}$. The precedence statistic $P_{(r)} = \sum\limits_{i=1}^{r} M_i$ is simply the number of X-failures that precede the rth Y-failure. The exceedance statistic $P'_{(r)} = \sum\limits_{i=r+1}^{n_2+1} M_i$ is similarly the number of X-failures that exceed the rth Y-failure. It is clear that

$$P_{(r)} + P'_{(r)} = \sum_{i=1}^{n_2+1} M_i = n_1,$$

for any $r = 1, 2, \cdots, n_2$, and therefore the precedence and the exceedance tests are equivalent. Katzenbeisser (1985, 1989) has discussed some properties of

Table 3.9: Times to insulating fluid breakdown data from Nelson (1982) for Samples 2 and 3

X-sample (Sample 3)	0.49	0.64	0.82	0.93	1.08	*	*	*	*	*
Y-sample (Sample 2)	0.00	0.18	0.55	0.66	0.71	1.30	1.63	*	*	*

exceedance tests.

It is important to mention here that the precedence statistic $P_{(r)}$, which represents the number of X-failures that precede $Y_{r:n_2}$, has also been referred to as the *placement* of $Y_{r:n_2}$ among the observations in the X-sample. Fligner and Wolfe (1976) and Orban and Wolfe (1982) have discussed various properties of these placements, including the first two moments under $H_0 : F_X = F_Y$.

The precedence or exceedance probability distributions, under the situation when both samples have the same distributions, have been used by Chakraborti and van der Laan (2000) to construct nonparametric prediction intervals for the number of Y-observations that exceed a specified order statistic from the X-sample.

3.8   ILLUSTRATIVE EXAMPLES

Nelson (1982, p. 462, Table 4.1) reported data on times to breakdown in minutes of an insulating fluid subjected to high voltage stress. The times are divided into six groups in the observed order. We shall use these data to construct the following two illustrative examples.

Example 3.3. For the first example, let us take our X- and Y-samples to be Samples 3 and 2 in Nelson (1982, p. 462), respectively. The observations are presented in Table 3.9.

For this example, if we had observed up to the 7th breakdown from the Y-sample (viz., 1.63), then we would have observed only the first five observations from the X-sample. In this case, we have the total number of items on test for the X- and Y-samples as $n_1 = n_2 = 10$ and the number of observations as $r_1 = 5$ and $r_2 = 7$, respectively.

Parametric Procedure. In analyzing these data, Nelson (1982) carried out a parametric test by assuming exponential distributions for the two samples with means θ_X and θ_Y, respectively. He then constructed a two-sided confidence interval for the ratio θ_X/θ_Y. If 1 is not included in the $100(1-\alpha)\%$ confidence interval for the ratio θ_X/θ_Y, it is concluded that the two exponential means differ significantly at $\alpha\%$ level of significance. Similarly, a one-sided $100(1-\alpha)\%$ confidence interval for the ratio θ_X/θ_Y can be used to test the hypothesis

$$H_0 : \theta_X = \theta_Y \text{ against } H_1 : \theta_X < \theta_Y. \tag{3.6}$$

If 1 is not included in the one-sided $100(1-\alpha)\%$ confidence interval for the ratio θ_X/θ_Y, then we could conclude that there is enough evidence in support of H_1 in (3.6). Note that testing the hypotheses in (3.6) is equivalent to testing the hypotheses in (3.1) under the exponential model.

Now, under the assumption of exponentiality for the two Type-II censored samples, we find the maximum likelihood estimates (MLEs) of θ_X and θ_Y to be (see Section 2.3.2)

$$\hat{\theta}_X = \frac{1}{r_1}\left[\sum_{i=1}^{r_1} X_{i:10} + (n_1 - r_1)X_{r_1:10}\right] = 1.872$$

$$\text{and} \quad \hat{\theta}_Y = \frac{1}{r_2}\left[\sum_{i=1}^{r_2} Y_{i:10} + (n_2 - r_2)Y_{r_2:10}\right] = 1.417,$$

respectively. Since the sampling distributions of $2r_1\hat{\theta}_X/\theta_X$ and $2r_2\hat{\theta}_Y/\theta_Y$ are chi-square with $2r_1$ and $2r_2$ degrees of freedom (see Theorem 2.7), respectively, the upper 95% confidence limit for the ratio θ_X/θ_Y is given by

$$\frac{\hat{\theta}_X}{\hat{\theta}_Y} \times \frac{1}{F_{2r_1,2r_2}(0.05)} = 3.437,$$

where $F_{\nu_1,\nu_2}(p)$ denotes the pth percentage point of the F-distribution with (ν_1, ν_2) degrees of freedom. Since the one-sided 95% confidence interval $(0, 3.437)$ for θ_X/θ_Y includes the value 1, we conclude that the observed data provide enough evidence toward H_0 in (3.6). In fact, with the value of $\hat{\theta}_X/\hat{\theta}_Y = 1.321$, we find the p-value to be 0.692.

Nonparametric Procedure. Considering the same data, we can carry out a nonparametric test for the hypotheses in (3.1) through a precedence test.

Table 3.10: Times to insulating fluid breakdown data from Nelson (1982) for Samples 3 and 6

X-sample (Sample 3)	0.49	0.64	0.82	0.93	1.08	1.99	2.06	2.15	2.57	*
Y-sample (Sample 6)	1.34	1.49	1.56	2.10	2.12	3.83	*	*	*	*

In this case, we specifically have $n_1 = n_2 = 10$, $r = 7$, $m_1 = m_2 = 0$, $m_3 = m_4 = 1$, $m_5 = m_6 = 0$, and $m_7 = 3$. Then, the precedence test statistic is

$$P_{(7)} = \sum_{i=1}^{7} m_i = 5.$$

From Table 3.1, the near 5% critical value for $n_1 = n_2 = 10$ and $r = 7$ is 10 (exact level of significance $= 0.0433$). Since in this example, $P_{(7)} < 10$, we do not reject the null hypothesis that the two samples are identically distributed at 5% level of significance. The p-value for this test is 0.498, which agrees with the conclusion of the parametric procedure above.

Example 3.4. For the second example, let us consider X- and Y-samples to be Samples 3 and 6 in Nelson (1982), respectively. The observations are presented in Table 3.10.

For these data, if we had observed only up to the 6th breakdown from the Y-sample, then the first nine observations from the X-sample would have been observed.

In this case, we have $n_1 = n_2 = 10$, $r_1 = 9$ and $r_2 = 6$, respectively.

Parametric Procedure. Following the same lines as in Example 3.3, under the assumption of exponentiality for the two Type-II censored samples, we find the MLEs of θ_X and θ_Y to be $\hat{\theta}_X = 1.7$, $\hat{\theta}_Y = 4.627$ and $\hat{\theta}_X/\hat{\theta}_Y = 0.367$, and the 95% one-sided confidence interval for θ_X/θ_Y as $(0, 0.861)$. Since this interval does not include the value 1, we conclude that the observed data provide enough evidence to reject H_0. In fact, the p-value of the test is 0.027.

Nonparametric Procedure. Considering the same data, we can carry out a nonparametric test for the hypotheses in (3.1) through a precedence test. In this case, we specifically have $n_1 = n_2 = 10$, $r = 6$, $m_1 = 5$, $m_2 = m_3 = 0$, $m_4 = 2$, $m_5 = 0$, and $m_6 = 2$. The value of the precedence test statistic is $P_{(6)} = 9$ and the corresponding p-value is 0.065. From this p-value, the precedence test would lead to accepting H_0 at 5% and rejecting at 10% level of significance.

It is important to keep in mind that the precedence test discussed here is applicable even though only a few early failures from the two samples are available as data. Furthermore, the decisions are reached without making any assumption on the lifetime distributions underlying the two samples.

Chapter 4

Maximal Precedence Test

4.1 INTRODUCTION

As mentioned in the previous chapter, the precedence test suffers from a masking effect. To begin with, we shall explain the masking effect of the precedence test by a simple example.

Example 4.1. Suppose we have $n_1 = n_2 = 20$ and we are using the precedence test with $r = 3$ and $s = 8$; then the null hypothesis in (3.1) will be rejected at 5% level of significance if there were at least 8 failures from the X-sample before the third failure from the Y-sample. If only 7 failures had occurred from the X-sample before the third failure from the Y-sample, then we will not reject the null hypothesis by the use of $P_{(3)}$. Nevertheless, if all 7 failures had occurred before the first failure from the Y-sample (the probability of this happening under H_0 is less than 1%), we would have suspected that there is a location-shift between the two populations. In fact, if we had used $P_{(1)}$ with $s = 4$ or $P_{(2)}$ with $s = 6$, we would have rejected the null hypothesis. Information given by $r = 3$ is thus getting masked in this case.

The masking effect of the precedence test can also be seen from the critical values for the precedence test presented earlier in Tables 3.1 and 3.2. When $r \geq 2$, the critical value of the precedence test is increasing with r, which usually results in such a masking effect.

In order to avoid this masking problem, a maximal precedence test was proposed by Balakrishnan and Frattina (2000) and Balakrishnan and Ng (2001). The maximal precedence test is a test procedure for testing the hypotheses in (3.1) and the test statistic is defined as the maximum number of failures occurring from the X-sample before the first, between the first and the second, $\cdots$, between the $(r-1)$th and rth failures from the Y-sample, that is, $M_{(r)} = \max(M_1, M_2, \cdots, M_r)$. For example, if we refer to Figure 3.1, with $r = 4$, the maximal precedence test statistic is $M_{(4)} = \max(0, 3, 4, 1) = 4$. Large values of $M_{(r)}$ lead to the rejection of H_0 and in favor of H_1 in (3.1). The null distribution of $M_{(r)}$ for the special case when $r = 2$ was derived by Balakrishnan and Frattina (2000) and extended by Balakrishnan and Ng (2001) for any general value of r.

We derive in Section 4.2 the null distribution of the maximal precedence test statistic $M_{(r)}$. Next, in Section 4.3, the exact power function under Lehmann alternative is derived. Monte Carlo simulations are used in Section 4.4 to discuss the power properties of the maximal precedence test under a location-shift alternative. Some comparative remarks are made in Section 4.5. Finally, some illustrative examples are presented in Section 4.6.

4.2 EXACT NULL DISTRIBUTION

Here, in order to derive the null distribution function of the maximal precedence test statistic $M_{(r)}$, we first obtain the joint probability mass function of $M_1, M_2, \cdots, M_r$ under $H_0 : F_X = F_Y$.

Theorem 4.1. The joint probability mass function of $M_1, M_2, \cdots, M_r$ is given by, under $H_0 : F_X = F_Y$,

$$\Pr\{M_1 = m_1, M_2 = m_2, \cdots, M_r = m_r \mid H_0 : F_X = F_Y\}$$
$$= \frac{\left(\begin{array}{c} n_1 + n_2 - \sum\limits_{i=1}^{r} m_i - r \\ n_2 - r \end{array} \right)}{\left(\begin{array}{c} n_1 + n_2 \\ n_2 \end{array} \right)}. \tag{4.1}$$

Proof. First, conditional on the Y-failures, the probability that there are m_1 X-failures before $y_{1:n_2}$ and m_i X-failures between $y_{i-1:n_2}$ and $y_{i:n_2}$, $i = 2, 3, \cdots, r$, is given by the multinomial probability as follows:

$$\Pr\left\{ m_1 X's \le y_{1:n_2}, \ m_2 X's \in (y_{1:n_1}, y_{2:n_2}], \cdots, \ m_r X's \in (y_{r-1:n_2}, y_{r:n_2}], \right.$$
$$\left. \left(n_1 - \sum_{i=1}^{r} m_i \right) X's > y_{r:n_2} \middle| Y_{1:n_2} = y_{1:n_2}, Y_{2:n_2} = y_{1:n_2}, \cdots, Y_{r:n_2} = y_{r:n_2} \right\}$$

$$= \frac{n_1!}{m_1! m_2! \cdots m_r! \left(n_1 - \sum\limits_{i=1}^{r} m_i\right)!} [F_X(y_{1:n_2})]^{m_1}$$

$$\times \left\{ \prod_{i=2}^{r} [F_X(y_{i:n_2}) - F_X(y_{i-1:n_2})]^{m_i} \right\} [1 - F_X(y_{r:n_2})]^{\left(n_1 - \sum\limits_{i=1}^{r} m_i\right)},$$

$$y_{1:n_2} < y_{2:n_2} < \cdots < y_{r:n_2}.$$

We also have the joint density of the first r order statistics from the Y-sample as (see Section 2.2.1)

$$f_{1,2,\cdots,r:n_2}(y_{1:n_2}, y_{2:n_2}, \cdots, y_{r:n_2})$$

$$= \frac{n_2!}{(n_2 - r)!} f_Y(y_{1:n_2}) f_Y(y_{2:n_2}) \cdots f_Y(y_{r:n_2})[1 - F_Y(y_{r:n_2})]^{n_2 - r},$$

$$y_{1:n_2} < y_{2:n_2} < \cdots < y_{r:n_2}.$$

Consequently, we obtain the unconditional probability of $\{M_1 = m_1, M_2 = m_2, \cdots, M_r = m_r\}$ as

$$\Pr\{M_1 = m_1, M_2 = m_2, \cdots, M_r = m_r\}$$

$$= \frac{n_1!}{m_1! m_2! \cdots m_r! \left(n_1 - \sum\limits_{i=1}^{r} m_i\right)!}$$

$$\times \int_{-\infty}^{\infty} \int_{-\infty}^{y_{r:n_2}} \cdots \int_{-\infty}^{y_{2:n_2}} [F_X(y_{1:n_2})]^{m_1}$$

$$\times \left\{ \prod_{i=2}^{r} [F_X(y_{i:n_2}) - F_X(y_{i-1:n_2})]^{m_i} \right\} [1 - F_X(y_{r:n_2})]^{\left(n_1 - \sum\limits_{i=1}^{r} m_i\right)}$$

$$\times f_{1,2,\cdots,r:n_2}(y_{1:n_2}, y_{2:n_2}, \cdots, y_{r:n_2}) dy_{1:n_2} dy_{2:n_2} \cdots dy_{r-1:n_2} dy_{r:n_2}$$

$$= C \int_{-\infty}^{\infty} \int_{-\infty}^{y_{r:n_2}} \cdots \int_{-\infty}^{y_{2:n_2}} [F_X(y_{1:n_2})]^{m_1}$$

$$\times \left\{ \prod_{i=2}^{r} [F_X(y_{i:n_2}) - F_X(y_{i-1:n_2})]^{m_i} \right\} [1 - F_X(y_{r:n_2})]^{\left(n_1 - \sum\limits_{i=1}^{r} m_i\right)}$$

$$\times \left\{ \prod_{i=1}^{r} f_Y(y_{i:n_2}) \right\} [1 - F_Y(y_{r:n_2})]^{n_2 - r} dy_{1:n_2} dy_{2:n_2} \cdots dy_{r-1:n_2} dy_{r:n_2},$$

$$\tag{4.2}$$

where

$$C = \frac{n_1! n_2!}{m_1! m_2! \cdots m_r! \left(n_1 - \sum\limits_{i=1}^{r} m_i\right)! (n_2 - r)!}.$$

Under the null hypothesis, $H_0 : F_X = F_Y$, the expression in (4.2) becomes

$$\Pr\{M_1 = m_1, M_2 = m_2, \cdots, M_r = m_r \mid H_0 : F_X = F_Y\}$$
$$= C \int_{-\infty}^{\infty} \int_{-\infty}^{y_{r:n_2}} \cdots \int_{-\infty}^{y_{2:n_2}} [F_X(y_{1:n_2})]^{m_1}$$
$$\times \left\{ \prod_{i=2}^{r} [F_X(y_{i:n_2}) - F_X(y_{i-1:n_2})]^{m_i} \right\}$$
$$\times [1 - F_X(y_{r:n_2})]^{\left(n_1 + n_2 - \sum\limits_{i=1}^{r} m_i - r\right)}$$
$$\times \left\{ \prod_{i=1}^{r} f_X(y_{i:n_2}) \right\} dy_{1:n_2} dy_{2:n_2} \cdots dy_{r-1:n_2} dy_{r:n_2}.$$

Upon setting $u_i = F_X(y_{i:n_2})$ for $i = 1, 2, \cdots, r$, the above expression for the unconditional probability becomes

$$\Pr\{M_1 = m_1, M_2 = m_2, \cdots, M_r = m_r \mid H_0 : F_X = F_Y\}$$
$$= C \int_0^1 \int_0^{u_r} \cdots \int_0^{u_2} u_1^{m_1} \left\{ \prod_{i=2}^{r} (u_i - u_{i-1})^{m_i} \right\}$$
$$\times (1 - u_r)^{\left(n_1 + n_2 - \sum\limits_{i=1}^{r} m_i - r\right)} du_1 \cdots du_r.$$

Using the transformation $w_1 = u_1/u_2$, we obtain

$$\Pr\{M_1 = m_1, M_2 = m_2, \cdots, M_r = m_r \mid F_X = F_Y\}$$
$$= C \int_0^1 w_1^{m_1} (1 - w_1)^{m_2} dw_1 \int_0^1 \int_0^{u_r} \cdots \int_0^{u_3} u_2^{m_1 + m_2 + 1}$$
$$\times \left\{ \prod_{i=3}^{r} (u_i - u_{i-1})^{m_i} \right\} (1 - u_r)^{\left(n_1 + n_2 - \sum\limits_{i=1}^{r} m_i - r\right)} du_2 \cdots du_r$$
$$= C \times B(m_1, m_2 + 1) \int_0^1 \int_0^{u_r} \cdots \int_0^{u_3} u_2^{m_1 + m_2 + 1}$$
$$\times \left\{ \prod_{i=3}^{r} (u_i - u_{i-1})^{m_i} \right\} (1 - u_r)^{\left(n_1 + n_2 - \sum\limits_{i=1}^{r} m_i - r\right)} du_2 \cdots du_r,$$

where $B(a, b) = \int_0^1 x^{a-1}(1 - x)^{b-1}dx$ denotes the complete beta function.

Similarly, upon using the transformations $w_l = u_l/u_{l+1}$ for $l = 2, \cdots, r-1$, we obtain

$$\Pr\{M_1 = m_1, M_2 = m_2, \cdots, M_r = m_r \mid H_0 : F_X = F_Y\}$$

$$= C\left\{\prod_{j=1}^{r-1} B\left(\sum_{i=1}^{j} m_i + j, m_{j+1} + 1\right)\right\}$$

$$\times \int_0^1 u_r^{\left(\sum_{i=1}^{r} m_i + r - 1\right)}(1 - u_r)^{\left(n_1 + n_2 - \sum_{i=1}^{r} m_i - r\right)} du_r$$

$$= C\left\{\prod_{j=1}^{r-1} B\left(\sum_{i=1}^{j} m_i + j, m_{j+1} + 1\right)\right\}$$

$$\times B\left(\sum_{i=1}^{r} m_i + r, n_1 + n_2 - \sum_{i=1}^{r} m_i - r + 1\right)$$

$$= \frac{n_1! n_2! \left(n_1 + n_2 - \sum_{i=1}^{r} m_i - r\right)!}{\left(n_1 - \sum_{i=1}^{r} m_i\right)!(n_2 - r)!(n_1 + n_2)!}$$

$$= \frac{\dbinom{n_1 + n_2 - \sum_{i=1}^{r} m_i - r}{n_2 - r}}{\dbinom{n_1 + n_2}{n_2}},$$

which completes the proof of the theorem.

$\odot$

From Theorem 4.1, the cumulative distribution function of $M_{(r)} = \max(M_1, M_2, \cdots, M_r)$, under $H_0 : F_X = F_Y$, can be readily expressed as

$$\Pr(M_{(r)} \leq m \mid H_0 : F_X = F_Y)$$
$$= \Pr(M_1 \leq m, M_2 \leq m, \cdots, M_r \leq m \mid H_0 : F_X = F_Y)$$
$$= \sum_{\substack{m_i(i=1,2,\cdots,r)=0 \\ \sum_{i=1}^{r} m_i \leq n_1}}^{m} \frac{\dbinom{n_1 + n_2 - \sum_{i=1}^{r} m_i - r}{n_2 - r}}{\dbinom{n_1 + n_2}{n_2}}. \tag{4.3}$$

For specified values of n_1, n_2, r and the level of significance α, the critical value s (corresponding to a level closest to α) for the maximal precedence test can be found from (4.3) as

$$
\begin{aligned}
\alpha &= \Pr(M_{(r)} \geq s \mid H_0 : F_X = F_Y) \\
&= 1 - \Pr(M_{(r)} \leq s - 1 \mid H_0 : F_X = F_Y) \\
&= 1 - \sum_{\substack{m_i (i=1,2,\cdots,r)=0 \\ \sum_{i=1}^{r} m_i \leq n_1}}^{s-1} \frac{\dbinom{n_1 + n_2 - \sum_{i=1}^{r} m_i - r}{n_2 - r}}{\dbinom{n_1 + n_2}{n_2}}.
\end{aligned}
\tag{4.4}
$$

In Tables 4.1 and 4.2, we have presented the critical value s and the exact level of significance α as close as possible to 5% and 10% for some sample sizes n_1 and n_2 and $r = 2(1)10$, respectively. (Note that when $r = 1$, the maximal precedence test is identical to the precedence test.)

4.3 EXACT POWER FUNCTION UNDER LEHMANN ALTERNATIVE

In this section, we present an explicit expression for the power function of the maximal precedence test under the Lehmann alternative $H_1 : [F_X]^\gamma = F_Y$ for some γ, as derived by Balakrishnan and Ng (2001).

As mentioned earlier in Section 3.3, the Lehmann alternative $H_1 : [F_X]^\gamma = F_Y$ is a subclass of the alternative $H_1 : F_X > F_Y$ when $\gamma > 1$. In the following theorem, we derive an explicit expression for the power function of the maximal precedence test under the Lehmann alternative.

Theorem 4.2. Under the Lehmann alternative $H_1 : [F_X]^\gamma = F_Y$, the probability mass function of $M_1, M_2, \cdots, M_r$ is given by

$$
\Pr\{M_1 = m_1, M_2 = m_2, \cdots, M_r = m_r \mid H_1 : [F_X]^\gamma = F_Y\}
$$

$$
= \frac{n_1! n_2! \gamma^r}{m_1!(n_2 - r)!} \left\{ \prod_{j=1}^{r-1} \frac{\Gamma\left(\sum_{i=1}^{j} m_i + j\gamma\right)}{\Gamma\left(\sum_{i=1}^{j+1} m_i + j\gamma + 1\right)} \right\}
$$

Table 4.1: Near 5% critical values and exact levels of significance for the maximal precedence test statistic $M_{(r)}$

n_1	n_2	$r = 2$		$r = 3$		$r = 4$		$r = 5$		$r = 6$	
		s	α	s	α	s	α	s	α	s	α
10	10	5	0.0325	5	0.0487	5	0.0650	6	0.0271	6	0.0325
10	15	4	0.0332	4	0.0497	4	0.0662	5	0.0237	5	0.0285
15	15	5	0.0420	5	0.0629	6	0.0337	6	0.0421	6	0.0505
15	20	4	0.0519	5	0.0277	5	0.0369	5	0.0461	5	0.0553
20	20	5	0.0469	6	0.0302	6	0.0403	6	0.0503	6	0.0602
20	25	4	0.0645	5	0.0379	5	0.0504	5	0.0629	6	0.0285
25	25	5	0.0498	6	0.0333	6	0.0443	6	0.0553	6	0.0662
25	30	5	0.0304	5	0.0455	5	0.0604	6	0.0304	6	0.0365
30	30	5	0.0518	6	0.0354	6	0.0471	6	0.0587	7	0.0315
30	50	4	0.0345	4	0.0514	4	0.0681	5	0.0295	5	0.0353
		$r = 7$		$r = 8$		$r = 9$		$r = 10$			
n_1	n_2	s	α	s	α	s	α	s	α		
10	10	6	0.0379	6	0.0433	6	0.0488	6	0.0542		
10	15	5	0.0332	5	0.0379	5	0.0427	5	0.0474		
15	15	6	0.0589	6	0.0673	7	0.0284	7	0.0316		
15	20	5	0.0644	6	0.0247	6	0.0277	6	0.0308		
20	20	6	0.0702	7	0.0332	7	0.0374	7	0.0415		
20	25	6	0.0332	6	0.0379	6	0.0427	6	0.0474		
25	25	7	0.0336	7	0.0384	7	0.0431	7	0.0479		
25	30	6	0.0425	6	0.0485	6	0.0546	6	0.0606		
30	30	7	0.0367	7	0.0419	7	0.0471	7	0.0523		
30	50	5	0.0411	5	0.0469	5	0.0527	5	0.0585		

Table 4.2: Near 10% critical values and exact levels of significance for the maximal precedence test statistic $M_{(r)}$

n_1	n_2	s	α	s	α	s	α	s	α	s	α
		\multicolumn{2}{c}{$r=2$}	\multicolumn{2}{c}{$r=3$}	\multicolumn{2}{c}{$r=4$}	\multicolumn{2}{c}{$r=5$}	\multicolumn{2}{c}{$r=6$}
10	10	4	0.0863	4	0.1290	5	0.0650	5	0.0812	5	0.0974
10	15	3	0.1032	4	0.0497	4	0.0662	4	0.0826	4	0.0990
15	15	4	0.0985	5	0.0629	5	0.0837	5	0.1044	5	0.1249
15	20	3	0.1360	4	0.0774	4	0.1026	4	0.1276	5	0.0553
20	20	4	0.1044	5	0.0700	5	0.0929	5	0.1156	5	0.1381
20	25	4	0.0645	4	0.0958	4	0.1266	5	0.0629	5	0.0753
25	25	4	0.1078	5	0.0743	5	0.0984	5	0.1222	6	0.0662
25	30	4	0.0733	4	0.1086	5	0.0604	5	0.0752	5	0.0900
30	30	4	0.1101	5	0.0771	5	0.1020	5	0.1265	6	0.0702
30	50	3	0.0969	3	0.1424	4	0.0681	4	0.0846	4	0.1010

n_1	n_2	s	α	s	α	s	α	s	α
		\multicolumn{2}{c}{$r=7$}	\multicolumn{2}{c}{$r=8$}	\multicolumn{2}{c}{$r=9$}	\multicolumn{2}{c}{$r=10$}
10	10	5	0.1137	5	0.1299	5	0.1461	6	0.0542
10	15	4	0.1153	4	0.1316	4	0.1479	5	0.0474
15	15	6	0.0589	6	0.0673	6	0.0757	6	0.0841
15	20	5	0.0644	5	0.0735	5	0.0827	5	0.0918
20	20	6	0.0702	6	0.0802	6	0.0901	6	0.1000
20	25	5	0.0876	5	0.0999	5	0.1121	5	0.1243
25	25	6	0.0771	6	0.0880	6	0.0988	6	0.1095
25	30	5	0.1046	5	0.1191	5	0.1335	5	0.1477
30	30	6	0.0817	6	0.0932	6	0.1045	6	0.1158
30	50	4	0.1171	4	0.1330	5	0.0527	5	0.0585

$$\times \sum_{k=0}^{n_2-r} \binom{n_2 - r}{k} (-1)^k \frac{\Gamma\left(\sum\limits_{i=1}^{r} m_i + (r+k)\gamma\right)}{\Gamma\left(n_1 + (r+k)\gamma + 1\right)}. \tag{4.5}$$

Proof. Under the Lehmann alternative $H_1 : [F_X]^\gamma = F_Y, \gamma > 1$, the expression in (4.2) can be simplified as follows:

$$\Pr\{M_1 = m_1, M_2 = m_2, \cdots, M_r = m_r \mid H_1 : [F_X]^\gamma = F_Y\}$$

$$= C \int_{-\infty}^{\infty} \int_{-\infty}^{y_{r:n_2}} \cdots \int_{-\infty}^{y_{2:n_2}} [F_X(y_{1:n_2})]^{m_1}$$

$$\times \left\{ \prod_{i=2}^{r} [F_X(y_{i:n_2}) - F_X(y_{i-1:n_2})]^{m_i} \right\} [1 - F_X(y_{r:n_2})]^{\left(n_1 - \sum\limits_{i=1}^{r} m_i\right)}$$

$$\times \left\{ \prod_{i=1}^{r} \gamma [F_X(y_{i:n_2})]^{\gamma-1} f_X(y_{i:n_2}) \right\} \{1 - [F_X(y_{r:n_2})]^\gamma\}^{n_2-r}$$

$$\times dy_{1:n_2} dy_{2:n_2} \cdots dy_{r-1:n_2} dy_{r:n_2}$$

$$= C\gamma^r \int_0^\infty \int_0^{y_{r:n_2}} \cdots \int_0^{y_{2:n_2}} [F_X(y_{1:n_2})]^{m_1+\gamma-1}$$

$$\times \left\{ \prod_{i=2}^{r} [F_X(y_{i:n_2})]^{\gamma-1} [F_X(y_{i:n_2}) - F_X(y_{i-1:n_2})]^{m_i} \right\}$$

$$\times [1 - F_X(y_{r:n_2})]^{\left(n_1 - \sum\limits_{i=1}^{r} m_i\right)} \left\{ \prod_{i=1}^{r} f_X(y_{i:n_2}) \right\}$$

$$\times \left\{ \sum_{k=0}^{n_2-r} (-1)^k \binom{n_2 - r}{k} [F_X(y_{r:n_2})]^{\gamma k} \right\} dy_{1:n_2} \cdots dy_{r:n_2}$$

$$= C\gamma^r \sum_{k=0}^{n_2-r} (-1)^k \binom{n_2 - r}{k} \int_0^\infty \int_{-\infty}^{y_{r:n_2}} \cdots \int_{-\infty}^{y_{2:n_2}} [F_X(y_{1:n_2})]^{m_1+\gamma-1}$$

$$\times \left\{ \prod_{i=1}^{r} f_X(y_{i:n_2}) \right\} \left\{ \prod_{i=2}^{r} [F_X(y_{i:n_2})]^{\gamma-1} [F_X(y_{i:n_2}) - F_X(y_{i-1:n_2})]^{m_i} \right\}$$

$$\times [F_X(y_{r:n_2})]^{\gamma k} [1 - F_X(y_{r:n_2})]^{\left(n_1 - \sum\limits_{i=1}^{r} m_i\right)} dy_{1:n_2} \cdots dy_{r:n_2}.$$

Upon setting $u_i = F_X(y_{i:n_2})$, $i = 1, 2, \cdots, r$, the above expression for the unconditional probability becomes

$$\Pr\{M_1 = m_1, M_2 = m_2, \cdots, M_r = m_r \mid H_1 : [F_X]^\gamma = F_Y\}$$

$$= C\gamma^r \sum_{k=0}^{n_2-r} (-1)^k \binom{n_2 - r}{k} \int_0^1 \int_0^{u_r} \cdots \int_0^{u_2} u_1^{m_1+\gamma-1}$$

$$\times \left\{ \prod_{i=2}^{r} u_i^{\gamma-1} (u_i - u_{i-1})^{m_i} \right\} u_r^{\gamma k} (1 - u_r)^{\left(n_1 - \sum_{i=1}^{r} m_i \right)} du_1 \cdots du_r.$$

Using the transformation $w_1 = u_1/u_2$, we obtain

$$\Pr\{M_1 = m_1, M_2 = m_2, \cdots, M_r = m_r \mid H_1 : [F_X]^\gamma = F_Y\}$$

$$= C\gamma^r \sum_{k=0}^{n_2-r} (-1)^k \binom{n_2 - r}{k} \int_0^1 w_1^{m_1+\gamma-1} (1 - w_1)^{m_2} dw_1$$

$$\times \int_{-\infty}^1 \int_{-\infty}^{u_r} \cdots \int_0^{u_3} u_2^{m_1+m_2+2\gamma-1} \left\{ \prod_{i=3}^{r} u_i^{\gamma-1} (u_i - u_{i-1})^{m_i} \right\}$$

$$\times u_r^{\gamma k} (1 - u_r)^{\left(n_1 - \sum_{i=1}^{r} m_i \right)} du_2 \cdots du_r$$

$$= C\gamma^r \; B(m_1 + \gamma, m_2 + 1) \sum_{k=0}^{n_2-r} (-1)^k \binom{n_2 - r}{k}$$

$$\times \int_0^1 \int_0^{u_r} \cdots \int_0^{u_3} u_2^{m_1+m_2+2\gamma-1} \left\{ \prod_{i=3}^{r} u_i^{\gamma-1} (u_i - u_{i-1})^{m_i} \right\}$$

$$\times u_r^{\gamma k} (1 - u_r)^{\left(n_1 - \sum_{i=1}^{r} m_i \right)} du_2 \cdots du_r,$$

where $B(a, b)$ denotes the complete beta function as before.

Similarly, upon using the transformations $w_l = u_l/u_{l+1}$ for $l = 2, \cdots, r-1$, we obtain

$$\Pr\{M_1 = m_1, M_2 = m_2, \cdots, M_r = m_r \mid H_1 : [F_X]^\gamma = F_Y\}$$

$$= C\gamma^r \left\{ \prod_{j=1}^{r-1} B \left(\sum_{i=1}^{j} m_i + j\gamma, m_{j+1} + 1 \right) \right\}$$

$$\times \sum_{k=0}^{n_2-r} (-1)^k \binom{n_2 - r}{k} \int_0^1 u_r^{\sum_{i=1}^{r} m_i + (r+k)\gamma - 1} (1 - u_r)^{\left(n_1 - \sum_{i=1}^{r} m_i \right)} du_r$$

$$= C\gamma^r \left\{ \prod_{j=1}^{r-1} B \left(\sum_{i=1}^{j} m_i + j\gamma, m_{j+1} + 1 \right) \right\}$$

$$\times \sum_{k=0}^{n_2-r} (-1)^k \binom{n_2 - r}{k} B \left(\sum_{i=1}^{r} m_i + (r+k)\gamma, n_1 - \sum_{i=1}^{r} m_i + 1 \right)$$

$$= \frac{n_1! n_2! \gamma^r}{m_1!(n_2-r)!} \left\{ \prod_{j=1}^{r-1} \frac{\Gamma\left(\sum\limits_{i=1}^{j} m_i + j\gamma\right)}{\Gamma\left(\sum\limits_{i=1}^{j+1} m_i + j\gamma + 1\right)} \right\}$$

$$\times \sum_{k=0}^{n_2-r} (-1)^k \binom{n_2-r}{k} \frac{\Gamma\left(\sum\limits_{i=1}^{r} m_i + (r+k)\gamma\right)}{\Gamma\left(n_1 + (r+k)\gamma + 1\right)}, \qquad (4.6)$$

which completes the proof of the theorem.

⊙

From Theorem 4.2, the cumulative distribution function of $M_{(r)} = \max(M_1, M_2, \cdots, M_r)$ under the Lehmann alternative is readily given by

$$\Pr\left\{ M_{(r)} \leq m \mid H_1 : [F_X]^\gamma = F_Y \right\}$$

$$= \Pr\left\{ M_1 \leq m, M_2 \leq m, \cdots, M_r \leq m \mid H_1 : [F_X]^\gamma = F_Y \right\}$$

$$= \sum_{\substack{m_i(i=1,2,\cdots,r)=0 \\ \sum_{i=1}^{r} m_i \leq n_1}}^{m} \Pr\left\{ M_1 = m_1, \cdots, M_r = m_r \mid H_1 : [F_X]^\gamma = F_Y \right\}$$

$$= \sum_{\substack{m_i(i=1,2,\cdots,r)=0 \\ \sum_{i=1}^{r} m_i \leq n_1}}^{m} \frac{n_1! n_2! \gamma^r}{m_1!(n_2-r)!} \left\{ \prod_{j=1}^{r-1} \frac{\Gamma\left(\sum\limits_{i=1}^{j} m_i + j\gamma\right)}{\Gamma\left(\sum\limits_{i=1}^{j+1} m_i + j\gamma + 1\right)} \right\}$$

$$\times \sum_{k=0}^{n_2-r} \binom{n_2-r}{k} (-1)^k \frac{\Gamma\left(\sum\limits_{i=1}^{r} m_i + (r+k)\gamma\right)}{\Gamma\left(n_1 + (r+k)\gamma + 1\right)}.$$

Therefore, under the Lehmann alternative $H_1 : [F_X]^\gamma = F_Y$, the power function of the maximal precedence test is given by

$$1 - \Pr\left\{ M_{(r)} \leq s-1 \mid H_1 : [F_X]^\gamma = F_Y \right\}$$

$$= 1 - \sum_{\substack{m_i(i=1,2,\cdots,r)=0 \\ \sum_{i=1}^{r} m_i \leq n_1}}^{s-1} \frac{n_1! n_2! \gamma^r}{m_1!(n_2-r)!} \left\{ \prod_{j=1}^{r-1} \frac{\Gamma\left(\sum\limits_{i=1}^{j} m_i + j\gamma\right)}{\Gamma\left(\sum\limits_{i=1}^{j+1} m_i + j\gamma + 1\right)} \right\}$$

$$\times \sum_{k=0}^{n_2-r} \binom{n_2-r}{k} (-1)^k \frac{\Gamma\left(\sum\limits_{i=1}^{r} m_i + (r+k)\gamma\right)}{\Gamma\left(n_1 + (r+k)\gamma + 1\right)}. \qquad (4.7)$$

Under the Lehmann alternative and for $n_1 = n_2 = 10, r = 2(1)3$, and $\gamma = 2(1)6$, the power values computed from the expression in (4.7) are presented in Table 4.3. For comparative purposes, the corresponding power values of the precedence test (taken from Table 3.3) are also included here. From Table 4.3, we observe that the maximal precedence test has nearly the same power as the precedence test for all values of r.

4.4   MONTE CARLO SIMULATION UNDER LOCATION-SHIFT ALTERNATIVE

In order to assess the power properties of the maximal precedence test, Balakrishnan and Ng (2001) considered the location-shift alternative $H_1 :$ $F_X(x) = F_Y(x + \theta)$ for some $\theta > 0$, where θ is a shift in location. The power of the precedence test with $r = 1(1)6$, the maximal precedence test with $r = 2(1)6$, and the Wilcoxon's rank-sum test W_R (based on complete samples) were all estimated by these authors through Monte Carlo simulations when $\theta = 0.5$ and $\theta = 1.0$. The following lifetime distributions were used in the Monte Carlo simulations in order to demonstrate the power performance of the maximal precedence test under this location-shift alternative:

1. Standard normal distribution

2. Standard exponential distribution

3. Gamma distribution with shape parameter a and standardized by mean a and standard deviation $\sqrt{a}$

4. Lognormal distribution with shape parameter σ and standardized by mean $e^{\sigma^2/2}$ and standard deviation $\sqrt{e^{\sigma^2}(e^{\sigma^2} - 1)}$

A brief description of these distributions and their properties has been provided in Section 2.5.

Table 4.3: Power values of precedence and maximal precedence tests under Lehmann alternative for $n_1 = n_2 = 10, r = 2(1)5$, and $\gamma = 2(1)6$

| r | γ | Power values of | |
		$P_{(r)}$	$M_{(r)}$
2	1	0.02864	0.03250
	2	0.23970	0.22095
	3	0.49048	0.45342
	4	0.67299	0.63634
	5	0.78963	0.76059
	6	0.86216	0.84119
3	1	0.03489	0.04875
	2	0.24581	0.24178
	3	0.48087	0.46802
	4	0.65212	0.64561
	5	0.76460	0.76641
	6	0.83731	0.84490
4	1	0.03489	0.06498
	2	0.21679	0.25407
	3	0.41996	0.47465
	4	0.57708	0.64914
	5	0.68834	0.76837
	6	0.76610	0.84603
5	1	0.02864	0.02709
	2	0.16106	0.11714
	3	0.31528	0.27327
	4	0.44601	0.43572
	5	0.54833	0.57255
	6	0.62715	0.67828

4.5 EVALUATION AND COMPARATIVE REMARKS

In Tables 4.4–4.8, we have presented the estimated power values of the precedence tests with $r = 1(1)10$, the maximal precedence tests with $r = 2(1)10$, and the Wilcoxon's rank-sum test for the underlying standard normal, standard exponential, standardized gamma, and standardized lognormal distributions, with location-shift being equal to 0.5 and 1.0. For comparison purposes, the corresponding critical values and the exact levels of significance are also presented.

From Tables 4.4–4.8, we see that the power values of all tests increase with increasing sample sizes as well as with increasing location-shift. When we compare the power values of the maximal precedence tests with those of the Wilcoxon's rank-sum test, we find that the Wilcoxon's rank-sum test performs better than the precedence tests and the maximal precedence tests if the underlying distributions are close to symmetry, such as the normal distribution, gamma distribution with large values of shape parameter a, and lognormal distribution with small values of shape parameter σ. However, under some right-skewed distributions such as the exponential distribution, gamma distribution with shape parameter $a = 2.0$, and lognormal distribution with shape parameter $\sigma = 0.5$, the maximal precedence tests have higher power values than the Wilcoxon's rank-sum test. For example, in Table 4.8, under gamma distribution with shape parameter $a = 2.0$ when $n_1 = n_2 = 20$ and the location-shift equals 0.5, the simulated power of the maximal precedence test with $r = 2$ is 0.6572 (exact level of significance is 0.0469) while the power of the Wilcoxon's rank-sum test (based on complete samples) is 0.5701 (exact level of significance is 0.0482). From Tables 4.4–4.8, it is evident that the more right-skewed the underlying distribution the more powerful the maximal precedence test.

Moreover, we can compare the power values of the precedence tests and the maximal precedence tests with the same value of r since they are both test procedures based on failures from the X-sample occurring before the rth failure from the Y-sample. The power values presented in Tables 4.4–4.8 show that the precedence test is more powerful than the maximal precedence test under the normal distribution. However, under the exponential distribution, the maximal precedence test performs better than the precedence test. From

Table 4.4: Power of precedence tests $(P_{(r)})$, maximal precedence tests $(M_{(r)})$, and Wilcoxon's rank-sum test (W_R) when $n_1 = n_2 = 10$

Test	r	Critical value	Exact l.o.s.	Location shift	Distribution N(0,1)	Exp(1)	LN(0.1)	LN(0.5)
$P_{(r)}$	1	4	0.0433	0.5	0.1736	0.7197	0.1936	0.3944
				1.0	0.4208	0.9779	0.4817	0.8709
	2	6	0.0286	0.5	0.1496	0.3916	0.1603	0.2693
				1.0	0.4243	0.8376	0.4571	0.7253
	3	7	0.0349	0.5	0.1779	0.3073	0.1828	0.2571
				1.0	0.4813	0.7239	0.4964	0.6676
	4	8	0.0349	0.5	0.1771	0.2251	0.1777	0.2149
				1.0	0.4799	0.5767	0.4783	0.5656
	5	9	0.0286	0.5	0.1473	0.1453	0.1444	0.1527
				1.0	0.4245	0.3960	0.4038	0.4144
	6	9	0.0704	0.5	0.2659	0.2385	0.2567	0.2531
				1.0	0.5832	0.4991	0.5545	0.5314
	7	10	0.0433	0.5	0.1721	0.1280	0.1614	0.1405
				1.0	0.4240	0.2757	0.3791	0.3074
	8	10	0.1053	0.5	0.3032	0.2253	0.2841	0.2427
				1.0	0.5777	0.3902	0.5265	0.4243
	9	10	0.2368	0.5	0.4880	0.3828	0.4598	0.4002
				1.0	0.7394	0.5407	0.6910	0.5720
	10	10	0.5000	0.5	0.7290	0.6267	0.7020	0.6395
				1.0	0.8887	0.7415	0.8546	0.7598
$M_{(r)}$	2	5	0.0325	0.5	0.1417	0.4923	0.1479	0.2543
				1.0	0.3768	0.9177	0.4083	0.7327
	3	5	0.0488	0.5	0.1753	0.4949	0.1781	0.2678
				1.0	0.4171	0.9178	0.4395	0.7360
	4	5	0.0650	0.5	0.1993	0.4974	0.1990	0.2768
				1.0	0.4359	0.9181	0.4538	0.7376
	5	6	0.0271	0.5	0.0901	0.2684	0.0905	0.1297
				1.0	0.2482	0.7791	0.2586	0.5284
	6	6	0.0325	0.5	0.0938	0.2690	0.0938	0.1309
				1.0	0.2496	0.7791	0.2596	0.5286
	7	6	0.0379	0.5	0.0965	0.2695	0.0960	0.1315
				1.0	0.2502	0.7792	0.2600	0.5286
	8	6	0.0433	0.5	0.0982	0.2700	0.0977	0.1322
				1.0	0.2504	0.7792	0.2601	0.5286
	9	6	0.0488	0.5	0.0992	0.2705	0.0989	0.1328
				1.0	0.2505	0.7793	0.2602	0.5287
	10	6	0.0542	0.5	0.0999	0.2711	0.0994	0.1331
				1.0	0.2506	0.7793	0.2602	0.5287
W_R		82	0.0446	0.5	0.2536	0.4332	0.2605	0.3508
				1.0	0.6480	0.8163	0.6554	0.7819

Table 4.4: (Continued)

Test	r	Critical value	Exact l.o.s.	Location shift	Distribution Gamma(2)	Gamma(5)	Gamma(10)
$P_{(r)}$	1	4	0.0433	0.5	0.4139	0.2644	0.2249
				1.0	0.8806	0.6731	0.5744
	2	6	0.0286	0.5	0.2519	0.1936	0.1731
				1.0	0.6877	0.5590	0.5046
	3	7	0.0349	0.5	0.2330	0.2013	0.1893
				1.0	0.6157	0.5462	0.5181
	4	8	0.0349	0.5	0.1935	0.1796	0.1755
				1.0	0.5103	0.4847	0.4764
	5	9	0.0286	0.5	0.1378	0.1371	0.1388
				1.0	0.3730	0.3760	0.3853
	6	9	0.0704	0.5	0.2349	0.2400	0.2436
				1.0	0.4962	0.5131	0.5280
	7	10	0.0433	0.5	0.1325	0.1411	0.1475
				1.0	0.2912	0.3205	0.3390
	8	10	0.1053	0.5	0.2345	0.2529	0.2616
				1.0	0.4146	0.4536	0.4794
	9	10	0.2368	0.5	0.3959	0.4218	0.4342
				1.0	0.5719	0.6148	0.6425
	10	10	0.5000	0.5	0.6422	0.6653	0.6777
				1.0	0.7657	0.8016	0.8228
$M_{(r)}$	2	5	0.0325	0.5	0.2461	0.1757	0.1611
				1.0	0.7318	0.5216	0.4585
	3	5	0.0488	0.5	0.2567	0.1964	0.1870
				1.0	0.7344	0.5336	0.4776
	4	5	0.0650	0.5	0.2638	0.2097	0.2040
				1.0	0.7361	0.5391	0.4860
	5	6	0.0271	0.5	0.1184	0.0924	0.0895
				1.0	0.5146	0.3285	0.2829
	6	6	0.0325	0.5	0.1197	0.0943	0.0921
				1.0	0.5147	0.3290	0.2834
	7	6	0.0379	0.5	0.1206	0.0957	0.0940
				1.0	0.5148	0.3292	0.2837
	8	6	0.0433	0.5	0.1215	0.0968	0.0952
				1.0	0.5149	0.3294	0.2838
	9	6	0.0488	0.5	0.1222	0.0976	0.0961
				1.0	0.5149	0.3295	0.2839
	10	6	0.0542	0.5	0.1228	0.0983	0.0968
				1.0	0.5150	0.3296	0.2839
W_R		82	0.0446	0.5	0.3337	0.2822	0.2663
				1.0	0.7513	0.6964	0.6719

Table 4.5: Power of precedence tests $(P_{(r)})$, maximal precedence tests $(M_{(r)})$, and Wilcoxon's rank-sum test (W_R) when $n_1 = 10, n_2 = 15$

Test	r	Critical value	Exact l.o.s.	Location shift	Distribution			
					N(0,1)	Exp(1)	LN(0.1)	LN(0.5)
$P_{(r)}$	1	3	0.0522	0.5	0.1971	0.8733	0.2290	0.5146
				1.0	0.4648	0.9952	0.5431	0.9395
	2	4	0.0640	0.5	0.2607	0.7638	0.2886	0.5243
				1.0	0.5942	0.9821	0.6500	0.9309
	3	5	0.0618	0.5	0.2741	0.6340	0.2956	0.4775
				1.0	0.6335	0.9480	0.6704	0.8933
	4	6	0.0533	0.5	0.2574	0.4933	0.2745	0.4037
				1.0	0.6260	0.8779	0.6488	0.8308
	5	7	0.0416	0.5	0.2246	0.3567	0.2324	0.3137
				1.0	0.5864	0.7587	0.5958	0.7311
	6	8	0.0287	0.5	0.1750	0.2323	0.1756	0.2189
				1.0	0.5096	0.5854	0.5041	0.5844
	7	8	0.0576	0.5	0.2704	0.3091	0.2687	0.3052
				1.0	0.6357	0.6524	0.6217	0.6658
	8	9	0.0339	0.5	0.1839	0.1789	0.1783	0.1847
				1.0	0.5056	0.4375	0.4769	0.4679
	9	9	0.0654	0.5	0.2759	0.2499	0.2646	0.2603
				1.0	0.6195	0.5137	0.5858	0.5509
	10	10	0.0283	0.5	0.1430	0.1084	0.1313	0.1160
				1.0	0.3927	0.2526	0.3510	0.2840
$M_{(r)}$	2	4	0.0332	0.5	0.1530	0.6919	0.1697	0.3464
				1.0	0.4225	0.9748	0.4676	0.8507
	3	4	0.0497	0.5	0.1947	0.6946	0.2098	0.3676
				1.0	0.4819	0.9750	0.5168	0.8559
	4	4	0.0662	0.5	0.2261	0.6971	0.2386	0.3805
				1.0	0.5160	0.9751	0.5435	0.8584
	5	5	0.0237	0.5	0.0990	0.4519	0.1032	0.1811
				1.0	0.2904	0.9078	0.3104	0.6707
	6	5	0.0285	0.5	0.1052	0.4525	0.1087	0.1830
				1.0	0.2948	0.9079	0.3136	0.6709
	7	5	0.0332	0.5	0.1102	0.4531	0.1127	0.1845
				1.0	0.2973	0.9079	0.3155	0.6711
	8	5	0.0379	0.5	0.1139	0.4537	0.1156	0.1856
				1.0	0.2989	0.9080	0.3164	0.6713
	9	5	0.0427	0.5	0.1168	0.4543	0.1180	0.1866
				1.0	0.2997	0.9080	0.3171	0.6713
	10	5	0.0474	0.5	0.1189	0.4552	0.1198	0.1875
				1.0	0.3001	0.9080	0.3173	0.6714
W_R		99	0.0455	0.5	0.2934	0.5020	0.2998	0.4103
				1.0	0.7306	0.8637	0.7331	0.8399

Table 4.5: (Continued)

Test	r	Critical value	Exact l.o.s.	Location shift	Distribution Gamma(2)	Gamma(5)	Gamma(10)
$P_{(r)}$	1	3	0.0522	0.5	0.2783	0.3384	0.2783
				1.0	0.6656	0.7798	0.6656
	2	4	0.0640	0.5	0.3335	0.3780	0.3335
				1.0	0.7327	0.8022	0.7327
	3	5	0.0618	0.5	0.3266	0.3575	0.3266
				1.0	0.7262	0.7736	0.7262
	4	6	0.0533	0.5	0.2926	0.3102	0.2926
				1.0	0.6840	0.7179	0.6840
	5	7	0.0416	0.5	0.2414	0.2500	0.2414
				1.0	0.6122	0.6307	0.6122
	6	8	0.0287	0.5	0.1781	0.1811	0.1781
				1.0	0.5048	0.5066	0.5048
	7	8	0.0576	0.5	0.2679	0.2690	0.2679
				1.0	0.6138	0.6115	0.6138
	8	9	0.0339	0.5	0.1720	0.1703	0.1720
				1.0	0.4533	0.4443	0.4533
	9	9	0.0654	0.5	0.2549	0.2505	0.2549
				1.0	0.5568	0.5410	0.5568
	10	10	0.0283	0.5	0.1229	0.1167	0.1229
				1.0	0.3146	0.2953	0.3146
$M_{(r)}$	2	4	0.0332	0.5	0.1959	0.2263	0.1959
				1.0	0.5533	0.6419	0.5533
	3	4	0.0497	0.5	0.2303	0.2556	0.2303
				1.0	0.5852	0.6614	0.5852
	4	4	0.0662	0.5	0.2552	0.2760	0.2552
				1.0	0.6021	0.6725	0.6021
	5	5	0.0237	0.5	0.1094	0.1193	0.1094
				1.0	0.3602	0.4290	0.3602
	6	5	0.0285	0.5	0.1136	0.1223	0.1136
				1.0	0.3619	0.4303	0.3619
	7	5	0.0332	0.5	0.1170	0.1247	0.1170
				1.0	0.3630	0.4310	0.3630
	8	5	0.0379	0.5	0.1197	0.1268	0.1197
				1.0	0.3638	0.4316	0.3638
	9	5	0.0427	0.5	0.1219	0.1286	0.1219
				1.0	0.3643	0.4319	0.3643
	10	5	0.0474	0.5	0.1235	0.1301	0.1235
				1.0	0.3647	0.4321	0.3647
W_R		99	0.0455	0.5	0.3142	0.3323	0.3142
				1.0	0.7495	0.7685	0.7495

Table 4.6: Power of precedence tests $(P_{(r)})$, maximal precedence tests $(M_{(r)})$, and Wilcoxon's rank-sum test (W_R) when $n_1 = n_2 = 15$

Test	r	Critical value	Exact l.o.s.	Location shift	Distribution			
					N(0,1)	Exp(1)	LN(0,1)	LN(0,5)
$P_{(r)}$	1	4	0.0498	0.5	0.2117	0.9360	0.2443	0.5676
				1.0	0.5025	0.9996	0.5867	0.9729
	2	6	0.0400	0.5	0.2206	0.7682	0.2464	0.4908
				1.0	0.5719	0.9927	0.6379	0.9468
	3	7	0.0543	0.5	0.2847	0.6963	0.3074	0.5158
				1.0	0.6719	0.9819	0.7212	0.9421
	4	8	0.0641	0.5	0.3254	0.6230	0.3434	0.5095
				1.0	0.7268	0.9621	0.7606	0.9276
	5	10	0.0328	0.5	0.2264	0.3687	0.2321	0.3226
				1.0	0.6251	0.8318	0.6441	0.7975
	6	11	0.0328	0.5	0.2272	0.3014	0.2262	0.2834
				1.0	0.6248	0.7393	0.6293	0.7312
	7	11	0.0697	0.5	0.3514	0.4024	0.3451	0.3944
				1.0	0.7525	0.8016	0.7488	0.8073
	8	12	0.0641	0.5	0.3292	0.3258	0.3174	0.3356
				1.0	0.7265	0.6990	0.7103	0.7258
	9	13	0.0543	0.5	0.2858	0.2494	0.2694	0.2625
				1.0	0.6726	0.5657	0.6406	0.6075
	10	14	0.0400	0.5	0.2226	0.1668	0.2024	0.1792
				1.0	0.5721	0.3968	0.5256	0.4437
$M_{(r)}$	2	5	0.0421	0.5	0.2016	0.8403	0.2233	0.4492
				1.0	0.5176	0.9973	0.5787	0.9397
	3	5	0.0629	0.5	0.2563	0.8421	0.2731	0.4744
				1.0	0.5864	0.9973	0.6342	0.9434
	4	6	0.0337	0.5	0.1578	0.6851	0.1675	0.3014
				1.0	0.4347	0.9885	0.4694	0.8584
	5	6	0.0421	0.5	0.1724	0.6858	0.1801	0.3055
				1.0	0.4481	0.9886	0.4780	0.8589
	6	6	0.0505	0.5	0.1835	0.6865	0.1892	0.3088
				1.0	0.4550	0.9886	0.4831	0.8592
	7	6	0.0589	0.5	0.1922	0.6872	0.1964	0.3115
				1.0	0.4593	0.9886	0.4860	0.8594
	8	6	0.0673	0.5	0.1986	0.6878	0.2020	0.3134
				1.0	0.4618	0.9886	0.4876	0.8595
	9	7	0.0284	0.5	0.0969	0.4901	0.0991	0.1705
				1.0	0.2898	0.9603	0.3118	0.7280
	10	7	0.0316	0.5	0.0983	0.4903	0.1001	0.1709
				1.0	0.2901	0.9603	0.3119	0.7280
W_R		192	0.0488	0.5	0.3600	0.5938	0.3637	0.4927
				1.0	0.8240	0.9406	0.8317	0.9216

Table 4.6: (Continued)

Test	r	Critical value	Exact l.o.s.	Location shift	Distribution Gamma(2)	Gamma(5)	Gamma(10)
$P_{(r)}$	1	4	0.0498	0.5	0.6390	0.3713	0.2991
				1.0	0.9844	0.8362	0.7214
	2	6	0.0400	0.5	0.4893	0.3311	0.2852
				1.0	0.9425	0.8081	0.7286
	3	7	0.0543	0.5	0.4900	0.3768	0.3414
				1.0	0.9275	0.8336	0.7794
	4	8	0.0641	0.5	0.4755	0.3938	0.3683
				1.0	0.9036	0.8326	0.7976
	5	10	0.0328	0.5	0.2899	0.2518	0.2412
				1.0	0.7473	0.6840	0.6622
	6	11	0.0328	0.5	0.2533	0.2331	0.2285
				1.0	0.6756	0.6389	0.6317
	7	11	0.0697	0.5	0.3630	0.3463	0.3433
				1.0	0.7636	0.7469	0.7463
	8	12	0.0641	0.5	0.3079	0.3046	0.3089
				1.0	0.6790	0.6820	0.6919
	9	13	0.0543	0.5	0.2426	0.2501	0.2588
				1.0	0.5646	0.5875	0.6079
	10	14	0.0400	0.5	0.1706	0.1805	0.1881
				1.0	0.4138	0.4533	0.4788
$M_{(r)}$	2	5	0.0421	0.5	0.4733	0.2962	0.2567
				1.0	0.9521	0.7676	0.6714
	3	5	0.0629	0.5	0.4890	0.3326	0.3008
				1.0	0.9537	0.7867	0.7060
	4	6	0.0337	0.5	0.3037	0.1946	0.1776
				1.0	0.8785	0.6215	0.5332
	5	6	0.0421	0.5	0.3075	0.2024	0.1876
				1.0	0.8790	0.6245	0.5384
	6	6	0.0505	0.5	0.3104	0.2087	0.1949
				1.0	0.8793	0.6262	0.5414
	7	6	0.05889	0.5	0.3130	0.2130	0.2005
				1.0	0.8795	0.6272	0.5431
	8	6	0.0673	0.5	0.3151	0.2166	0.2048
				1.0	0.8797	0.6277	0.5442
	9	7	0.0284	0.5	0.1687	0.1059	0.1001
				1.0	0.7511	0.4473	0.3645
	10	7	0.0316	0.5	0.1692	0.1067	0.1010
				1.0	0.7511	0.4474	0.3647
W_R		192	0.0488	0.5	0.4707	0.3987	0.3797
				1.0	0.9042	0.8640	0.8481

Table 4.7: Power of precedence tests ($P_{(r)}$), maximal precedence tests ($M_{(r)}$), and Wilcoxon's rank-sum test (W_R) when $n_1 = 15, n_2 = 20$

Test	r	Critical value	Exact l.o.s.	Location shift	N(0,1)	Exp(1)	LN(0.1)	LN(0.5)
$P_{(r)}$	1	3	0.0695	0.5	0.2610	0.9803	0.3004	0.6898
				1.0	0.5661	0.9999	0.6559	0.9902
	2	5	0.0401	0.5	0.2280	0.8712	0.2587	0.5644
				1.0	0.5893	0.9978	0.6678	0.9710
	3	6	0.0461	0.5	0.2698	0.7941	0.2992	0.5654
				1.0	0.6691	0.9937	0.7284	0.9648
	4	7	0.0472	0.5	0.2868	0.7041	0.3130	0.5347
				1.0	0.7056	0.9829	0.7494	0.9499
	5	8	0.0449	0.5	0.2889	0.6064	0.3109	0.4867
				1.0	0.7182	0.9603	0.7495	0.9250
	6	9	0.0404	0.5	0.2796	0.5049	0.2932	0.4289
				1.0	0.7120	0.9183	0.7343	0.8864
	7	10	0.0345	0.5	0.2572	0.4012	0.2651	0.3622
				1.0	0.6912	0.8495	0.6987	0.8285
	8	10	0.0646	0.5	0.3644	0.4854	0.3708	0.4608
				1.0	0.7913	0.8822	0.7932	0.8754
	9	11	0.0521	0.5	0.3220	0.3813	0.3212	0.3772
				1.0	0.7528	0.7938	0.7438	0.8020
	10	12	0.0392	0.5	0.2683	0.2580	0.2626	0.2890
				1.0	0.6943	0.6463	0.6717	0.6950
$M_{(r)}$	2	4	0.0519	0.5	0.2393	0.9326	0.2671	0.5711
				1.0	0.5770	0.9993	0.6523	0.9751
	3	5	0.0277	0.5	0.1594	0.8247	0.1757	0.3851
				1.0	0.4613	0.9967	0.5211	0.9259
	4	5	0.0369	0.5	0.1851	0.8256	0.1990	0.3957
				1.0	0.4949	0.9968	0.5465	0.9274
	5	5	0.0461	0.5	0.2049	0.8264	0.2171	0.4029
				1.0	0.5162	0.9968	0.5613	0.9281
	6	5	0.0553	0.5	0.2209	0.8271	0.2316	0.4083
				1.0	0.5298	0.9968	0.5713	0.9286
	7	5	0.0644	0.5	0.2337	0.8277	0.2430	0.4123
				1.0	0.5388	0.9968	0.5775	0.9290
	8	6	0.0247	0.5	0.1103	0.6588	0.1150	0.2279
				1.0	0.3326	0.9867	0.3698	0.8269
	9	6	0.0277	0.5	0.1134	0.6590	0.1175	0.2288
				1.0	0.3341	0.9867	0.3710	0.8270
	10	6	0.0308	0.5	0.1161	0.6593	0.1192	0.2293
				1.0	0.3350	0.9867	0.3716	0.8271
W_R		220	0.0497	0.5	0.3990	0.6438	0.4078	0.5445
				1.0	0.8710	0.9560	0.8735	0.9426

Table 4.7: (Continued)

| Test | r | Critical value | Exact l.o.s. | Location shift | Distribution | | |
					Gamma(2)	Gamma(5)	Gamma(10)
$P_{(r)}$	1	3	0.0695	0.5	0.7777	0.4727	0.3804
				1.0	0.9951	0.9071	0.7998
	2	5	0.0401	0.5	0.5814	0.3737	0.3143
				1.0	0.9723	0.8599	0.7747
	3	6	0.0461	0.5	0.5495	0.3945	0.3459
				1.0	0.9579	0.8662	0.8071
	4	7	0.0472	0.5	0.5033	0.3873	0.3494
				1.0	0.9351	0.8554	0.8081
	5	8	0.0449	0.5	0.4474	0.3637	0.3359
				1.0	0.9012	0.8274	0.7913
	6	9	0.0404	0.5	0.3878	0.3295	0.3092
				1.0	0.8519	0.7881	0.7625
	7	10	0.0345	0.5	0.3233	0.2858	0.2745
				1.0	0.7833	0.7317	0.7166
	8	10	0.0646	0.5	0.4161	0.3869	0.3768
				1.0	0.8407	0.8085	0.8004
	9	11	0.0521	0.5	0.3382	0.3238	0.3208
				1.0	0.7599	0.7397	0.7409
	10	12	0.0392	0.5	0.2798	0.2552	0.2564
				1.0	0.6682	0.6456	0.6543
$M_{(r)}$	2	4	0.0519	0.5	0.6159	0.3766	0.3182
				1.0	0.9824	0.8562	0.7602
	3	5	0.0277	0.5	0.4138	0.2339	0.2022
				1.0	0.9443	0.7235	0.6149
	4	5	0.0369	0.5	0.4210	0.2501	0.2216
				1.0	0.9450	0.7318	0.6308
	5	5	0.0461	0.5	0.4263	0.2618	0.2360
				1.0	0.9456	0.7363	0.6392
	6	5	0.0553	0.5	0.4305	0.2710	0.2482
				1.0	0.9460	0.7390	0.6445
	7	5	0.0644	0.5	0.4342	0.2781	0.2574
				1.0	0.9463	0.7410	0.6484
	8	6	0.0247	0.5	0.2413	0.1351	0.1217
				1.0	0.8587	0.5453	0.4416
	9	6	0.0277	0.5	0.2423	0.1368	0.1236
				1.0	0.8588	0.5456	0.4422
	10	6	0.0308	0.5	0.2430	0.1383	0.1253
				1.0	0.8589	0.5457	0.4425
W_R		220	0.0497	0.5	0.5180	0.4451	0.4229
				1.0	0.9298	0.8994	0.8845

Table 4.8: Power of precedence tests $(P_{(r)})$, maximal precedence tests $(M_{(r)})$, and Wilcoxon's rank-sum test (W_R) when $n_1 = n_2 = 20$

Test	r	Critical value	Exact l.o.s.	Location shift	N(0,1)	Exp(1)	LN(0.1)	LN(0.5)
$P_{(r)}$	1	4	0.0530	0.5	0.2386	0.9887	0.2804	0.6855
				1.0	0.5553	1.0000	0.6543	0.9943
	2	6	0.0457	0.5	0.2662	0.9355	0.3038	0.6453
				1.0	0.6529	0.9998	0.7338	0.9901
	3	7	0.0637	0.5	0.3480	0.9008	0.3851	0.6841
				1.0	0.7625	0.9994	0.8199	0.9910
	4	9	0.0412	0.5	0.2947	0.7499	0.3219	0.5614
				1.0	0.7352	0.9932	0.7877	0.9716
	5	10	0.0479	0.5	0.3317	0.6904	0.3525	0.5555
				1.0	0.7787	0.9864	0.8163	0.9648
	6	11	0.0527	0.5	0.3565	0.6291	0.3706	0.5369
				1.0	0.8055	0.9733	0.8295	0.9540
	7	12	0.0555	0.5	0.3686	0.5670	0.3782	0.5107
				1.0	0.8186	0.9532	0.8337	0.9367
	8	13	0.0564	0.5	0.3736	0.5045	0.3780	0.4777
				1.0	0.8241	0.9214	0.8297	0.9135
	9	14	0.0555	0.5	0.3708	0.4410	0.3665	0.4347
				1.0	0.8190	0.8732	0.8173	0.8792
	10	15	0.0527	0.5	0.3558	0.3784	0.3455	0.3846
				1.0	0.8066	0.8064	0.7939	0.8283
$M_{(r)}$	2	5	0.0469	0.5	0.2434	0.9634	0.2741	0.5983
				1.0	0.5994	1.0000	0.6788	0.9882
	3	6	0.0302	0.5	0.1847	0.9051	0.2016	0.4460
				1.0	0.5235	0.9997	0.5896	0.9654
	4	6	0.0403	0.5	0.2138	0.9053	0.2273	0.4572
				1.0	0.5605	0.9997	0.6170	0.9662
	5	6	0.0503	0.5	0.2360	0.9058	0.2472	0.4642
				1.0	0.5833	0.9997	0.6327	0.9666
	6	6	0.0603	0.5	0.2545	0.9062	0.2626	0.4688
				1.0	0.5977	0.9997	0.6423	0.9668
	7	6	0.0702	0.5	0.2687	0.9065	0.2749	0.4726
				1.0	0.6067	0.9997	0.6488	0.9670
	8	7	0.0332	0.5	0.1522	0.8014	0.1539	0.3066
				1.0	0.4287	0.9981	0.4738	0.9156
	9	7	0.0374	0.5	0.1566	0.8016	0.1573	0.3076
				1.0	0.4305	0.9981	0.4751	0.9156
	10	7	0.0415	0.5	0.1601	0.8018	0.1600	0.3083
				1.0	0.4316	0.9981	0.4758	0.9156
W_R		348	0.0482	0.5	0.4403	0.7011	0.4462	0.5941
				1.0	0.9136	0.9802	0.9179	0.9716

Table 4.8: (Continued)

Test	r	Critical value	Exact l.o.s.	Location shift	Distribution Gamma(2)	Gamma(5)	Gamma(10)
$P_{(r)}$	1	4	0.0530	0.5	0.7882	0.4525	0.3561
				1.0	0.9982	0.9145	0.8027
	2	6	0.0457	0.5	0.6709	0.4367	0.3646
				1.0	0.9914	0.9126	0.8373
	3	7	0.0637	0.5	0.6795	0.4998	0.4390
				1.0	0.9888	0.9341	0.8877
	4	9	0.0412	0.5	0.5308	0.4014	0.3575
				1.0	0.9616	0.8876	0.8422
	5	10	0.0479	0.5	0.5165	0.4189	0.3813
				1.0	0.9488	0.8876	0.8539
	6	11	0.0527	0.5	0.4967	0.4208	0.3920
				1.0	0.9319	0.8794	0.8570
	7	12	0.0555	0.5	0.4663	0.4129	0.3930
				1.0	0.9083	0.8649	0.8498
	8	13	0.0564	0.5	0.4324	0.3968	0.3850
				1.0	0.8779	0.8447	0.8353
	9	14	0.0555	0.5	0.3933	0.3736	0.3675
				1.0	0.8373	0.8154	0.8119
	10	15	0.0527	0.5	0.3489	0.3422	0.3418
				1.0	0.7822	0.7753	0.7771
$M_{(r)}$	2	5	0.0469	0.5	0.6572	0.3896	0.3247
				1.0	0.9934	0.8856	0.7900
	3	6	0.0302	0.5	0.4874	0.2723	0.2329
				1.0	0.9782	0.7935	0.6879
	4	6	0.0403	0.5	0.4946	0.2895	0.2546
				1.0	0.9785	0.8008	0.7034
	5	6	0.0503	0.5	0.4997	0.3021	0.2712
				1.0	0.9789	0.8047	0.7122
	6	6	0.0603	0.5	0.5040	0.3120	0.2835
				1.0	0.9791	0.8073	0.7174
	7	6	0.0702	0.5	0.5075	0.3200	0.2933
				1.0	0.9792	0.8092	0.7208
	8	7	0.0332	0.5	0.3307	0.1843	0.1647
				1.0	0.9415	0.6640	0.5520
	9	7	0.0374	0.5	0.3318	0.1864	0.1670
				1.0	0.9415	0.6644	0.5526
	10	7	0.0415	0.5	0.3325	0.1881	0.1690
				1.0	0.9416	0.6646	0.5530
W_R		348	0.0482	0.5	0.5701	0.4905	0.4646
				1.0	0.9623	0.9385	0.9275

the power values of the precedence tests, we observe that the masking effect in the precedence test becomes more obvious when r becomes larger. The maximal precedence test eliminates the masking effect present in the precedence test. The power values reveal that the larger the value of r, the more superior the maximal precedence test becomes as compared to the precedence test under an exponential distribution. For example, in Table 4.8, under exponential distribution when $n_1 = n_2 = 20$ and the location-shift equals 0.5, we find the simulated power of the precedence test to be 0.9355 (exact level of significance is 0.0457) while the power of the maximal precedence test is 0.9634 (exact level of significance is 0.0469) for $r = 2$; and the power of the precedence test is 0.6291 (exact level of significance is 0.0527) while the power of the maximal precedence test is 0.9062 (exact level of significance is 0.0602) for $r = 6$.

Through the simulation work, we have seen that the maximal precedence test does avoid the masking effect, but still suffers from a loss of power when compared to the classical Wilcoxon's rank-sum test (based on complete samples). So, we will try next to extend the precedence test in another way. We can see that even though we have suggested an alternative to the precedence test in the form of maximal precedence test, it too is based on frequencies of failures preceding the rth failure. One possible extension may be to construct a Wilcoxon-type rank-test that takes into account the magnitude of the failure times.

4.6 ILLUSTRATIVE EXAMPLES

For the purpose of illustration, let us reconsider the data in Examples 3.3 and 3.4.

Example 4.2. Consider Example 3.3 wherein we had observed only the first five observations from the X-sample that occurred up to the 7th breakdown from the Y-sample. In this case, we have $n_1 = n_2 = 10$, $r = 7$, $m_1 = m_2 = 0$, $m_3 = m_4 = 1$, $m_5 = m_6 = 0$, and $m_7 = 3$. The maximal precedence test statistic is

$$M_{(7)} = \max\{m_1, m_2, \cdots, m_7\} = \max\{0, 0, 1, 1, 0, 0, 3\} = 3.$$

From Table 4.1, the near 5% critical value for $n_1 = n_2 = 10$ and $r = 7$ is 6 (exact level of significance = 0.0379). In this example, $M_{(7)} < 6$, we do not reject the null hypothesis that the two samples are identically distributed at 5% level of significance. The p-value for this test is in fact 0.625.

Example 4.3. Consider Example 3.4 wherein we had observed only the first nine observations from the X-sample that occurred up to the 6th breakdown from the Y-sample. In this case, we have $n_1 = n_2 = 10$, $r = 6$, $m_1 = 5$, $m_2 = m_3 = 0$, $m_4 = 2$, $m_5 = 0$, and $m_6 = 2$. The maximal precedence test statistic is

$$M_{(6)} = \max\{m_1, m_2, \cdots, m_6\} = \max\{5, 0, 0, 2, 0, 2\} = 5$$

and the corresponding p-value is 0.097. From this p-value, the maximal precedence test would lead to accepting H_0 at 5% and rejecting it at 10% level of significance.

Chapter 5

Weighted Precedence and Weighted Maximal Precedence Tests

5.1 INTRODUCTION

In this chapter, we describe weighted precedence and weighted maximal precedence tests, introduced by Ng and Balakrishnan (2005), for testing the hypotheses in (3.1). This is another logical extension of the precedence and maximal precedence tests. The motivation is explained by the following two cases in the example mentioned earlier in Example 3.1.

Example 5.1. A manufacturer of electronic components wishes to compare two designs A and B with respect to life. Specifically, he or she wants to abandon design A if there is significant evidence that it produces components with shorter life. Suppose $n_1 = 10$ components from design A and $n_2 = 10$ components from design B are placed simultaneously on a life-test. The test will be terminated when the 5th failure from the sample from design B occurs. Figures 5.1 and 5.2 show two possible outcomes of the life-testing experiment in this example.

The precedence and maximal precedence test statistics are equal in both cases. The test statistics are $P_{(5)} = 8$ and $M_{(5)} = 5$. In Tables 3.1 and 4.1, we have the critical values with $n_1 = n_2 = 10$ and $r = 5$ as 9 (with level of significance 0.02864) and 6 (with level of significance 0.02709) for the precedence and maximal precedence tests, respectively. Therefore, we

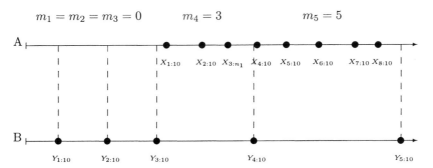

Figure 5.1: Case 1 life-test in Example 5.1

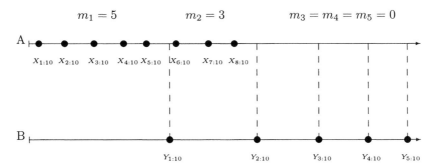

Figure 5.2: Case 2 life-test in Example 5.1

will not reject the null hypothesis that two distributions are equal in both cases at the same level of significance. However, we feel that Case 2 provides much more evidence that the components from design B are better than the components from design A. This suggests that we should try to develop a test procedure that distinguishes between Case 1 and Case 2, and the weighted precedence and weighted maximal precedence tests is one such attempt in this direction.

5.2 TEST STATISTICS AND EXACT NULL DISTRIBUTIONS

In this section, we describe the weighted precedence and weighted maximal precedence tests, introduced by Ng and Balakrishnan (2005), by giving a decreasing weight to m_i as i increases. The weighted precedence test statistic $P^*_{(r)}$ is thus defined as

$$P^*_{(r)} = \sum_{i=1}^{r} (n_2 - i + 1)\, m_i, \tag{5.1}$$

and the weighted maximal precedence test statistic $M^*_{(r)}$ as

$$M^*_{(r)} = \max_{1 \le i \le r} \left\{ (n_2 - i + 1)\, m_i \right\}. \tag{5.2}$$

It is clear that large values of $P^*_{(r)}$ or $M^*_{(r)}$ would lead to the rejection of H_0 and in favor of H_1 in (3.1).

For a fixed level of significance α, the critical region for the weighted precedence test will be $\{s, s+1, \cdots, n_1 n_2\}$, where

$$\alpha = \Pr(P^*_{(r)} \ge s | H_0 : F_X = F_Y). \tag{5.3}$$

It follows from the result of Theorem 4.1 that, for specified values of n_1, n_2, s and r, an expression for α in (5.3) is given by

$$\alpha = \sum_{\substack{m_i (i=1,2,\cdots,r)=0 \\ s \le \sum_{i=1}^{r} (n_2-i+1)m_i \le n_1 n_2}}^{n_1} \Pr(M_1 = m_1, \cdots, M_r = m_r | H_0 : F_X = F_Y). \tag{5.4}$$

Similarly, for a fixed level of significance α, the critical region for the weighted maximal precedence test will be $\{t, t+1, \cdots, n_1 n_2\}$, where

$$\alpha = \Pr(M^*_{(r)} \geq t | H_0 : F_X = F_Y)$$

$$= \sum_{\substack{m_i(i=1,2,\cdots,r)=0 \\ t \leq \max_{1 \leq i \leq r}(n_2-i+1)m_i \leq n_1 n_2}}^{n_1} \Pr(M_1 = m_1, \cdots, M_r = m_r | H_0 : F_X = F_Y),$$

$$(5.5)$$

where

$$\Pr(M_1 = m_1, \cdots, M_r = m_r | H_0 : F_X = F_Y) = \frac{\binom{n_1 + n_2 - \sum_{i=1}^{r} m_i - r}{n_2 - r}}{\binom{n_1 + n_2}{n_2}}.$$

The critical values s and t and the exact level of significance α (as close to 5% as possible) for different choices of the sample sizes n_1 and n_2 and $r = 2(1)5$ are presented in Tables 5.1 and 5.2, respectively. It should be noted that for $r = 1$, the test statistics $P^*_{(1)}$ and $M^*_{(1)}$ become $P_{(1)}$ and $M_{(1)}$, respectively.

For example, refer to Case 1 of Example 5.1 (see Figure 5.1); in this case, the weighted precedence and weighted maximal precedence test statistics are $P^*_{(5)} = (7 \times 3) + (6 \times 5) = 51$ and $M^*_{(5)} = \max\{(7 \times 3), (6 \times 5)\} = 30$. Similarly, in Case 2 (see Figure 5.2), the weighted precedence and weighted maximal precedence test statistics are $P^*_{(5)} = (10 \times 5) + (9 \times 3) = 77$ and $M^*_{(5)} = \max\{(10 \times 5), (9 \times 3)\} = 50$. Comparing these with the critical values in Tables 5.1 and 5.2, we will not reject the null hypothesis that two distributions are equal in Case 1, but we will conclude that design B is better than design A in Case 2 at 5% level of significance.

5.3 EXACT POWER FUNCTION UNDER LEHMANN ALTERNATIVE

Here, we present the explicit expression for the power function of the weighted precedence and weighted maximal precedence tests under the Lehmann alternative $H_1 : [F_X]^\gamma = F_Y$ for some γ, as derived by Ng and Balakrishnan (2005); see Section 3.4 for pertinent expressions.

Table 5.1: Near 5% critical values and exact levels of significance for the weighted precedence test statistic $P_{(r)}^*$

n_1	n_2	$r = 2$	$r = 3$	$r = 4$	$r = 5$
		s (exact l.o.s.)	s (exact l.o.s.)	s (exact l.o.s.)	s (exact l.o.s.)
10	10	48(0.04954)	56(0.05096)	62(0.05065)	66(0.05108)
15	15	75(0.04741)	95(0.04945)	107(0.05094)	118(0.04991)
20	20	100(0.05326)	133(0.04979)	152(0.05023)	171(0.04967)
30	30	174(0.05139)	205(0.05082)	250(0.04964)	280(0.04996)
30	50	198(0.04803)	245(0.04945)	290(0.05040)	333(0.04965)

Table 5.2: Near 5% critical values and exact levels of significance for the weighted maximal precedence test statistic $M_{(r)}^*$

n_1	n_2	$r = 2$	$r = 3$	$r = 4$	$r = 5$
		t (exact l.o.s.)	t (exact l.o.s.)	t (exact l.o.s.)	t (exact l.o.s.)
10	10	40(0.05954)	45(0.03792)	42(0.04334)	42(0.04489)
15	15	70(0.04205)	70(0.05042)	72(0.04627)	72(0.04942)
20	20	95(0.04691)	100(0.04359)	100(0.05357)	102(0.04439)
30	30	145(0.05179)	150(0.04943)	162(0.04707)	162(0.05225)
30	50	150(0.06610)	192(0.05138)	196(0.04604)	196(0.05181)

The power functions under the Lehmann alternative for weighted precedence and weighted maximal precedence tests are given by

Power

$$
= \sum_{\substack{m_i(i=1,2,\cdots,r)=0 \\ s\leq\sum_{i=1}^{r}(n_2-i+1)m_i\leq n_1 n_2}}^{n_1} \Pr(M_1 = m_1,\cdots, M_r = m_r | H_1 : [F_X]^\gamma = F_Y),
$$

$$(5.6)$$

and

Power

$$
= \sum_{\substack{m_i(i=1,2,\cdots,r)=0 \\ t\leq \max_{1\leq i\leq r}(n_2-i+1)m_i\leq n_1 n_2}}^{n_1} \Pr(M_1 = m_1,\cdots, M_r = m_r | H_1 : [F_X]^\gamma = F_Y),
$$

$$(5.7)$$

respectively, where

$$
\Pr(M_1 = m_1,\cdots, M_r = m_r | H_1 : [F_X]^\gamma = F_Y)
$$

$$
= \frac{n_1! n_2! \gamma^r}{m_1!(n_2-r)!} \left\{ \prod_{j=1}^{r-1} \frac{\Gamma\left(\sum_{i=1}^{j} m_i + j\gamma\right)}{\Gamma\left(\sum_{i=1}^{j+1} m_i + j\gamma + 1\right)} \right\}
$$

$$
\times \sum_{k=0}^{n_2-r} \binom{n_2-r}{k} (-1)^k \frac{\Gamma\left(\sum_{i=1}^{r} m_i + (r+k)\gamma\right)}{\Gamma\left(n_1 + (r+k)\gamma + 1\right)}, \qquad (5.8)
$$

as derived earlier in Eq. (3.5).

Now, we compare the power under the Lehmann alternative for the precedence test (PT), the maximal precedence test (MPT), the weighted precedence test (WPT), and the weighted maximal precedence test (WMPT). For $n_1 = n_2 = 10$, $r = 2, 3$, and $\gamma = 2(1)6$, the exact power values computed from (5.6)–(5.8) are presented in Table 5.3.

From Table 5.3, we see clearly that the weighted precedence and weighted maximal precedence tests yield better power than the precedence and maximal precedence tests under the Lehmann alternative, with the weighted precedence test being overall the best.

Table 5.3: Power comparison of precedence, maximal precedence, weighted precedence, and weighted maximal precedence tests under Lehmann alternative for $n_1 = n_2 = 10$, $r = 2, 3$, and $\gamma = 2(1)6$

r	Test	$\gamma = 1$	$\gamma = 2$	$\gamma = 3$	$\gamma = 4$	$\gamma = 5$	$\gamma = 6$
2	$P_{(2)}$	0.02864	0.23970	0.49048	0.67299	0.78963	0.86216
	$M_{(2)}$	0.03250	0.22095	0.45342	0.63634	0.76059	0.84119
	$P_{(2)}^*$	0.04954	0.35681	0.64389	0.80970	0.89619	0.94126
	$M_{(2)}^*$	0.05954	0.35732	0.62980	0.79435	0.88451	0.93342
3	$P_{(2)}$	0.03489	0.24581	0.48087	0.65212	0.76460	0.83731
	$M_{(2)}$	0.04875	0.24178	0.46802	0.64561	0.76641	0.84490
	$P_{(2)}^*$	0.05096	0.35100	0.63139	0.79597	0.88402	0.93137
	$M_{(2)}^*$	0.03792	0.22811	0.45815	0.63914	0.76225	0.84220

5.4 MONTE CARLO SIMULATION UNDER LOCATION-SHIFT ALTERNATIVE

In order to further examine the power properties of the precedence and maximal precedence tests with the corresponding weighted versions, we consider the location-shift alternative $H_1 : F_X(x) = F_Y(x + \theta)$ for some $\theta > 0$, where θ is a shift in location, as before.

Power values of all four tests were estimated by Ng and Balakrishnan (2005) through Monte Carlo simulations when $\theta = 0.5$ and 1.0. The following lifetime distributions were used:

1. Standard normal distribution

2. Standard exponential distribution

3. Gamma distribution with shape parameter a and standardized by mean a and standard deviation $\sqrt{a}$

4. Lognormal distribution with shape parameter σ and standardized by mean $e^{\sigma^2/2}$ and standard deviation $\sqrt{e^{\sigma^2}(e^{\sigma^2} - 1)}$

5. Standard extreme-value distribution standardized by mean $-0.5772\ldots$ (Euler's constant) and standard deviation $\pi/\sqrt{6}$

A brief description of these distributions and their properties has been provided in Section 2.5. For different choices of sample sizes, 10,000 sets of data were generated in order to obtain the estimated rejection rates. In Tables 5.4–5.7, we have presented the estimated power values of the precedence and maximal precedence tests and their weighted versions for $r = 2(1)6$ for the distributions listed above with location-shift being equal to 0.5 and 1.0. For comparative purposes, the corresponding exact levels of significance are also presented.

When we compare the power values of the weighted precedence and weighted maximal precedence tests with those of precedence and maximal precedence tests, we find that the former tests perform better than the latter tests if the underlying distributions are right-skewed, such as the exponential distribution, gamma distribution with shape parameter $a = 2.0$, and lognormal distribution with shape parameter $\sigma = 0.5$. In addition, even if the underlying distributions are nearly symmetric, the power values of the weighted precedence and weighted maximal precedence tests are close to those of the precedence and maximal precedence tests. Overall, we find that the weighted precedence test outperforms the other tests in almost all cases considered.

5.5   ILLUSTRATIVE EXAMPLES

We illustrate here the weighted precedence and weighted maximal precedence tests with the data presented earlier in Example 3.3.

Example 5.2. As in Example 3.3, let us take the X- and Y-samples to be Samples 3 and 2 in Nelson (1982, p. 462), respectively. In this example, if we had observed only up to the 6th breakdown from the Y-sample, then we would have observed only the first five X-failures. The resulting observations are presented in Table 5.8.

In this case, we have $n_1 = n_2 = 10$, $r_1 = 5$, and $r_2 = 6$, respectively.

Parametric Procedure. Following Example 3.3, based on the assumption of exponentiality for the two samples, we find the MLEs of θ_X and θ_Y to be (see Section 2.3.2)

$$\hat{\theta}_X = \frac{1}{r_1} \left[\sum_{i=1}^{r_1} X_{i:10} + (n_1 - r_1) X_{r_1:10} \right] = 1.872$$

Table 5.4: Power of precedence, maximal precedence, weighted precedence, and weighted maximal precedence tests when $n_1 = n_2 = 10$ and $\theta = 0.5$

r	Dist.	$P_{(r)}$	$M_{(r)}$	$P^*_{(r)}$	$M^*_{(r)}$
2	Exact l.o.s.	0.0286	0.0325	0.0495	0.0595
	N(0,1)	0.1496	0.1417	0.2275	0.2309
	Exp(1)	0.3916	0.4923	0.6145	0.7197
	Gamma(2)	0.2525	0.2474	0.4080	0.4344
	Gamma(10)	0.1731	0.1611	0.2708	0.2735
	LN(0.1)	0.1603	0.1479	0.2379	0.2457
	LN(0.5)	0.2693	0.2543	0.4206	0.4301
	EV	0.1463	0.1388	0.2002	0.2034
3	Exact l.o.s.	0.0349	0.0487	0.0510	0.0379
	N(0,1)	0.1779	0.1753	0.2400	0.1542
	Exp(1)	0.3073	0.4949	0.5124	0.4886
	Gamma(2)	0.2343	0.2579	0.3707	0.2483
	Gamma(10)	0.1893	0.1870	0.2724	0.1739
	LN(0.1)	0.1828	0.1781	0.2518	0.1562
	LN(0.5)	0.2571	0.2678	0.3839	0.2559
	EV	0.1941	0.1869	0.2414	0.1584
4	Exact l.o.s.	0.0348	0.0650	0.0507	0.0449
	N(0,1)	0.1771	0.1993	0.2492	0.1715
	Exp(1)	0.2251	0.4974	0.4210	0.4854
	Gamma(2)	0.1936	0.2653	0.3360	0.2522
	Gamma(10)	0.1755	0.2040	0.2684	0.1791
	LN(0.1)	0.1777	0.1990	0.2560	0.1709
	LN(0.5)	0.2149	0.2768	0.3549	0.2671
	EV	0.2221	0.2228	0.2720	0.1799
5	Exact l.o.s.	0.0271	0.0271	0.0511	0.0449
	N(0,1)	0.1473	0.0901	0.2558	0.1638
	Exp(1)	0.1453	0.2684	0.4246	0.4857
	Gamma(2)	0.1380	0.1197	0.3372	0.2512
	Gamma(10)	0.1388	0.0895	0.2669	0.1739
	LN(0.1)	0.1444	0.0905	0.2609	0.1684
	LN(0.5)	0.1527	0.1297	0.3541	0.2702
	EV	0.2143	0.1131	0.2935	0.1699
6	Exact l.o.s.	0.0704	0.0325	0.0490	0.0449
	N(0,1)	0.2659	0.0938	0.2589	0.1715
	Exp(1)	0.2385	0.2690	0.4105	0.4854
	Gamma(2)	0.2344	0.1208	0.3255	0.2522
	Gamma(10)	0.2436	0.0921	0.2794	0.1791
	LN(0.1)	0.2567	0.0938	0.2597	0.1709
	LN(0.5)	0.2531	0.1309	0.3466	0.2671
	EV	0.3716	0.1203	0.3021	0.1799

Table 5.5: Power of precedence, maximal precedence, weighted precedence, and weighted maximal precedence tests when $n_1 = n_2 = 10$ and $\theta = 1.0$

r	Dist.	$P_{(r)}$	$M_{(r)}$	$P^*_{(r)}$	$M^*_{(r)}$
2	Exact l.o.s.	0.0286	0.0325	0.0495	0.0595
	N(0,1)	0.4243	0.3768	0.5278	0.5067
	Exp(1)	0.8376	0.9177	0.9468	0.9784
	Gamma(2)	0.6879	0.7303	0.8525	0.8881
	Gamma(10)	0.5060	0.4569	0.6487	0.6313
	LN(0.1)	0.4571	0.4083	0.5803	0.5535
	LN(0.5)	0.7253	0.7327	0.8727	0.8801
	EV	0.3969	0.3573	0.4691	0.4504
3	Exact l.o.s.	0.0349	0.0487	0.0510	0.0379
	N(0,1)	0.4813	0.4171	0.5839	0.3942
	Exp(1)	0.7239	0.9178	0.8824	0.9164
	Gamma(2)	0.6148	0.7327	0.7987	0.7338
	Gamma(10)	0.5177	0.4765	0.6540	0.4610
	LN(0.1)	0.4964	0.4395	0.6087	0.4145
	LN(0.5)	0.6676	0.7360	0.8221	0.7289
	EV	0.5204	0.4398	0.5769	0.4007
4	Exact l.o.s.	0.0348	0.0650	0.0507	0.0449
	N(0,1)	0.4799	0.4359	0.6083	0.3984
	Exp(1)	0.5767	0.9181	0.7921	0.9155
	Gamma(2)	0.5114	0.7344	0.7172	0.7446
	Gamma(10)	0.4758	0.4853	0.6435	0.4711
	LN(0.1)	0.4783	0.4538	0.6165	0.4170
	LN(0.5)	0.5656	0.7376	0.7765	0.7381
	EV	0.5935	0.4862	0.6499	0.4330
5	Exact l.o.s.	0.0271	0.0271	0.0511	0.0449
	N(0,1)	0.4245	0.2482	0.6332	0.4047
	Exp(1)	0.3960	0.7791	0.8193	0.9111
	Gamma(2)	0.3725	0.5136	0.7373	0.7409
	Gamma(10)	0.3833	0.2833	0.6558	0.4684
	LN(0.1)	0.4038	0.2586	0.6334	0.4262
	LN(0.5)	0.4144	0.5284	0.7821	0.7464
	EV	0.6073	0.2945	0.7073	0.4223
6	Exact l.o.s.	0.0704	0.0325	0.0490	0.0449
	N(0,1)	0.5832	0.2496	0.6372	0.3984
	Exp(1)	0.4991	0.7791	0.7965	0.9155
	Gamma(2)	0.4954	0.5138	0.7200	0.7446
	Gamma(10)	0.5266	0.2839	0.6643	0.4711
	LN(0.1)	0.5545	0.2596	0.6380	0.4170
	LN(0.5)	0.5314	0.5286	0.7620	0.7381
	EV	0.7785	0.3000	0.7407	0.4330

Table 5.6: Power of precedence, maximal precedence, weighted precedence, and weighted maximal precedence tests when $n_1 = n_2 = 20$ and $\theta = 0.5$

r	Dist.	$P_{(r)}$	$M_{(r)}$	$P^*_{(r)}$	$M^*_{(r)}$
2	Exact l.o.s.	0.0457	0.0469	0.0533	0.0469
	N(0,1)	0.2662	0.2434	0.2906	0.2401
	Exp(1)	0.9355	0.9634	0.9700	0.9642
	Gamma(2)	0.6711	0.6580	0.7352	0.6563
	Gamma(10)	0.3638	0.3235	0.3929	0.3240
	LN(0.1)	0.3038	0.2741	0.3201	0.2664
	LN(0.5)	0.6453	0.5983	0.6917	0.6048
	EV	0.1992	0.1865	0.2154	0.1845
3	Exact l.o.s.	0.0637	0.0302	0.0498	0.0436
	N(0,1)	0.3485	0.1811	0.3096	0.2328
	Exp(1)	0.9034	0.9061	0.9015	0.9626
	Gamma(2)	0.6767	0.4842	0.6741	0.6488
	Gamma(10)	0.4391	0.2333	0.4102	0.3169
	LN(0.1)	0.3833	0.2009	0.3486	0.2699
	LN(0.5)	0.6838	0.4468	0.6769	0.5879
	EV	0.2864	0.1549	0.2493	0.1933
4	Exact l.o.s.	0.0412	0.0403	0.0502	0.0536
	N(0,1)	0.2947	0.2138	0.3301	0.2576
	Exp(1)	0.7499	0.9053	0.8582	0.9617
	Gamma(2)	0.5317	0.4949	0.6453	0.6603
	Gamma(10)	0.3583	0.2546	0.4231	0.3442
	LN(0.1)	0.3219	0.2273	0.3695	0.2964
	LN(0.5)	0.5614	0.4572	0.6640	0.5993
	EV	0.2642	0.1963	0.2864	0.2304
5	Exact l.o.s.	0.0479	0.0503	0.0497	0.0444
	N(0,1)	0.3317	0.2360	0.3501	0.2251
	Exp(1)	0.6904	0.9058	0.8043	0.9038
	Gamma(2)	0.5173	0.5004	0.5978	0.4968
	Gamma(10)	0.3808	0.2709	0.4189	0.2631
	LN(0.1)	0.3525	0.2472	0.3816	0.2471
	LN(0.5)	0.5555	0.4642	0.6250	0.4574
	EV	0.3182	0.2276	0.3128	0.2010
6	Exact l.o.s.	0.0527	0.0602	0.0495	0.0485
	N(0,1)	0.3565	0.2545	0.3598	0.2252
	Exp(1)	0.6291	0.9062	0.7498	0.9023
	Gamma(2)	0.4966	0.5044	0.5678	0.4969
	Gamma(10)	0.3920	0.2834	0.4210	0.2667
	LN(0.1)	0.3706	0.2626	0.3777	0.2428
	LN(0.5)	0.5369	0.4688	0.5982	0.4581
	EV	0.3620	0.2560	0.3500	0.2254

Table 5.7: Power of precedence, maximal precedence, weighted precedence, and weighted maximal precedence tests when $n_1 = n_2 = 20$ and $\theta = 1.0$

r	Dist.	$P_{(r)}$	$M_{(r)}$	$P_{(r)}^*$	$M_{(r)}^*$
2	Exact l.o.s.	0.0457	0.0469	0.0533	0.0469
	N(0,1)	0.6529	0.5994	0.6780	0.5985
	Exp(1)	0.9998	1.0000	0.9998	0.9998
	Gamma(2)	0.9914	0.9934	0.9958	0.9932
	Gamma(10)	0.8371	0.7893	0.8599	0.7905
	LN(0.1)	0.7338	0.6788	0.7575	0.6743
	LN(0.5)	0.9901	0.9882	0.9940	0.9885
	EV	0.4774	0.4412	0.5001	0.4481
3	Exact l.o.s.	0.0637	0.0302	0.0498	0.0436
	N(0,1)	0.7615	0.5249	0.7359	0.6088
	Exp(1)	0.9993	0.9996	0.9994	1.0000
	Gamma(2)	0.9886	0.9785	0.9886	0.9913
	Gamma(10)	0.8891	0.6880	0.8816	0.7941
	LN(0.1)	0.8210	0.5894	0.8047	0.6874
	LN(0.5)	0.9907	0.9644	0.9908	0.9857
	EV	0.6310	0.4285	0.5919	0.4823
4	Exact l.o.s.	0.0412	0.0403	0.0502	0.0536
	N(0,1)	0.7352	0.5605	0.7800	0.6381
	Exp(1)	0.9932	0.9997	0.9986	1.0000
	Gamma(2)	0.9615	0.9790	0.9829	0.9919
	Gamma(10)	0.8418	0.7030	0.8900	0.7923
	LN(0.1)	0.7877	0.6170	0.8298	0.7160
	LN(0.5)	0.9716	0.9662	0.9881	0.9872
	EV	0.6594	0.5061	0.6751	0.5459
5	Exact l.o.s.	0.0479	0.0503	0.0497	0.0444
	N(0,1)	0.7787	0.5833	0.7999	0.5641
	Exp(1)	0.9864	0.9997	0.9964	0.9993
	Gamma(2)	0.9487	0.9793	0.9749	0.9770
	Gamma(10)	0.8532	0.7115	0.8852	0.7071
	LN(0.1)	0.8163	0.6327	0.8401	0.6320
	LN(0.5)	0.9648	0.9666	0.9840	0.9641
	EV	0.7423	0.5564	0.7448	0.5358
6	Exact l.o.s.	0.0527	0.0602	0.0495	0.0485
	N(0,1)	0.8055	0.5977	0.8208	0.5729
	Exp(1)	0.9733	0.9997	0.9912	0.9997
	Gamma(2)	0.9323	0.9796	0.9642	0.9751
	Gamma(10)	0.8558	0.7171	0.8813	0.7015
	LN(0.1)	0.8295	0.6423	0.8604	0.6303
	LN(0.5)	0.9540	0.9668	0.9770	0.9668
	EV	0.8055	0.5974	0.8039	0.5609

Table 5.8: Times to insulating fluid breakdown data from Nelson (1982) for Samples 2 and 3

X-sample (Sample 3)	0.49	0.64	0.82	0.93	1.08	*	*	*	*	*
Y-sample (Sample 2)	0.00	0.18	0.55	0.66	0.71	1.30	*	*	*	*

$$\text{and} \quad \hat{\theta}_Y = \frac{1}{r_2}\left[\sum_{i=1}^{r_2} Y_{i:10} + (n_2 - r_2)Y_{r_2:10}\right] = 1.433.$$

Then, $\hat{\theta}_X/\hat{\theta}_Y = 1.306$ and the 95% one-sided confidence interval for θ_X/θ_Y is $(0, 3.804)$. Since this interval includes the value 1, we conclude that the data provide enough evidence to accept H_0. In fact the p-value of the test is 0.674, based on which we do not reject H_0 in (3.6).

Nonparametric Procedure. Considering the same data, we can carry out a nonparametric test for hypotheses in (3.1). In this case, we would have $n_1 = n_2 = 10$, $r = 5$, $m_1 = m_2 = 0$, $m_3 = m_4 = 1$, $m_5 = 0$, and $m_6 = 3$. From these, we have the following values for the test statistics and the corresponding p-values:

Test	Test statistic	p-value
Precedence test	$P_{(6)} = 5$	0.672
Maximal precedence test	$M_{(6)} = 3$	0.552
Weighted precedence test	$P_{(6)}^* = 30$	0.744
Weighted maximal precedence test	$M_{(6)}^* = 15$	0.756

Based on these p-values, we once again do not reject the null hypothesis that the two samples are identically distributed by any of the tests.

Chapter 6

Wilcoxon-type Rank-sum Precedence Tests

6.1 INTRODUCTION

First, recall from Section 2.6 that for testing the hypotheses in (3.1), if complete samples of size n_1 and n_2 were available, then the standard Wilcoxon's rank-sum statistic, proposed by Wilcoxon (1945), is

$$W_R = R_1 + R_2 + \cdots + R_{n_1}, \tag{6.1}$$

where $(R_1, R_2, \cdots, R_{n_1})$ are the ranks of the X-failures among all failures. Small values of W_R would lead to the rejection of H_0. If there are no ties, the exact mean and variance of W_R, under the null hypothesis of equal distributions, are given by

$$E(W_R) = \frac{n_1(n_1 + n_2 + 1)}{2}, \quad Var(W_R) = \frac{n_1 n_2(n_1 + n_2 + 1)}{12}, \tag{6.2}$$

and the statistic is symmetric about its mean.

In this chapter, we describe Wilcoxon-type rank-sum precedence tests, introduced by Ng and Balakrishnan (2004), for testing the hypotheses in (3.1) when the Y-sample is observed only until its rth failure (i.e., Type-II censored sample). This test is a variation of the precedence life-test described in Section 3.2 and is a generalization of the Wilcoxon rank-sum test. Three Wilcoxon-type rank-sum precedence test statistics—the minimal, maximal, and expected rank-sum statistics—are presented in Section 6.2. We also present the null distributions of these three Wilcoxon-type rank-sum precedence test statistics. Critical values for some choices of sample sizes are presented. After noting that the large-sample normal approximation for the null distribution is not satisfactory in the case of small or moderate sample sizes, we describe in Section 6.3 an Edgeworth approximation to the significance probabilities. Next, in Section 6.4, we present the exact power function under the Lehmann alternative. We also examine the power properties of the Wilcoxon-type rank-sum precedence tests under a location-shift alternative through Monte Carlo simulations in Section 6.5. Next, in Section 6.6, some comparative remarks are made among the precedence test, the maximal precedence test, and the Wilcoxon's rank-sum test (based on complete samples). Finally, in Section 6.7, two examples are used to illustrate the Wilcoxon-type rank-sum precedence tests.

6.2   TEST STATISTICS AND EXACT NULL DISTRIBUTIONS

In this section, we describe three new test statistics of Wilcoxon-type, introduced by Ng and Balakrishnan (2004), for the precedence testing situation described in Section 3.2. Their null distributions are also presented.

6.2.1   Wilcoxon-type Rank-sum Precedence Test Statistics

In order to test the hypotheses in (3.1), instead of using the maximum of the frequencies of failures from the first sample between the first r failures of the second sample (which avoids masking effects, but suffers from lack of power), one could use the sum of the ranks of those failures and construct Wilcoxon-type analogs. These tests will be particularly useful when data are not completely observed. Further, these tests, in addition to avoiding the masking effect, will also retain high power as compared with the complete-sample Wilcoxon rank-sum test.

More specifically, suppose $m_1, m_2, \cdots, m_r$ denote the number of X-failures that occurred before the first, between the first and the second, $\cdots$, between the $(r-1)$th and rth Y-failures, respectively; see Figure 3.1. For notational convenience, let us denote $S_r = \sum_{i=1}^{r} M_i$ and $S_r^* = \sum_{i=1}^{r} iM_i$, and their realizations by s_r and s_r^*, respectively. Let W_r be the rank-sum of the X-failures that occurred before the rth Y-failure. The Wilcoxon's test statistic will be smallest when all the remaining $(n_1 - s_r)$ X-failures occur between the rth and $(r+1)$th Y-failures. The test statistic in this case would then become

$$\begin{aligned} W_{\min,r} &= W_r + (S_r + r + 1) + (S_r + r + 2) + \cdots + (n_1 + r) \\ &= \frac{n_1(n_1 + 2r + 1)}{2} - (r+1)S_r + S_r^*. \end{aligned}$$

This is referred to as the *minimal rank-sum statistic.*

The Wilcoxon's test statistic will be the largest when all the remaining $(n_1 - s_r)$ X-failures occur after the n_2th Y-failure. Such a test statistic is referred to as the *maximal rank-sum statistic* and is given by

$$\begin{aligned} W_{\max,r} &= W_r + (S_r + n_2 + 1) + (S_r + n_2 + 2) + \cdots + (n_1 + n_2) \\ &= \frac{n_1(n_1 + 2n_2 + 1)}{2} - (n_2 + 1)S_r + S_r^*. \end{aligned}$$

We could similarly propose a rank-sum statistic using the expected rank sums of failures from the X-sample between the rth and the $(r+1)$th, $\cdots$, after the n_2th failures of the Y-sample, denoted by $W_{E,r}$. It can be shown that $W_{E,r}$ is simply the average of $W_{\min,r}$ and $W_{\max,r}$, and is given by

$$\begin{aligned} W_{E,r} &= W_r + \frac{1}{2}[(S_r + r + 1) + (S_r + r + 2) + \cdots + (n_1 + r)] \\ &\quad + \frac{1}{2}[(S_r + n_2 + 1) + (S_r + n_2 + 2) + \cdots + (n_1 + n_2)] \\ &= \frac{n_1(n_1 + n_2 + r + 1)}{2} - \left(\frac{n_2 + r}{2} + 1\right)S_r + S_r^*. \end{aligned}$$

For example, from Figure 3.1, when $n_1 = n_2 = 10$ with $r = 4$, we have

$$\begin{aligned} W_{\min,4} &= 2 + 3 + 4 + 6 + 7 + 8 + 9 + 11 + 13 + 14 = 77, \\ W_{\max,4} &= 2 + 3 + 4 + 6 + 7 + 8 + 9 + 11 + 17 + 18 = 89, \\ W_{E,4} &= \frac{77 + 89}{2} = 83. \end{aligned}$$

It is evident that small values of $W_{\min,r}$, $W_{\max,r}$, and $W_{E,r}$ lead to the rejection of H_0 and in favor of H_1 in (3.1). Moreover, in the special case when $r = n_2$ (that is, when we observe all the failures from the second sample), we have

$W_{\min,n_2} = W_{\max,n_2} = W_{E,n_2}$ and, in this case, they are all equivalent to the classical Wilcoxon's rank-sum statistic defined in (6.1).

As seen above, these forms are natural extensions of Wilcoxon's rank-sum statistic as they accommodate the unobserved $n_2 - r$ last Y-failures. It will, therefore, be of interest to compare the relative merits of these three Wilcoxon-type rank-sum statistics even though they are motivated by a similar idea. As will be shown later, the minimal rank-sum test turns out to be overall slightly better than the other two.

6.2.2   Null Distributions

The joint probability mass function of $M_1, M_2, \cdots, M_r$ under the null hypothesis $H_0 : F_X = F_Y$ is given in Theorem 3.1 as

$$\Pr\left\{M_1 = m_1, M_2 = m_2, \cdots, M_r = m_r \mid H_0 : F_X = F_Y\right\}$$
$$= \frac{\dbinom{n_1 + n_2 - s_r - r}{n_2 - r}}{\dbinom{n_1 + n_2}{n_2}}.$$

For the minimal rank-sum test statistic, we have

$$\Pr\left(W_{\min,r} = w \mid M_1 = m_1, M_2 = m_2, \cdots, M_r = m_r\right)$$
$$= \begin{cases} 1 & \text{if } w = \frac{n_1(n_1 + 2r + 1)}{2} - (r+1)s_r + s_r^*, \\ 0 & \text{otherwise.} \end{cases}$$

Therefore, the null distribution of $W_{\min,r}$ is simply given by

$$\Pr(W_{\min,r} = w \mid H_0 : F_X = F_Y)$$
$$= \Pr\left(\frac{n_1(n_1 + 2r + 1)}{2} - (r+1)S_r + S_r^* = w \,\middle|\, H_0 : F_X = F_Y\right)$$
$$= \sum_{\substack{m_i(i=1,2,\cdots,r)=0,\ s_r \leq n_1 \\ \frac{n_1(n_1+2r+1)}{2} - (r+1)s_r + s_r^* = w}}^{n_1} \frac{\dbinom{n_1 + n_2 - s_r - r}{n_2 - r}}{\dbinom{n_1 + n_2}{n_2}}. \tag{6.3}$$

For specified values of n_1, n_2, r and the level of significance α, the critical value c (corresponding to a level closest to α) for the minimal rank-sum precedence test can then be found from (6.3) as

$$\alpha = \Pr(W_{\min,r} \le c \mid H_0 : F_X = F_Y)$$

$$= \Pr\left(\frac{n_1(n_1 + 2r + 1)}{2} - (r+1)S_r + S_r^* \le c \;\middle|\; F_X = F_Y\right)$$

$$= \sum_{m_i(i=1,2,\cdots,r)=0}^{n_1} \frac{\binom{n_1 + n_2 - s_r - r}{n_2 - r}}{\binom{n_1 + n_2}{n_2}} I_{\min,c}(m_1, m_2, \cdots, m_r),$$

where $I_{\min,c}$ is the indicator function defined by

$$I_{\min,c}(m_1, \cdots, m_r) = \begin{cases} 1 & \text{if } \frac{n_1(n_1+2r+1)}{2} - (r+1)s_r + s_r^* \le c \\ & \text{and } s_r \le n_1, \\ 0 & \text{otherwise.} \end{cases} \tag{6.4}$$

The null distributions of the maximal rank-sum statistic and the expected rank-sum statistic can be derived in a similar manner. The critical value c for the maximal rank-sum statistic and the expected rank-sum statistic can then be shown to be

$$\alpha = \Pr(W_{\max,r} \le c \mid H_0 : F_X = F_Y)$$

$$= \Pr\left(\frac{n_1(n_1 + 2n_2 + 1)}{2} - (n_2 + 1)S_r + S_r^* \le c \;\middle|\; F_X = F_Y\right)$$

$$= \sum_{m_i(i=1,2,\cdots,r)=0}^{n_1} \frac{\binom{n_1 + n_2 - s_r - r}{n_2 - r}}{\binom{n_1 + n_2}{n_2}} I_{\max,c}(m_1, m_2, \cdots, m_r),$$

and

$$\alpha = \Pr(W_{E,r} \le c \mid H_0 : F_X = F_Y)$$

$$= \Pr\left(\frac{n_1(n_1 + n_2 + r + 1)}{2} - \left(\frac{n_2 + r}{2} + 1\right)S_r + S_r^* \le c \;\middle|\; F_X = F_Y\right)$$

$$= \sum_{m_i(i=1,2,\cdots,r)=0}^{n_1} \frac{\binom{n_1 + n_2 - s_r - r}{n_2 - r}}{\binom{n_1 + n_2}{n_2}} I_{E,c}(m_1, m_2, \cdots, m_r),$$

respectively, where $I_{\max,c}$ is the indicator function defined by

$$
I_{\max,c}(m_1, \cdots, m_r) = \begin{cases} 1 & \text{if } \frac{n_1(n_1+2n_2+1)}{2} - (n_2+1)s_r + s_r^* \leq c \\ & \text{and } s_r \leq n_1 \\ \\ 0 & \text{otherwise,} \end{cases}
\tag{6.5}
$$

and $I_{E,c}$ is the indicator function defined by

$$
I_{E,c}(m_1, \cdots, m_r) = \begin{cases} 1 & \text{if } \frac{n_1(n_1+n_2+r+1)}{2} - \left(\frac{n_2+r}{2} + 1\right)s_r + s_r^* \leq c \\ & \text{and } s_r \leq n_1 \\ \\ 0 & \text{otherwise.} \end{cases}
\tag{6.6}
$$

In Tables 6.1 and 6.2, we have presented the critical value c and the exact level of significance α (as close as possible to 5% and 10%), respectively, for some sample sizes n_1 and n_2 and $r = 2(1)7$ for the minimal, maximal, and expected rank-sum precedence test statistics.

Instead of testing the hypotheses in (3.1), we can also use these three test statistics to test a two-sided alternative $H_1 : F_X \neq F_Y$. We will reject the null hypothesis for large and small values of the test statistics $W_{\min,r}$, $W_{\max,r}$, and $W_{E,r}$; that is, we reject the null hypothesis if the test statistic is at most c_1 or at least c_2, where $c_1 < c_2$. The critical values (c_1, c_2) for specified values of n_1, n_2, r and the level of significance α can be computed in a straightforward manner from the null distributions presented above.

However, for a specified value of α, there may not be a critical region for which the probabilities on both sides are close to $\alpha/2$. For example, when $n_1 = 30$, $n_2 = 50$, and $r = 5$, the maximum possible value of $W_{\min,5}$ is 615 and the probability that $W_{\min,5} = 615$ is 0.0881 under the null hypothesis. Thus, it will be impossible to find a critical region corresponding to $\alpha = 10\%$ that has probability close to 5% on both sides. In this case, we may choose $c_1 = 594$ $[\Pr(W_{\min,5} \leq 594|H_0) = 0.0632]$ and $c_2 = 615$ to attain a critical region with 15.13% level of significance. Similarly, we can construct the critical regions for $W_{\max,5}$ and $W_{E,5}$ in this example, which turn out to be $c_1 = 1671, c_2 = 1965$ (with 15.26% level of significance) and $c_1 = 1131, c_2 = 1290$ (with 15.26% level of significance), respectively.

Table 6.1: Near 5% upper critical values and exact levels of significance for the Wilcoxon's rank-sum precedence test statistics

n_1	n_2	$r = 2$	$r = 3$	$r = 4$
		$W_{\min,r}$		
10	10	67 (0.060)	71 (0.046)	75 (0.054)
10	15	68 (0.033)	74 (0.045)	79 (0.047)
15	15	141 (0.042)	150 (0.047)	158 (0.048)
15	20	143 (0.052)	153 (0.053)	162 (0.050)
20	20	241 (0.047)	255 (0.053)	267 (0.047)
30	30	516 (0.052)	539 (0.046)	561 (0.046)
30	50	518 (0.034)	544 (0.047)	569 (0.052)
		$W_{\max,r}$		
10	10	107 (0.050)	99 (0.051)	93 (0.051)
10	15	147 (0.047)	136 (0.050)	129 (0.050)
15	15	270 (0.047)	250 (0.049)	238 (0.051)
15	20	340 (0.051)	320 (0.049)	302 (0.050)
20	20	510 (0.053)	477 (0.050)	458 (0.050)
30	30	1191 (0.051)	1160 (0.051)	1115 (0.050)
30	50	1767 (0.048)	1720 (0.049)	1675 (0.050)
		$W_{E,r}$		
10	10	87 (0.050)	85 (0.047)	84 (0.049)
10	15	108 (0.047)	106 (0.050)	105 (0.052)
15	15	205 (0.047)	201 (0.049)	199 (0.051)
15	20	241 (0.051)	236 (0.051)	232 (0.052)
20	20	375 (0.053)	367 (0.050)	363 (0.051)
30	30	855 (0.051)	850 (0.051)	841 (0.050)
30	50	1143 (0.048)	1133 (0.049)	1123 (0.050)

Table 6.1: (Continued)

n_1	n_2	$r = 5$	$r = 6$	$r = 7$
		$W_{\min,r}$		
10	10	77 (0.045)	80 (0.055)	81 (0.050)
10	15	84 (0.056)	87 (0.048)	90 (0.048)
15	15	165 (0.050)	171 (0.051)	176 (0.050)
15	20	170 (0.047)	178 (0.051)	185 (0.053)
20	20	279 (0.052)	289 (0.050)	298 (0.048)
30	30	582 (0.047)	602 (0.049)	621 (0.052)
30	50	593 (0.052)	616 (0.052)	638 (0.049)
		$W_{\max,r}$		
10	10	89 (0.051)	86 (0.049)	84 (0.047)
10	15	122 (0.051)	116 (0.051)	111 (0.049)
15	15	227 (0.050)	218 (0.049)	211 (0.049)
15	20	287 (0.050)	274 (0.049)	264 (0.050)
20	20	439 (0.050)	423 (0.050)	410 (0.051)
30	30	1085 (0.050)	1057 (0.051)	1029 (0.050)
30	50	1632 (0.050)	1595 (0.050)	1578 (0.050)
		$W_{E,r}$		
10	10	84 (0.052)	83 (0.049)	83 (0.049)
10	15	103 (0.051)	102 (0.049)	101 (0.049)
15	15	197 (0.052)	195 (0.048)	194 (0.049)
15	20	229 (0.050)	227 (0.051)	225 (0.049)
20	20	360 (0.051)	357 (0.050)	355 (0.049)
30	30	835 (0.050)	830 (0.050)	826 (0.050)
30	50	1115 (0.050)	1110 (0.050)	1106 (0.050)

Table 6.2: Near 10% upper critical values and exact levels of significance for the Wilcoxon's rank-sum precedence test statistics

n_1	n_2	$r = 2$	$r = 3$	$r = 4$
		$W_{\min,r}$		
10	10	68 (0.086)	73 (0.091)	78 (0.111)
10	15	70 (0.103)	76 (0.097)	82 (0.106)
15	15	143 (0.099)	153 (0.109)	161 (0.091)
15	20	144 (0.095)	155 (0.096)	165 (0.101)
20	20	243 (0.104)	257 (0.088)	271 (0.102)
30	30	518 (0.110)	542 (0.096)	565 (0.095)
30	50	520 (0.097)	546 (0.094)	572 (0.104)
		$W_{\max,r}$		
10	10	116 (0.103)	107 (0.102)	100 (0.100)
10	15	161 (0.111)	148 (0.096)	138 (0.099)
15	15	285 (0.100)	264 (0.100)	251 (0.098)
15	20	344 (0.093)	330 (0.101)	318 (0.100)
20	20	530 (0.107)	497 (0.100)	479 (0.102)
30	30	1220 (0.097)	1190 (0.096)	1160 (0.098)
30	50	1816 (0.104)	1768 (0.094)	1723 (0.100)
		$W_{E,r}$		
10	10	92 (0.103)	90 (0.095)	89 (0.100)
10	15	116 (0.111)	112 (0.096)	111 (0.099)
15	15	214 (0.100)	209 (0.096)	207 (0.099)
15	20	245 (0.093)	245 (0.101)	242 (0.102)
20	20	386 (0.107)	378 (0.100)	375 (0.101)
30	30	870 (0.097)	866 (0.096)	862 (0.101)
30	50	1168 (0.104)	1157 (0.094)	1148 (0.100)

Table 6.2: (Continued)

n_1	n_2	$r = 5$	$r = 6$	$r = 7$
		$W_{\min,r}$		
10	10	81 (0.104)	83 (0.095)	85 (0.098)
10	15	87 (0.109)	91 (0.104)	94 (0.094)
15	15	169 (0.099)	176 (0.104)	181 (0.095)
15	20	174 (0.100)	182 (0.097)	190 (0.104)
20	20	283 (0.097)	295 (0.107)	305 (0.104)
30	30	587 (0.096)	608 (0.099)	628 (0.102)
30	50	596 (0.092)	620 (0.094)	644 (0.105)
		$W_{\max,r}$		
10	10	95 (0.098)	92 (0.101)	90 (0.103)
10	15	132 (0.101)	125 (0.098)	120 (0.098)
15	15	240 (0.098)	231 (0.099)	223 (0.098)
15	20	305 (0.101)	292 (0.102)	281 (0.102)
20	20	460 (0.098)	444 (0.100)	430 (0.100)
30	30	1121 (0.100)	1093 (0.101)	1066 (0.099)
30	50	1680 (0.099)	1668 (0.100)	1629 (0.099)
		$W_{E,r}$		
10	10	88 (0.094)	88 (0.103)	88 (0.103)
10	15	109 (0.097)	109 (0.105)	108 (0.103)
15	15	205 (0.099)	204 (0.099)	203 (0.101)
15	20	239 (0.098)	237 (0.099)	236 (0.100)
20	20	372 (0.099)	370 (0.101)	368 (0.099)
30	30	856 (0.099)	852 (0.101)	848 (0.099)
30	50	1140 (0.099)	1140 (0.100)	1135 (0.099)

6.3   LARGE-SAMPLE APPROXIMATION FOR THE NULL DISTRIBUTIONS

For small values of r, n_1, and n_2, the critical values and the exact significance probabilities of the Wilcoxon-type rank-sum precedence test statistics can be computed without any difficulty, as presented in Tables 6.1 and 6.2. However, for large values of r, n_1, or n_2, this would require a heavy computational effort and time. For this reason, we describe in this section some large-sample approximations for the null distributions of the Wilcoxon-type rank-sum precedence statistics, as developed by Ng and Balakrishnan (2002).

By the distributional properties of a linear rank statistic T, we know that $\frac{T-E(T)}{\sqrt{Var(T)}}$ is asymptotically normally distributed with zero mean and unit variance; see, for example, Lehmann (1975). Hence, the distribution function of $\frac{T-E(T)}{\sqrt{Var(T)}}$ can be written in a series form as

$$
\Pr\left\{\frac{T-E(T)}{\sqrt{Var(T)}} \leq t\right\}
$$

$$
= \Phi(t) + \phi(t)\left\{\frac{-\sqrt{\beta_1}}{6}(t^2-1)\right\}
$$

$$
+ \phi(t)(-t)\left\{\frac{(\beta_2-3)}{24}(t^2-3) + \frac{\beta_1}{72}(t^4-10t^2+15)\right\} + \cdots,
$$

$$\tag{6.7}$$

where $\sqrt{\beta_1}$ and β_2 are the coefficients of skewness and kurtosis of T, respectively, and $\Phi(t)$ is the cdf of the standard normal distribution with corresponding pdf $\phi(t)$. Formula (6.7) is known as the Edgeworth expansion; see, for example, Hall (1992) and Johnson, Kotz, and Balakrishnan (1994). The approximation to the distribution of the classical Wilcoxon's rank-sum statistic by using Edgeworth expansion has been discussed earlier by many authors including Fix and Hodges (1955) and Bickel (1974).

Though $W_{\min,r}$, $W_{\max,r}$, and $W_{E,r}$ are not linear rank statistics, yet their distributions seem to be approximately normal even for values of n_1 and n_2 as small as 10 when r is close to n_2; see Figure 6.1, where $q = \frac{r}{n_2}$. In other cases, however, the normal approximation does not seem to be reasonable; see Figure 6.2, for example. So, we consider the use of the second-order Edgeworth expansion to approximate the distribution of these test statistics, i.e.,

$$F_W(w) \approx \Phi(w') - \phi(w') \left\{ \frac{\sqrt{\beta_1}}{6}(w'^2 - 1) + \frac{(\beta_2 - 3)}{24}(w'^3 - 3w') \right.$$
$$\left. + \frac{\beta_1}{72}(w'^5 - 10w'^3 + 15w') \right\}, \tag{6.8}$$

where $w' = \frac{w - E(W)}{\sqrt{Var(W)}}$. In order to use the second-order Edgeworth approximation, we require the knowledge of the first four moments of $W_{\min,r}$, $W_{\max,r}$, and $W_{E,r}$. The exact explicit formulas for these moments are presented in Appendix A.

In addition, we can modify the approximation by applying the continuity correction as all three test statistics are discrete. In other words, w' in formula (6.8) should be modified as

$$w' = \frac{w - E(W) + 0.5}{\sqrt{Var(W)}}. \tag{6.9}$$

In Tables 6.3–6.11, we provide a comparison of the normal approximation and the Edgeworth approximation (both with continuity correction) to the exact significance probabilities for the Wilcoxon-type rank-sum precedence test statistics $W_{\min,r}$, $W_{\max,r}$, and $W_{E,r}$. Specifically, the approximate values corresponding to the exact significance probabilities of 1%, 5% and 10% of the Wilcoxon-type rank-sum precedence test statistics (for some combinations of n_1 and n_2, and $r = 2(1)7$) are presented in these tables.

From Tables 6.3–6.11, we find that even though the Edgeworth approximation may not always be better than the normal approximation, it improves the normal approximation considerably in most of the cases considered. We also note that the accuracy of the approximation does not in general increase with n_1 and n_2 for fixed values of r, and that the accuracy of the approximation usually depends on the value of r. Intuitively, this may be due to the fact that the null distributions of the Wilcoxon-type rank-sum precedence statistics are highly skewed when r is too small relative to n_2. From our extensive simulation results, we have observed that the normal approximation is quite reasonable in the practical range of sample sizes (like those considered here, say 10 to 50) as long as r is at least 50% of n_2. For example, in Figure 6.1, we have presented the simulated histogram of the distribution of $W_{E,r}$ for $n_1 = n_2 = 10$ and 30 when $q = \frac{r}{n_2}$ equals 60%, 80%, and 90%. These plots support this observation with the normal approximation being good even for sample sizes as small as 10.

Even though the normal approximation does not work for small or moderate values of n_2 for small values of q, we have observed that even when q is small, the normal approximation works quite well for large values of n_1 and n_2. In Figure 6.2, this point is illustrated by the case when $q = 20\%$ for $n_1 = n_2 = 10, 30$, and 100. While the distributions are quite skewed (to the left for sample sizes 10 and 30, the normal approximation seems to be quite good for sample size 100.

6.4 EXACT POWER FUNCTION UNDER LEHMANN ALTERNATIVE

In this section, we present an expression for the power functions of the minimal, maximal, and expected rank-sum precedence tests under the Lehmann alternative $H_1 : [F_X]^\gamma = F_Y$, as derived by Ng and Balakrishnan (2002, 2004).

Earlier in Chapter 2, we presented the joint probability mass function of $M_1, M_2, \cdots, M_r$ under the Lehmann alternative as

$$\Pr\left\{M_1 = m_1, M_2 = m_2, \cdots, M_r = m_r \mid H_1 : [F_X]^\gamma = F_Y\right\}$$

$$= \frac{n_1! n_2! \gamma^r}{m_1!(n_2 - r)!} \left\{\prod_{j=1}^{r-1} \frac{\Gamma\left(\sum\limits_{i=1}^{j} m_i + j\gamma\right)}{\Gamma\left(\sum\limits_{i=1}^{j+1} m_i + j\gamma + 1\right)}\right\}$$

$$\times \sum_{k=0}^{n_2-r} \binom{n_2 - r}{k} (-1)^k \frac{\Gamma\left(\sum\limits_{i=1}^{r} m_i + (r+k)\gamma\right)}{\Gamma\left(n_1 + (r+k)\gamma + 1\right)}.$$

So under the Lehmann alternative, the power function of the minimal rank-sum precedence test is given by

$$\Pr\left\{W_{\min,r} \le s \mid H_1 : [F_X]^\gamma = F_Y\right\}$$

$$= \sum_{m_i(i=1,2,\cdots,r)=0}^{n_1} [\Pr\left\{M_1 = m_1, \cdots, M_r = m_r \mid H_1 : [F_X]^\gamma = F_Y\right\}$$

$$\times I_{\min,s}(m_1, m_2, \cdots, m_r)], \qquad (6.10)$$

where $I_{\min,s}(m_1, m_2, \cdots, m_r)$ is as defined in Eq. (6.4).

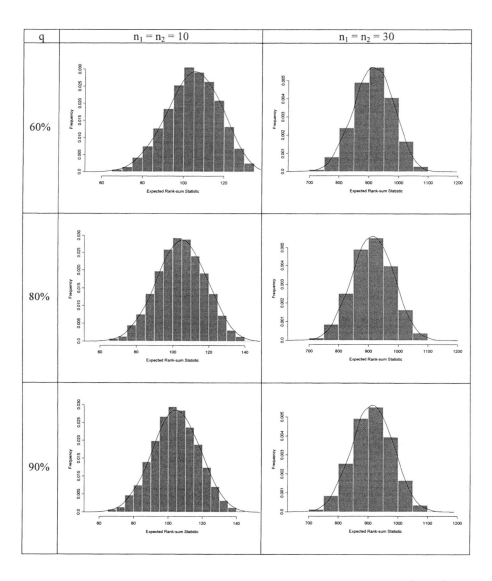

Figure 6.1: Null distribution of the expected rank-sum statistic ($W_{E,r}$) for $n_1 = n_2 = 10$ and $n_1 = n_2 = 30$ when $q = \frac{r}{n_2}$ equals 60%, 80%, and 90%

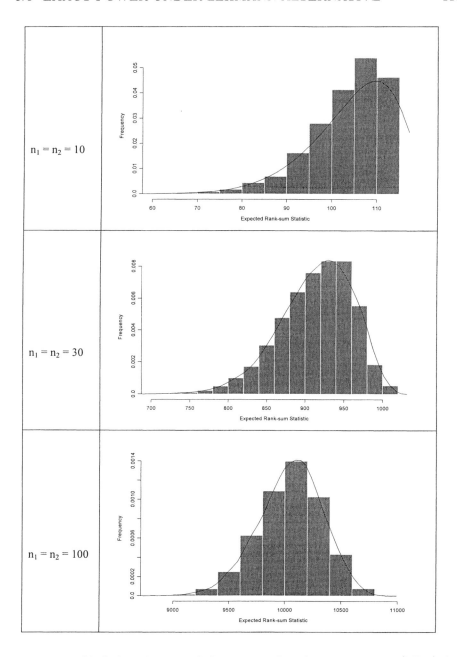

Figure 6.2: Null distribution of the expected rank-sum statistic $(W_{E,r})$ for $n_1 = n_2 = 30$ and $n_1 = n_2 = 100$ when $q = \frac{r}{n_2}$ equals 20%

Table 6.3: Exact and approximate values of $\Pr(W_{\min,r} \leq s)$ by normal approximation and Edgeworth approximation (near 1% critical values)

n_1	n_2		$r = 2$	$r = 3$	$r = 4$	$r = 5$	$r = 6$	$r = 7$
10	10	s	64	67	69	71	73	73
		(a)	0.010836	0.010349	0.008530	0.009228	0.011258	0.009006
		(b)	0.014773	0.011450	0.009140	0.009244	0.010870	0.008860
		(c)	0.001291	0.001774	0.002220	0.003688	0.006387	0.006372
10	15	s	66	70	74	77	80	82
		(a)	0.009486	0.008457	0.008770	0.008721	0.009723	0.009371
		(b)	0.014540	0.010208	0.010666	0.009646	0.010111	0.009501
		(c)	0.000648	0.000669	0.001392	0.001881	0.003026	0.003676
15	15	s	138	145	152	157	161	165
		(a)	0.011589	0.009270	0.011336	0.010106	0.009076	0.009551
		(b)	0.015129	0.011616	0.012825	0.010958	0.009533	0.009650
		(c)	0.000805	0.000982	0.002224	0.002542	0.002831	0.003840
15	20	s	140	148	156	163	169	174
		(a)	0.012331	0.008702	0.009992	0.010201	0.010015	0.009375
		(b)	0.016876	0.011348	0.012122	0.011802	0.011025	0.009998
		(c)	0.000729	0.000634	0.001299	0.001932	0.002408	0.002683
20	20	s	237	249	260	270	278	286
		(a)	0.008314	0.009383	0.010053	0.010823	0.009721	0.010263
		(b)	0.011628	0.012010	0.012262	0.012344	0.010684	0.010826
		(c)	0.000356	0.000772	0.001419	0.002228	0.002357	0.003180
30	30	s	512	533	553	572	590	606
		(a)	0.010534	0.009225	0.009415	0.009837	0.010456	0.009884
		(b)	0.016301	0.012620	0.012037	0.012045	0.012210	0.011151
		(c)	0.000565	0.000631	0.000951	0.001433	0.002040	0.002212
30	50	s	516	540	563	585	606	627
		(a)	0.011837	0.011955	0.011020	0.010481	0.009825	0.010636
		(b)	0.021007	0.017302	0.015263	0.013701	0.012350	0.012840
		(c)	0.000678	0.000835	0.001031	0.001195	0.001306	0.001844

Remark: (a) Exact probability; (b) Edgeworth approximation; (c) normal approximation.

Table 6.4: Exact and approximate values of $\Pr(W_{\min,r} \leq s)$ by normal approximation and Edgeworth approximation (near 5% critical values)

n_1	n_2		$r = 2$	$r = 3$	$r = 4$	$r = 5$	$r = 6$	$r = 7$
10	10	s	67	71	75	77	80	81
		(a)	0.059538	0.046386	0.053774	0.044616	0.054678	0.050077
		(b)	0.050721	0.042527	0.049805	0.042107	0.052925	0.049029
		(c)	0.032129	0.025736	0.035555	0.030737	0.043429	0.042195
10	15	s	68	74	79	84	87	90
		(a)	0.033160	0.044523	0.047266	0.055544	0.048403	0.047942
		(b)	0.037555	0.041784	0.043322	0.052846	0.045866	0.045875
		(c)	0.012363	0.021999	0.026107	0.037666	0.032446	0.034171
15	15	s	141	150	158	165	171	176
		(a)	0.042046	0.046543	0.048259	0.050005	0.050702	0.050462
		(b)	0.042547	0.043279	0.045437	0.047417	0.048583	0.048702
		(c)	0.019464	0.023841	0.028529	0.032495	0.035422	0.037221
15	20	s	143	153	162	170	178	185
		(a)	0.051866	0.053019	0.049504	0.046842	0.051492	0.053333
		(b)	0.049987	0.048369	0.046683	0.044490	0.049218	0.051304
		(c)	0.029718	0.028469	0.028264	0.027675	0.034065	0.037530
20	20	s	241	255	267	279	289	298
		(a)	0.046906	0.053347	0.046714	0.052337	0.049868	0.048164
		(b)	0.046579	0.049516	0.044720	0.050050	0.047903	0.046464
		(c)	0.023795	0.029956	0.026263	0.033710	0.032946	0.032910
30	30	s	516	539	561	582	602	621
		(a)	0.051787	0.045743	0.046115	0.047360	0.049444	0.051772
		(b)	0.050083	0.045444	0.044751	0.045925	0.047917	0.050188
		(c)	0.028797	0.023195	0.024603	0.027663	0.031195	0.034758
30	50	s	518	544	569	593	616	638
		(a)	0.034453	0.047051	0.052283	0.052450	0.051595	0.049401
		(b)	0.044409	0.046485	0.049387	0.050450	0.049820	0.048017
		(c)	0.012148	0.022901	0.028884	0.031308	0.031526	0.030443

Remark: (a) Exact probability; (b) Edgeworth approximation; (c) normal approximation.

Table 6.5: Exact and approximate values of $\Pr(W_{\min,r} \leq s)$ by normal approximation and Edgeworth approximation (near 10% critical values)

n_1	n_2		$r = 2$	$r = 3$	$r = 4$	$r = 5$	$r = 6$	$r = 7$
10	10	s	68	73	78	81	83	85
		(a)	0.086330	0.091277	0.110806	0.104191	0.094931	0.098124
		(b)	0.084350	0.085485	0.110303	0.102471	0.093664	0.097368
		(c)	0.071777	0.071680	0.099463	0.091675	0.083969	0.089600
10	15	s	70	76	82	87	91	94
		(a)	0.103162	0.097446	0.105516	0.109229	0.104138	0.093967
		(b)	0.104895	0.089679	0.105054	0.107394	0.102191	0.092556
		(c)	0.101234	0.078193	0.094805	0.096997	0.091375	0.081679
15	15	s	143	153	161	169	176	181
		(a)	0.098517	0.108542	0.091454	0.098716	0.103816	0.094681
		(b)	0.095878	0.103110	0.088323	0.096813	0.102358	0.093428
		(c)	0.090315	0.094957	0.076064	0.085783	0.092362	0.083330
15	20	s	144	155	165	174	182	190
		(a)	0.094631	0.096413	0.101107	0.099555	0.096648	0.103928
		(b)	0.080078	0.092818	0.097385	0.097363	0.094509	0.102528
		(c)	0.073204	0.084411	0.087776	0.086892	0.083497	0.092808
20	20	s	243	257	271	283	295	305
		(a)	0.104391	0.088148	0.102114	0.097170	0.107160	0.103803
		(b)	0.100900	0.083687	0.098877	0.094690	0.105569	0.102509
		(c)	0.100469	0.073092	0.089936	0.084150	0.096652	0.093236
30	30	s	518	542	565	587	608	628
		(a)	0.110112	0.095906	0.095081	0.096080	0.098619	0.101776
		(b)	0.105230	0.090470	0.090034	0.092755	0.096327	0.099980
		(c)	0.111186	0.084505	0.081110	0.083253	0.087018	0.091128
30	50	s	520	546	572	596	620	644
		(a)	0.096856	0.093982	0.103598	0.091977	0.093880	0.104767
		(b)	0.090392	0.082490	0.099533	0.087102	0.090588	0.102145
		(c)	0.096615	0.077199	0.095847	0.077620	0.081004	0.094616

Remark: (a) Exact probability; (b) Edgeworth approximation; (c) normal approximation.

Table 6.6: Exact and approximate values of $\Pr(W_{\max,r} \leq s)$ by normal approximation and Edgeworth approximation (near 1% critical values)

n_1	n_2		$r = 2$	$r = 3$	$r = 4$	$r = 5$	$r = 6$	$r = 7$
10	10	s	92	85	80	78	76	75
		(a)	0.009883	0.009818	0.009218	0.010338	0.009656	0.009526
		(b)	0.011075	0.009923	0.009118	0.010590	0.010239	0.010043
		(c)	0.001611	0.003838	0.005961	0.009580	0.011313	0.012466
10	15	s	130	118	109	103	98	95
		(a)	0.009424	0.009843	0.010400	0.009894	0.009531	0.009759
		(b)	0.014155	0.011534	0.009984	0.009747	0.009309	0.009777
		(c)	0.001332	0.002450	0.003585	0.005214	0.006521	0.008422
15	15	s	241	224	212	202	195	190
		(a)	0.009720	0.009752	0.010186	0.009800	0.009963	0.010330
		(b)	0.012842	0.010883	0.010272	0.009624	0.009773	0.010308
		(c)	0.001180	0.002308	0.003840	0.005142	0.006842	0.008772
15	20	s	302	284	267	253	242	234
		(a)	0.010071	0.010219	0.009895	0.009859	0.009777	0.010177
		(b)	0.012835	0.012589	0.011014	0.009982	0.009575	0.009952
		(c)	0.000789	0.002024	0.002870	0.003678	0.004665	0.006116
20	20	s	458	437	416	399	384	372
		(a)	0.009828	0.010023	0.009896	0.010221	0.009983	0.009988
		(b)	0.010342	0.011670	0.010733	0.010343	0.009855	0.009787
		(c)	0.000536	0.001812	0.002823	0.003962	0.004897	0.006004
30	30	s	1129	1080	1048	1013	982	955
		(a)	0.009844	0.009830	0.010140	0.010041	0.010031	0.009944
		(b)	0.012295	0.010701	0.011698	0.010863	0.010286	0.010027
		(c)	0.000568	0.001008	0.002178	0.002807	0.003410	0.004086
30	50	s	1671	1624	1578	1535	1495	1460
		(a)	0.009987	0.010260	0.009877	0.010001	0.010062	0.010015
		(b)	0.008197	0.010745	0.010793	0.010392	0.009950	0.009892
		(c)	0.000171	0.000635	0.001126	0.001571	0.001969	0.002479

Remark: (a) Exact probability; (b) Edgeworth approximation; (c) normal approximation.

Table 6.7: Exact and approximate values of $\Pr(W_{\max,r} \leq s)$ by normal approximation and Edgeworth approximation (near 5% critical values)

n_1	n_2		$r = 2$	$r = 3$	$r = 4$	$r = 5$	$r = 6$	$r = 7$
10	10	s	107	99	93	89	86	84
		(a)	0.049536	0.050965	0.050651	0.051078	0.049000	0.046618
		(b)	0.041075	0.045758	0.047745	0.050099	0.049470	0.047712
		(c)	0.024494	0.033200	0.038816	0.044524	0.047139	0.047949
10	15	s	147	136	129	122	116	111
		(a)	0.047431	0.050489	0.050350	0.051336	0.051481	0.049190
		(b)	0.037632	0.039618	0.047414	0.049451	0.049475	0.048303
		(c)	0.017946	0.024250	0.034535	0.039043	0.041535	0.042790
15	15	s	270	250	238	227	218	211
		(a)	0.047414	0.049450	0.050938	0.049910	0.049115	0.049094
		(b)	0.048188	0.042414	0.046987	0.047543	0.047651	0.048319
		(c)	0.029703	0.027123	0.034433	0.037573	0.040114	0.043021
15	20	s	340	320	302	287	274	264
		(a)	0.050737	0.048743	0.049698	0.050024	0.049116	0.049709
		(b)	0.052739	0.053981	0.051190	0.049313	0.047284	0.047998
		(c)	0.034351	0.037314	0.036562	0.036831	0.036881	0.039415
20	20	s	510	477	458	439	423	410
		(a)	0.053262	0.049792	0.050228	0.049673	0.049519	0.050532
		(b)	0.055142	0.044032	0.048551	0.047806	0.047730	0.049100
		(c)	0.037235	0.027125	0.034169	0.035603	0.037510	0.040652
30	30	s	1191	1160	1115	1085	1057	1029
		(a)	0.051395	0.050824	0.049641	0.049961	0.050515	0.050095
		(b)	0.040255	0.049806	0.043367	0.046748	0.049107	0.048795
		(c)	0.016852	0.031820	0.027140	0.032415	0.036336	0.037443
30	50	s	1767	1720	1675	1632	1595	1578
		(a)	0.048031	0.049454	0.050398	0.049652	0.050218	0.049752
		(b)	0.042354	0.043165	0.042962	0.041659	0.042130	0.054496
		(c)	0.016996	0.022494	0.024590	0.025007	0.026927	0.040620

Remark: (a) Exact probability; (b) Edgeworth approximation; (c) normal approximation.

Table 6.8: Exact and approximate values of $\Pr(W_{\max,r} \leq s)$ by normal approximation and Edgeworth approximation (near 10% critical values)

n_1	n_2		$r = 2$	$r = 3$	$r = 4$	$r = 5$	$r = 6$	$r = 7$
10	10	s	116	107	100	95	92	90
		(a)	0.102941	0.102048	0.100359	0.098411	0.101128	0.102974
		(b)	0.097726	0.100288	0.098048	0.096865	0.100974	0.103736
		(c)	0.083405	0.086711	0.086788	0.088465	0.095203	0.100246
10	15	s	161	148	138	132	125	120
		(a)	0.111462	0.095923	0.098707	0.101296	0.098274	0.098435
		(b)	0.099747	0.092948	0.090126	0.101186	0.096765	0.097613
		(c)	0.088165	0.078888	0.076926	0.089936	0.087166	0.089830
15	15	s	285	264	251	240	231	223
		(a)	0.100383	0.099800	0.098176	0.098025	0.099481	0.098008
		(b)	0.108353	0.089156	0.093616	0.095867	0.098248	0.097191
		(c)	0.098960	0.075500	0.081355	0.085255	0.089374	0.090057
15	20	s	344	330	318	305	292	281
		(a)	0.093346	0.100692	0.100019	0.101377	0.101836	0.101594
		(b)	0.062711	0.083472	0.100599	0.104788	0.101589	0.100156
		(c)	0.046540	0.069506	0.088750	0.093967	0.091504	0.091148
20	20	s	530	497	479	460	444	430
		(a)	0.106871	0.100244	0.102209	0.098465	0.099591	0.099996
		(b)	0.117278	0.088898	0.099892	0.098200	0.098593	0.098974
		(c)	0.112829	0.075981	0.088453	0.087342	0.088809	0.090356
30	30	s	1220	1190	1160	1121	1093	1066
		(a)	0.097255	0.096046	0.098183	0.100450	0.101126	0.099490
		(b)	0.068958	0.092703	0.105727	0.094046	0.097360	0.098011
		(c)	0.055407	0.081931	0.096404	0.083045	0.087268	0.088622
30	50	s	1816	1768	1723	1680	1668	1629
		(a)	0.104303	0.093926	0.100452	0.099100	0.100467	0.099477
		(b)	0.089094	0.090873	0.088134	0.083553	0.116357	0.108275
		(c)	0.084692	0.081381	0.076549	0.071031	0.109208	0.099792

Remark: (a) Exact probability; (b) Edgeworth approximation; (c) normal approximation.

Table 6.9: Exact and approximate values of $\Pr(W_{E,r} \leq s)$ by normal approximation and Edgeworth approximation (near 1% critical values)

n_1	n_2		$r = 2$	$r = 3$	$r = 4$	$r = 5$	$r = 6$	$r = 7$
10	10	s	79	77	75	75	75	75
		(a)	0.011669	0.010800	0.008914	0.009710	0.010944	0.010760
		(b)	0.012888	0.011120	0.008951	0.010050	0.010886	0.011385
		(c)	0.002124	0.004097	0.004900	0.007512	0.009672	0.011231
10	15	s	98	94	93	91	90	89
		(a)	0.009424	0.009565	0.010555	0.010164	0.009690	0.009452
		(b)	0.013669	0.010622	0.011155	0.009934	0.009765	0.009280
		(c)	0.001192	0.001934	0.003909	0.004708	0.005906	0.006651
15	15	s	189	185	183	181	179	178
		(a)	0.009720	0.009854	0.010448	0.010541	0.009515	0.009663
		(b)	0.011887	0.010768	0.010788	0.010382	0.009588	0.009514
		(c)	0.000966	0.002120	0.003792	0.005056	0.005767	0.006791
15	20	s	221	216	212	209	207	205
		(a)	0.010071	0.010112	0.010121	0.009787	0.010096	0.009537
		(b)	0.012675	0.011823	0.010768	0.010140	0.010023	0.009629
		(c)	0.000757	0.001702	0.002592	0.003481	0.004506	0.005212
20	20	s	352	344	339	335	332	331
		(a)	0.009828	0.009962	0.010100	0.009630	0.009633	0.010250
		(b)	0.014636	0.011856	0.011225	0.010049	0.009622	0.010271
		(c)	0.001137	0.001809	0.002707	0.003501	0.004295	0.005707
30	30	s	821	811	801	794	788	783
		(a)	0.009844	0.010077	0.010001	0.009981	0.010055	0.009927
		(b)	0.012523	0.012675	0.011331	0.010863	0.010466	0.010208
		(c)	0.000587	0.001432	0.001989	0.002707	0.003351	0.003968
30	50	s	1095	1084	1072	1063	1055	1048
		(a)	0.009987	0.010300	0.009877	0.010001	0.010100	0.009895
		(b)	0.009093	0.009152	0.011143	0.011073	0.011041	0.010795
		(c)	0.000207	0.000737	0.001193	0.001763	0.002308	0.002839

Remark: (a) Exact probability; (b) Edgeworth approximation; (c) normal approximation.

Table 6.10: Exact and approximate values of $\Pr(W_{E,r} \leq s)$ by normal approximation and Edgeworth approximation (near 5% critical values)

n_1	n_2		$r = 2$	$r = 3$	$r = 4$	$r = 5$	$r = 6$	$r = 7$
10	10	s	87	85	84	84	83	83
		(a)	0.049536	0.047100	0.048713	0.052031	0.049465	0.049205
		(b)	0.041706	0.043331	0.046235	0.053103	0.049005	0.051162
		(c)	0.024873	0.030409	0.036340	0.045524	0.044137	0.048107
10	15	s	108	106	105	103	102	101
		(a)	0.047431	0.050489	0.051937	0.050801	0.048957	0.048970
		(b)	0.038802	0.043743	0.050586	0.048014	0.048865	0.047877
		(c)	0.019073	0.027924	0.037100	0.036950	0.039815	0.040749
15	15	s	205	201	199	197	195	194
		(a)	0.047414	0.048909	0.051009	0.051789	0.048083	0.049019
		(b)	0.044765	0.045261	0.049174	0.049611	0.047589	0.048093
		(c)	0.025773	0.029633	0.036072	0.038776	0.038910	0.041207
15	20	s	241	236	232	229	227	225
		(a)	0.050737	0.051341	0.051610	0.050371	0.050944	0.048776
		(b)	0.050175	0.050302	0.048977	0.048250	0.048958	0.048032
		(c)	0.031158	0.033337	0.034147	0.035418	0.037876	0.038598
20	20	s	375	367	363	360	357	355
		(a)	0.053262	0.049792	0.051100	0.050705	0.050168	0.049410
		(b)	0.052148	0.046491	0.054833	0.049223	0.048360	0.048778
		(c)	0.033541	0.029514	0.033291	0.036584	0.037519	0.039514
30	30	s	855	850	841	835	830	826
		(a)	0.051395	0.050824	0.049718	0.049874	0.049882	0.049852
		(b)	0.042261	0.049617	0.047340	0.047751	0.048058	0.048672
		(c)	0.019457	0.031558	0.031096	0.033255	0.035061	0.036994
30	50	s	1143	1133	1123	1115	1110	1106
		(a)	0.048031	0.049500	0.050398	0.049652	0.049800	0.049955
		(b)	0.042559	0.062072	0.043910	0.044327	0.054503	0.049841
		(c)	0.017314	0.023519	0.025566	0.027668	0.031732	0.035745

Remark: (a) Exact probability; (b) Edgeworth approximation; (c) normal approximation.

Table 6.11: Exact and approximate values of $\Pr(W_{E,r} \leq s)$ by normal approximation and Edgeworth approximation (near 10% critical values)

n_1	n_2		$r = 2$	$r = 3$	$r = 4$	$r = 5$	$r = 6$	$r = 7$
10	10	s	92	90	89	88	88	88
		(a)	0.102941	0.094900	0.100435	0.093669	0.102719	0.102719
		(b)	0.094876	0.095276	0.098786	0.095984	0.102400	0.106092
		(c)	0.080528	0.081645	0.087380	0.086890	0.095409	0.100866
10	15	s	116	112	111	109	109	108
		(a)	0.111462	0.095923	0.098707	0.096702	0.104692	0.103481
		(b)	0.106748	0.091273	0.101294	0.094900	0.105362	0.102754
		(c)	0.096487	0.077225	0.088806	0.083398	0.095664	0.094392
15	15	s	214	209	207	205	204	203
		(a)	0.100383	0.096010	0.098905	0.098716	0.099350	0.100871
		(b)	0.106643	0.093894	0.098414	0.097286	0.100111	0.100206
		(c)	0.097154	0.080839	0.086556	0.086658	0.091016	0.092504
15	20	s	245	245	242	239	237	236
		(a)	0.093346	0.100692	0.102157	0.097702	0.099042	0.099956
		(b)	0.069277	0.101315	0.102143	0.098106	0.097552	0.100815
		(c)	0.054532	0.089931	0.090626	0.086840	0.087227	0.091700
20	20	s	386	378	375	372	370	368
		(a)	0.106871	0.100244	0.101000	0.099071	0.101144	0.099089
		(b)	0.111721	0.093063	0.098375	0.098121	0.099939	0.099302
		(c)	0.106146	0.080856	0.086747	0.087326	0.090219	0.090546
30	30	s	870	866	862	856	852	848
		(a)	0.097255	0.096046	0.101405	0.099315	0.101217	0.099133
		(b)	0.073423	0.091985	0.101369	0.098786	0.099947	0.098998
		(c)	0.061235	0.081149	0.091415	0.088466	0.090191	0.089711
30	50	s	1168	1157	1148	1140	1140	1135
		(a)	0.104303	0.093900	0.100452	0.099100	0.100000	0.099313
		(b)	0.089095	0.093199	0.089461	0.087548	0.101409	0.102564
		(c)	0.084723	0.080588	0.078170	0.075677	0.092896	0.093314

Remark: (a) Exact probability; (b) Edgeworth approximation; (c) normal approximation.

Similarly, under the Lehmann alternative, the power function of the maximal and expected rank-sum precedence tests are given by

$$
\Pr\left\{W_{\max,r} \le s \mid [F_X]^\gamma = F_Y\right\}
$$
$$
= \sum_{m_i(i=1,2,\cdots,r)=0}^{n_1} [\Pr\left\{M_1 = m_1, \cdots, M_r = m_r \mid H_1 : [F_X]^\gamma = F_Y\right\}
$$
$$
\times I_{\max,s}(m_1, m_2, \cdots, m_r)] \tag{6.11}
$$

and

$$
\Pr\left\{W_{E,r} \le s \mid [F_X]^\gamma = F_Y\right\}
$$
$$
= \sum_{m_i(i=1,2,\cdots,r)=0}^{n_1} [\Pr\left\{M_1 = m_1, \cdots, M_r = m_r \mid H_1 : [F_X]^\gamma = F_Y\right\}
$$
$$
\times I_{E,s}(m_1, m_2, \cdots, m_r)], \tag{6.12}
$$

respectively, where $I_{\max,s}(m_1, m_2, \cdots, m_r)$ and $I_{E,s}(m_1, m_2, \cdots, m_r)$ are as defined in Eqs. (6.5) and (6.6).

Now, we demonstrate the use of the formulas in (6.10)–(6.12) as well as the Monte Carlo simulation method for the computation of the power of the Wilcoxon-type rank-sum precedence tests under the Lehmann alternative. For this purpose, we generated 100,000 sets of data (from F_X and F_X^γ) and computed the test statistics $W_{\min,r}$, $W_{\max,r}$, and $W_{E,r}$ for each set. The power values were estimated by the rejection rates of the null hypothesis for different values of γ. For $n_1 = n_2 = 10$, $r = 2, 3$, and $\gamma = 2(1)6$, the power values computed from (6.10)–(6.12) and those estimated through the above-described Monte Carlo simulation method are presented in Table 6.12. We observe that the estimated values of the power determined from Monte Carlo simulations are quite close to the exact values. In addition to revealing the correctness of the formulas in (6.10)–(6.12), these results also suggest that the Monte Carlo simulation method provides a feasible and accurate way to estimate the power of the proposed test procedures.

Though the location-shift in some cases (like extreme-value distribution) becomes a Lehmann alternative and hence the exact power can be determined, that is not the case for other distributions such as normal and gamma. Therefore, we examine the power under location-shift alternative using Monte Carlo simulation in the next section.

Table 6.12: Power values under Lehmann alternative for $n_1 = n_2 = 10, r = 2, 3$, and $\gamma = 2(1)6$

r	γ	Power computed from formulas			Simulated power		
		$W_{\min,r}$	$W_{\max,r}$	$W_{E,r}$	$W_{\min,r}$	$W_{\max,r}$	$W_{E,r}$
r=2	1	0.05954	0.04954	0.04954	0.05852	0.04894	0.04894
	2	0.40396	0.35681	0.35681	0.40406	0.35538	0.35538
	3	0.69831	0.64389	0.64389	0.69941	0.64453	0.64453
	4	0.85319	0.80970	0.80970	0.85289	0.80999	0.80999
	5	0.92692	0.89619	0.89619	0.92651	0.89565	0.89565
	6	0.96212	0.94126	0.94126	0.96220	0.94057	0.94057
r=3	1	0.04639	0.05096	0.04707	0.04515	0.04959	0.04713
	2	0.36428	0.35100	0.34624	0.36379	0.35127	0.34852
	3	0.66331	0.63139	0.62922	0.66492	0.63201	0.63094
	4	0.82952	0.79597	0.79513	0.83027	0.79593	0.79547
	5	0.91190	0.88402	0.88368	0.91160	0.88427	0.88415
	6	0.95263	0.93137	0.93123	0.95230	0.93096	0.93093

6.5 MONTE CARLO SIMULATION UNDER LOCATION-SHIFT ALTERNATIVE

In order to assess the power properties of the proposed Wilcoxon-type rank-sum precedence tests, we consider now the location-shift alternative H_1 : $F_X(x) = F_Y(x + \theta)$ for some $\theta > 0$, where θ is a shift in location. The power of the precedence test with $r = 1(1)7$, the maximal precedence test with $r = 2(1)7$, the three Wilcoxon-type rank-sum precedence tests with $r = 2(1)7$, and the classical Wilcoxon's rank-sum test (based on complete samples) were all estimated by Ng and Balakrishnan (2002, 2004) through Monte Carlo simulations when $\theta = 0.5$ and $\theta = 1.0$. The lifetime distributions listed earlier in Section 3.5 were used in the Monte Carlo simulations in order to demonstrate the power performance of the Wilcoxon-type rank-sum precedence tests under this location-shift alternative.

In Tables 6.13–6.17, we have presented the estimated power values of the precedence tests with $r = 1(1)7$, the maximal precedence tests with $r = 2(1)7$, the three Wilcoxon-type rank-sum precedence tests with $r = 2(1)7$, and the classical Wilcoxon's rank-sum test for the underlying standard

Table 6.13: Power of precedence tests, maximal precedence tests, minimal, maximal, and expected rank-sum precedence tests, and Wilcoxon's rank-sum test (based on complete samples) when $n_1 = n_2 = 10$

Test	r	s	α	θ	Normal	Exp(1)	LN(0.25)	LN(0.5)
$P_{(r)}$	1	4	0.04334	0.5	0.17355	0.71973	0.23848	0.39436
				1.0	0.42078	0.97785	0.60775	0.87090
	2	6	0.02864	0.5	0.14958	0.39156	0.18522	0.26933
				1.0	0.42430	0.83764	0.53318	0.72532
	3	7	0.03489	0.5	0.17793	0.30731	0.19804	0.25708
				1.0	0.48132	0.72393	0.53939	0.66761
	4	8	0.03483	0.5	0.17708	0.22511	0.18323	0.21493
				1.0	0.47989	0.57667	0.49217	0.56561
	5	9	0.02864	0.5	0.14725	0.14531	0.14163	0.15265
				1.0	0.42447	0.39603	0.39065	0.41444
	6	9	0.07043	0.5	0.26586	0.23847	0.24790	0.25305
				1.0	0.58320	0.49909	0.52909	0.53141
	7	9	0.13545	0.5	0.42308	0.36325	0.38830	0.38359
				1.0	0.73305	0.61390	0.66626	0.64903
	8	9	0.24768	0.5	0.60199	0.51741	0.55841	0.54376
				1.0	0.85312	0.73126	0.79120	0.76368
	9	9	0.39474	0.5	0.77887	0.69617	0.73852	0.71894
				1.0	0.93564	0.84599	0.89279	0.86626
	10	9	0.52632	0.5	0.92282	0.87246	0.89630	0.88168
				1.0	0.98338	0.94081	0.96452	0.94961
$M_{(r)}$	2	5	0.03250	0.5	0.14174	0.49229	0.17004	0.25426
				1.0	0.37675	0.91770	0.48601	0.73270
	3	5	0.04875	0.5	0.17526	0.49485	0.19469	0.26781
				1.0	0.41709	0.91784	0.50380	0.73601
	4	5	0.06500	0.5	0.19933	0.49736	0.21137	0.27682
				1.0	0.43592	0.91810	0.51146	0.73755
	5	6	0.02700	0.5	0.09008	0.26844	0.09463	0.12970
				1.0	0.24818	0.77912	0.30676	0.52844
	6	6	0.03251	0.5	0.09378	0.26897	0.09703	0.13085
				1.0	0.24957	0.77914	0.30725	0.52857
	7	6	0.03793	0.5	0.09648	0.26951	0.09849	0.13153
				1.0	0.25023	0.77920	0.30743	0.52860
	8	6	0.04334	0.5	0.09819	0.27001	0.09984	0.13222
				1.0	0.25039	0.77923	0.30751	0.52861
	9	6	0.04876	0.5	0.09921	0.27052	0.10080	0.13276
				1.0	0.25050	0.77928	0.30761	0.52869
	10	6	0.05423	0.5	0.09986	0.27107	0.10134	0.13313
				1.0	0.25055	0.77930	0.30763	0.52872

Table 6.13: (Continued)

Test	r	s	α	θ	Gamma(2)	Gamma(5)	Gamma(10)
$P_{(r)}$	1	4	0.04334	0.5	0.41408	0.26426	0.22491
				1.0	0.88064	0.67354	0.57424
	2	6	0.02864	0.5	0.25252	0.19241	0.17311
				1.0	0.68794	0.55814	0.50597
	3	7	0.03489	0.5	0.23431	0.20245	0.18929
				1.0	0.61482	0.54595	0.51771
	4	8	0.03483	0.5	0.19357	0.18073	0.17548
				1.0	0.51144	0.48487	0.47581
	5	9	0.02864	0.5	0.13799	0.13726	0.13884
				1.0	0.37249	0.37815	0.38331
	6	9	0.07043	0.5	0.23441	0.24132	0.24361
				1.0	0.49543	0.51311	0.52664
	7	9	0.13545	0.5	0.36545	0.38085	0.38607
				1.0	0.62361	0.64824	0.66820
	8	9	0.24768	0.5	0.52933	0.54896	0.55762
				1.0	0.74884	0.77783	0.79587
	9	9	0.39474	0.5	0.70765	0.72728	0.73847
				1.0	0.86121	0.88478	0.89805
	10	9	0.52632	0.5	0.88038	0.89239	0.89989
				1.0	0.94853	0.96066	0.96735
$M_{(r)}$	2	5	0.03250	0.5	0.24736	0.17686	0.16108
				1.0	0.73033	0.52376	0.45692
	3	5	0.04875	0.5	0.25785	0.19838	0.18700
				1.0	0.73274	0.53569	0.47649
	4	5	0.06500	0.5	0.26527	0.21143	0.20397
				1.0	0.73435	0.54141	0.48526
	5	6	0.02700	0.5	0.11968	0.09218	0.08952
				1.0	0.51361	0.32886	0.28329
	6	6	0.03251	0.5	0.12081	0.09436	0.09210
				1.0	0.51379	0.32931	0.28385
	7	6	0.03793	0.5	0.12190	0.09564	0.09399
				1.0	0.51392	0.32961	0.28422
	8	6	0.04334	0.5	0.12278	0.09683	0.09520
				1.0	0.51397	0.32971	0.28438
	9	6	0.04876	0.5	0.12341	0.09764	0.09614
				1.0	0.51400	0.32983	0.28446
	10	6	0.05423	0.5	0.12408	0.09834	0.09676
				1.0	0.51406	0.32991	0.28448

Table 6.13: (Continued)

Test	r	s	α	θ	Normal	Exp(1)	LN(0.25)	LN(0.5)
$W_{\min,r}$	2	67	0.05954	0.5	0.23942	0.75179	0.31477	0.48316
				1.0	0.55242	0.98116	0.72338	0.91794
	3	71	0.04639	0.5	0.22288	0.62926	0.28257	0.42193
				1.0	0.55802	0.94955	0.70033	0.88107
	4	75	0.05377	0.5	0.25717	0.62267	0.31335	0.44410
				1.0	0.61907	0.94681	0.73585	0.88638
	5	77	0.04462	0.5	0.23764	0.52904	0.28236	0.39265
				1.0	0.60701	0.90048	0.70318	0.84718
	6	80	0.05468	0.5	0.27557	0.54185	0.31749	0.42206
				1.0	0.66094	0.90138	0.73689	0.85762
	7	81	0.05008	0.5	0.26530	0.49483	0.30080	0.39274
				1.0	0.65536	0.86798	0.71745	0.82970
	8	82	0.05056	0.5	0.27085	0.47701	0.30097	0.38469
				1.0	0.66483	0.85289	0.71506	0.81753
	9	82	0.04624	0.5	0.25841	0.44688	0.28482	0.36066
				1.0	0.65229	0.83064	0.69529	0.79418
	10	82	0.04460[†]	0.5	0.25363	0.43322	0.27797	0.35084
				1.0	0.64799	0.81627	0.68684	0.78191
$W_{\max,r}$	2	107	0.04954	0.5	0.21864	0.61386	0.27814	0.41342
				1.0	0.53289	0.94466	0.67657	0.86760
	3	99	0.05096	0.5	0.23908	0.50740	0.28068	0.38523
				1.0	0.58076	0.88496	0.67947	0.82888
	4	93	0.05065	0.5	0.24765	0.41622	0.27619	0.35310
				1.0	0.60476	0.79500	0.66268	0.76894
	5	89	0.05108	0.5	0.25641	0.42763	0.28094	0.35188
				1.0	0.62957	0.81969	0.67121	0.77665
	6	86	0.04900	0.5	0.25880	0.40750	0.27868	0.34450
				1.0	0.64022	0.79303	0.67066	0.75990
	7	84	0.04662	0.5	0.25600	0.40644	0.27624	0.34126
				1.0	0.64532	0.77765	0.67264	0.75783
	8	83	0.04650	0.5	0.25950	0.42334	0.27980	0.34871
				1.0	0.65223	0.80088	0.68384	0.77149
	9	83	0.05064	0.5	0.27432	0.45219	0.29770	0.37101
				1.0	0.67193	0.82583	0.70669	0.79468
	10	82	0.04460[†]	0.5	0.25363	0.43322	0.27797	0.35084
				1.0	0.64799	0.81627	0.68684	0.78191

† When $r = n_2 = 10$, $W_{\min,r} = W_{\max,r} = W_{E,r}$ and they are all equivalent to the Wilcoxon's rank sum test statistic.

Table 6.13: (Continued)

Test	r	s	α	θ	Gamma(2)	Gamma(5)	Gamma(10)
$W_{\min,r}$	2	67	0.05954	0.5	0.49375	0.34262	0.29921
				1.0	0.91494	0.77416	0.69533
	3	71	0.04639	0.5	0.41788	0.30283	0.26819
				1.0	0.86529	0.73787	0.67326
	4	75	0.05377	0.5	0.43420	0.33085	0.29946
				1.0	0.86784	0.76259	0.71060
	5	77	0.04462	0.5	0.37798	0.29481	0.26907
				1.0	0.81846	0.72485	0.67932
	6	80	0.05468	0.5	0.40483	0.32899	0.30405
				1.0	0.83001	0.75389	0.71749
	7	81	0.05008	0.5	0.37369	0.30976	0.28812
				1.0	0.79851	0.73116	0.69950
	8	82	0.05056	0.5	0.36709	0.30815	0.28815
				1.0	0.78641	0.72710	0.69840
	9	82	0.04624	0.5	0.34403	0.29076	0.27256
				1.0	0.76266	0.70628	0.67944
	10	82	0.04460[†]	0.5	0.33441	0.28330	0.26633
				1.0	0.75081	0.69701	0.67134
$W_{\max,r}$	2	107	0.04954	0.5	0.40659	0.29714	0.26448
				1.0	0.85125	0.71578	0.64977
	3	99	0.05096	0.5	0.36818	0.29194	0.26866
				1.0	0.79624	0.69985	0.65486
	4	93	0.05065	0.5	0.33127	0.28210	0.26392
				1.0	0.72522	0.66992	0.64209
	5	89	0.05108	0.5	0.32935	0.28533	0.26937
				1.0	0.73843	0.67789	0.65597
	6	86	0.04900	0.5	0.32376	0.28292	0.26830
				1.0	0.72400	0.67553	0.65529
	7	84	0.04662	0.5	0.32386	0.28095	0.26517
				1.0	0.72302	0.67849	0.65873
	8	83	0.04650	0.5	0.33213	0.28493	0.26965
				1.0	0.73981	0.69251	0.66899
	9	83	0.05064	0.5	0.35389	0.30321	0.28627
				1.0	0.76506	0.71584	0.69200
	10	82	0.04460[†]	0.5	0.33441	0.28330	0.26633
				1.0	0.75081	0.69701	0.67134

[†] When $r = n_2 = 10$, $W_{\min,r} = W_{\max,r} = W_{E,r}$ and they are all equivalent to the Wilcoxon's rank sum test statistic.

Table 6.13: (Continued)

Test	r	s	α	θ	Normal	Exp(1)	LN(0.25)	LN(0.5)
$W_{E,r}$	2	87	0.04954	0.5	0.21864	0.61386	0.27814	0.41342
				1.0	0.53289	0.94466	0.67657	0.86760
	3	85	0.04707	0.5	0.23359	0.50733	0.27695	0.38366
				1.0	0.57611	0.88496	0.67820	0.82879
	4	84	0.04871	0.5	0.24586	0.48267	0.28239	0.37451
				1.0	0.60944	0.87470	0.68524	0.81560
	5	84	0.05203	0.5	0.27860	0.48460	0.31251	0.40119
				1.0	0.65965	0.84538	0.71729	0.82098
	6	83	0.04947	0.5	0.26307	0.46053	0.29154	0.37322
				1.0	0.64946	0.84060	0.69872	0.80446
	7	83	0.04921	0.5	0.27844	0.46934	0.30601	0.38478
				1.0	0.67160	0.84411	0.71501	0.81044
	8	83	0.05207	0.5	0.27729	0.46363	0.30310	0.37985
				1.0	0.67320	0.83989	0.71369	0.80650
	9	83	0.05109	0.5	0.28397	0.47218	0.31036	0.38830
				1.0	0.68197	0.84352	0.72115	0.81153
	10	82	0.04460†	0.5	0.25363	0.43322	0.27797	0.35084
				1.0	0.64799	0.81627	0.68684	0.78191
W_R		82	0.04460†	0.5	0.25363	0.43322	0.27797	0.35084
				1.0	0.64799	0.81627	0.68684	0.78191

† When $r = n_2 = 10$, $W_{\min,r} = W_{\max,r} = W_{E,r}$ and they are all equivalent to the Wilcoxon's rank sum test statistic.

Table 6.13: (Continued)

Test	r	s	α	θ	Gamma(2)	Gamma(5)	Gamma(10)
$W_{E,r}$	2	87	0.04954	0.5	0.40659	0.29714	0.26448
				1.0	0.85125	0.71578	0.64977
	3	85	0.04707	0.5	0.36738	0.28915	0.26483
				1.0	0.79621	0.69928	0.65331
	4	84	0.04871	0.5	0.35581	0.29056	0.26953
				1.0	0.78214	0.69957	0.66307
	5	84	0.05203	0.5	0.38023	0.32038	0.29960
				1.0	0.78590	0.72757	0.70054
	6	83	0.04947	0.5	0.35292	0.29818	0.27934
				1.0	0.77051	0.70977	0.68180
	7	83	0.04921	0.5	0.36525	0.31221	0.29334
				1.0	0.77934	0.72432	0.69885
	8	83	0.05207	0.5	0.36281	0.30971	0.29138
				1.0	0.77557	0.72302	0.69748
	9	83	0.05109	0.5	0.36976	0.31706	0.29789
				1.0	0.78173	0.73106	0.70622
	10	82	0.04460[†]	0.5	0.33441	0.28330	0.26633
				1.0	0.75081	0.69701	0.67134
W_R		82	0.04460[†]	0.5	0.33441	0.28330	0.26633
				1.0	0.75081	0.69701	0.67134

† When $r = n_2 = 10$, $W_{\min,r} = W_{\max,r} = W_{E,r}$ and they are all equivalent to the Wilcoxon's rank sum test statistic.

Table 6.14: Power of precedence tests, maximal precedence tests, minimal, maximal, and expected rank-sum precedence tests, and Wilcoxon's rank-sum test (based on complete samples) when $n_1 = 10, n_2 = 15$

Test	r	s	α	θ	Normal	Exp(1)	LN(0.25)	LN(0.5)
$P_{(r)}$	1	3	0.05217	0.5	0.19712	0.87328	0.29466	0.51455
				1.0	0.46480	0.99524	0.69616	0.93954
	2	4	0.06403	0.5	0.26069	0.76379	0.34764	0.52427
				1.0	0.59416	0.98208	0.75896	0.93088
	3	5	0.06185	0.5	0.27407	0.63397	0.34122	0.47751
				1.0	0.63349	0.94802	0.75019	0.89333
	4	6	0.05327	0.5	0.25742	0.49333	0.30474	0.40369
				1.0	0.62596	0.87788	0.70580	0.83081
	5	7	0.04158	0.5	0.22462	0.35667	0.24962	0.31365
				1.0	0.58636	0.75865	0.62941	0.73112
	6	8	0.02861	0.5	0.17503	0.23232	0.18244	0.21891
				1.0	0.50964	0.58541	0.51525	0.58444
	7	8	0.05727	0.5	0.27041	0.30909	0.27224	0.30518
				1.0	0.63565	0.65240	0.62068	0.66583
$M_{(r)}$	2	4	0.03316	0.5	0.15295	0.69192	0.20721	0.34635
				1.0	0.42252	0.97482	0.58499	0.85073
	3	4	0.04968	0.5	0.19467	0.69459	0.24214	0.36758
				1.0	0.48194	0.97497	0.61467	0.85587
	4	4	0.06615	0.5	0.22612	0.69714	0.26619	0.38050
				1.0	0.51596	0.97509	0.62967	0.85837
	5	5	0.02371	0.5	0.09903	0.45191	0.11562	0.18108
				1.0	0.29035	0.90776	0.38368	0.67065
	6	5	0.02845	0.5	0.10515	0.45248	0.11965	0.18295
				1.0	0.29478	0.90785	0.38521	0.67090
	7	5	0.03320	0.5	0.11019	0.45314	0.12261	0.18449
				1.0	0.29732	0.90794	0.38603	0.67113
$W_{\min,r}$	2	68	0.03316	0.5	0.17258	0.75476	0.25464	0.44897
				1.0	0.48197	0.98174	0.68965	0.91761
	3	74	0.04452	0.5	0.22457	0.77202	0.31468	0.50833
				1.0	0.58015	0.98346	0.75992	0.93635
	4	79	0.04727	0.5	0.24651	0.75878	0.33307	0.51457
				1.0	0.62674	0.98167	0.78098	0.93482
	5	84	0.05554	0.5	0.28485	0.72092	0.36865	0.53499
				1.0	0.68328	0.96853	0.80999	0.93467
	6	87	0.04840	0.5	0.27134	0.67670	0.34367	0.49876
				1.0	0.67934	0.95867	0.79314	0.91918
	7	90	0.04794	0.5	0.27667	0.64783	0.34222	0.48416
				1.0	0.69276	0.95145	0.79098	0.90894

Table 6.14: (Continued)

Test	r	s	α	θ	Gamma(2)	Gamma(5)	Gamma(10)
$P_{(r)}$	1	3	0.05217	0.5	0.57176	0.33812	0.27897
				1.0	0.95367	0.78083	0.66569
	2	4	0.06403	0.5	0.52528	0.37730	0.33366
				1.0	0.92419	0.80314	0.73197
	3	5	0.06185	0.5	0.45446	0.35675	0.32678
				1.0	0.87349	0.77462	0.72702
	4	6	0.05327	0.5	0.37266	0.31205	0.29322
				1.0	0.79245	0.71811	0.68466
	5	7	0.04158	0.5	0.28524	0.25207	0.24153
				1.0	0.68107	0.63326	0.61399
	6	8	0.02861	0.5	0.19446	0.18170	0.17833
				1.0	0.53156	0.50855	0.50514
	7	8	0.05727	0.5	0.27912	0.27008	0.26864
				1.0	0.61900	0.61345	0.61475
$M_{(r)}$	2	4	0.03316	0.5	0.36049	0.22587	0.19675
				1.0	0.86261	0.64247	0.55219
	3	4	0.04968	0.5	0.37434	0.25466	0.23118
				1.0	0.86588	0.66203	0.58416
	4	4	0.06615	0.5	0.38539	0.27538	0.25568
				1.0	0.86789	0.67291	0.60049
	5	5	0.02371	0.5	0.18095	0.11995	0.10989
				1.0	0.68311	0.43025	0.35977
	6	5	0.02845	0.5	0.18263	0.12333	0.11415
				1.0	0.68336	0.43145	0.36168
	7	5	0.03320	0.5	0.18407	0.12615	0.11738
				1.0	0.68364	0.43224	0.36301
$W_{\min,r}$	2	68	0.03316	0.5	0.47238	0.28885	0.24050
				1.0	0.91591	0.75601	0.65882
	3	74	0.04452	0.5	0.52036	0.34847	0.29931
				1.0	0.93114	0.80993	0.73106
	4	79	0.04727	0.5	0.51749	0.36307	0.31739
				1.0	0.92722	0.82046	0.75432
	5	84	0.05554	0.5	0.52594	0.39376	0.35218
				1.0	0.92159	0.83810	0.78649
	6	87	0.04840	0.5	0.48492	0.36620	0.33001
				1.0	0.90213	0.81814	0.76999
	7	90	0.04794	0.5	0.46738	0.36189	0.32937
				1.0	0.88969	0.81182	0.76966

Table 6.14: (Continued)

Test	r	s	α	θ	Normal	Exp(1)	LN(0.25)	LN(0.5)
$W_{\max,r}$	2	147	0.04743	0.5	0.22052	0.76278	0.30920	0.49963
				1.0	0.55191	0.98204	0.73779	0.92864
	3	136	0.05049	0.5	0.24876	0.63347	0.32052	0.46750
				1.0	0.61105	0.94800	0.74192	0.89271
	4	109	0.05035	0.5	0.25240	0.49328	0.30132	0.40257
				1.0	0.62314	0.87788	0.70506	0.83079
	5	122	0.05134	0.5	0.27163	0.55230	0.32282	0.43772
				1.0	0.66561	0.90396	0.74897	0.86830
	6	116	0.05148	0.5	0.27844	0.48610	0.31678	0.40462
				1.0	0.67769	0.87089	0.72842	0.82587
	7	111	0.04919	0.5	0.27954	0.47535	0.31222	0.39955
				1.0	0.68871	0.84081	0.73148	0.82441
$W_{E,r}$	2	108	0.04743	0.5	0.22052	0.76278	0.30920	0.49963
				1.0	0.55191	0.98204	0.73779	0.92864
	3	106	0.05049	0.5	0.24876	0.63347	0.32052	0.46750
				1.0	0.61105	0.94800	0.74192	0.89271
	4	105	0.05194	0.5	0.27863	0.67346	0.34796	0.49356
				1.0	0.66594	0.95871	0.77758	0.91337
	5	103	0.05080	0.5	0.27487	0.56455	0.33235	0.45415
				1.0	0.67372	0.90624	0.76193	0.87477
	6	102	0.04896	0.5	0.28794	0.57176	0.33826	0.45150
				1.0	0.69595	0.91616	0.76827	0.87726
	7	101	0.04897	0.5	0.28496	0.53943	0.32842	0.43439
				1.0	0.70153	0.89715	0.76006	0.86199
W_R		99	0.04550	0.5	0.29337	0.50201	0.32272	0.41034
				1.0	0.73058	0.86374	0.76170	0.83985

Table 6.14: (Continued)

Test	r	s	α	θ	Gamma(2)	Gamma(5)	Gamma(10)
$W_{\text{max},r}$	2	147	0.04743	0.5	0.51160	0.34271	0.29439
				1.0	0.92325	0.79125	0.70872
	3	136	0.05049	0.5	0.44856	0.34054	0.30588
				1.0	0.87319	0.76987	0.71819
	4	109	0.05035	0.5	0.37192	0.30930	0.28999
				1.0	0.79244	0.71773	0.68385
	5	122	0.05134	0.5	0.41431	0.33315	0.31036
				1.0	0.83769	0.76415	0.72805
	6	116	0.05148	0.5	0.37775	0.32441	0.30703
				1.0	0.79086	0.73542	0.71425
	7	111	0.04919	0.5	0.37613	0.32015	0.30327
				1.0	0.78889	0.74024	0.71832
$W_{E,r}$	2	108	0.04743	0.5	0.51160	0.34271	0.29439
				1.0	0.92325	0.79125	0.70872
	3	106	0.05049	0.5	0.44856	0.34054	0.30588
				1.0	0.87319	0.76987	0.71819
	4	105	0.05194	0.5	0.47908	0.36642	0.33454
				1.0	0.89594	0.80260	0.75425
	5	103	0.05080	0.5	0.43060	0.34491	0.31919
				1.0	0.84490	0.77711	0.74010
	6	102	0.04896	0.5	0.42875	0.35030	0.32738
				1.0	0.85061	0.78256	0.75152
	7	101	0.04897	0.5	0.41136	0.34022	0.31777
				1.0	0.83313	0.77338	0.74527
W_R		99	0.04550	0.5	0.39241	0.33335	0.31417
				1.0	0.81379	0.76922	0.75008

Table 6.15: Power of precedence tests, maximal precedence tests, minimal, maximal, and expected rank-sum precedence tests, and Wilcoxon's rank-sum test (based on complete samples) when $n_1 = n_2 = 15$

Test	r	s	α	θ	Normal	Exp(1)	LN(0.25)	LN(0.5)
$P_{(r)}$	1	4	0.04981	0.5	0.21166	0.93598	0.31908	0.56758
				1.0	0.50252	0.99955	0.74977	0.97286
	2	6	0.04004	0.5	0.22061	0.76824	0.30350	0.49083
				1.0	0.57185	0.99273	0.75944	0.94680
	3	7	0.05432	0.5	0.28472	0.69628	0.35968	0.51584
				1.0	0.67187	0.98192	0.80630	0.94209
	4	9	0.03022	0.5	0.21126	0.44001	0.25202	0.34941
				1.0	0.60134	0.89959	0.69747	0.84304
	5	10	0.03280	0.5	0.22635	0.36866	0.25068	0.32262
				1.0	0.62508	0.83183	0.68471	0.79751
	6	11	0.03280	0.5	0.22719	0.30143	0.23544	0.28335
				1.0	0.62484	0.73930	0.64836	0.73122
	7	11	0.06971	0.5	0.35144	0.40243	0.35078	0.39436
				1.0	0.75253	0.80155	0.75525	0.80731
$M_{(r)}$	2	5	0.04205	0.5	0.20162	0.84028	0.27254	0.44924
				1.0	0.51762	0.99729	0.70499	0.93968
	3	5	0.06292	0.5	0.25626	0.84211	0.31613	0.47444
				1.0	0.58639	0.99732	0.73674	0.94336
	4	6	0.03400	0.5	0.15784	0.68508	0.19076	0.30140
				1.0	0.43474	0.98852	0.56833	0.85843
	5	6	0.04209	0.5	0.17235	0.68581	0.20037	0.30553
				1.0	0.44805	0.98857	0.57248	0.85892
	6	6	0.05050	0.5	0.18348	0.68654	0.20713	0.30876
				1.0	0.45504	0.98857	0.57515	0.85920
	7	6	0.05889	0.5	0.19223	0.68718	0.21236	0.31145
				1.0	0.45931	0.98859	0.57640	0.85936
$W_{\min,r}$	2	141	0.04205	0.5	0.22459	0.88518	0.32929	0.57076
				1.0	0.57820	0.99832	0.80054	0.97474
	3	150	0.04654	0.5	0.25997	0.87742	0.36651	0.59499
				1.0	0.65183	0.99821	0.84036	0.97793
	4	158	0.04826	0.5	0.28632	0.83497	0.38683	0.59747
				1.0	0.69937	0.99570	0.85993	0.97646
	5	165	0.05000	0.5	0.30607	0.81192	0.39997	0.59452
				1.0	0.73382	0.99463	0.86956	0.97379
	6	171	0.05070	0.5	0.32068	0.77397	0.40558	0.58443
				1.0	0.75832	0.99013	0.87448	0.96942
	7	176	0.05046	0.5	0.33025	0.74438	0.40584	0.57106
				1.0	0.77541	0.98709	0.87493	0.96445

Table 6.15: (Continued)

Test	r	s	α	θ	Gamma(2)	Gamma(5)	Gamma(10)
$P_{(r)}$	1	4	0.04981	0.5	0.64016	0.37132	0.29851
				1.0	0.98405	0.83848	0.72008
	2	6	0.04004	0.5	0.49059	0.32997	0.28478
				1.0	0.94228	0.80772	0.72752
	3	7	0.05432	0.5	0.49085	0.37752	0.34087
				1.0	0.92751	0.83281	0.77819
	4	9	0.03022	0.5	0.31724	0.25579	0.23896
				1.0	0.79981	0.70956	0.66764
	5	10	0.03280	0.5	0.29113	0.25256	0.24100
				1.0	0.74790	0.68407	0.66128
	6	11	0.03280	0.5	0.25289	0.23457	0.22839
				1.0	0.67456	0.63918	0.63055
	7	11	0.06971	0.5	0.36276	0.34668	0.34421
				1.0	0.76281	0.74719	0.74472
$M_{(r)}$	2	5	0.04205	0.5	0.47509	0.29568	0.31930
				1.0	0.95219	0.76861	0.76648
	3	5	0.06292	0.5	0.49078	0.33221	0.33340
				1.0	0.95378	0.78817	0.77670
	4	6	0.03400	0.5	0.30398	0.19437	0.34526
				1.0	0.87859	0.62283	0.78889
	5	6	0.04209	0.5	0.30797	0.20220	0.34436
				1.0	0.87898	0.62551	0.79358
	6	6	0.05050	0.5	0.31089	0.20820	0.34596
				1.0	0.87921	0.62745	0.79972
	7	6	0.05889	0.5	0.31355	0.21252	0.34827
				1.0	0.87949	0.62837	0.80394
$W_{\min,r}$	2	141	0.04205	0.5	0.60499	0.37449	0.25543
				1.0	0.97677	0.86393	0.67058
	3	150	0.04654	0.5	0.61237	0.40738	0.29928
				1.0	0.97668	0.88705	0.70524
	4	158	0.04826	0.5	0.59952	0.42024	0.17648
				1.0	0.97128	0.89459	0.53219
	5	165	0.05000	0.5	0.58975	0.42921	0.18621
				1.0	0.96718	0.89760	0.53755
	6	171	0.05070	0.5	0.57326	0.43190	0.19376
				1.0	0.96165	0.89771	0.54069
	7	176	0.05046	0.5	0.55487	0.42913	0.19943
				1.0	0.95492	0.89453	0.54239

Table 6.15: (Continued)

Test	r	s	α	θ	Normal	Exp(1)	LN(0.25)	LN(0.5)
$W_{\max,r}$	2	270	0.04741	0.5	0.24479	0.86046	0.33934	0.54855
				1.0	0.60349	0.99782	0.79679	0.96711
	3	250	0.04945	0.5	0.27457	0.69627	0.35277	0.51347
				1.0	0.66497	0.98192	0.80493	0.94203
	4	238	0.05094	0.5	0.29630	0.62237	0.36286	0.50003
				1.0	0.70813	0.96207	0.81423	0.92743
	5	227	0.04991	0.5	0.30688	0.59903	0.35903	0.47857
				1.0	0.73050	0.95887	0.81470	0.91932
	6	218	0.04912	0.5	0.31516	0.57724	0.35731	0.46871
				1.0	0.74960	0.94373	0.81906	0.91465
	7	211	0.04909	0.5	0.32803	0.54681	0.36000	0.46137
				1.0	0.76885	0.92760	0.81903	0.90232
$W_{E,r}$	2	205	0.04741	0.5	0.24479	0.86046	0.33934	0.54855
				1.0	0.60349	0.99782	0.79679	0.96711
	3	201	0.04891	0.5	0.27510	0.76629	0.35865	0.53395
				1.0	0.66937	0.99237	0.81678	0.95475
	4	199	0.05101	0.5	0.30958	0.73329	0.38500	0.54494
				1.0	0.72412	0.98637	0.84200	0.95521
	5	197	0.05179	0.5	0.32040	0.67778	0.38439	0.53117
				1.0	0.74916	0.97340	0.84639	0.94640
	6	195	0.04808	0.5	0.32528	0.64703	0.38154	0.51394
				1.0	0.76584	0.96720	0.84622	0.93778
	7	194	0.04902	0.5	0.33368	0.62344	0.38073	0.50532
				1.0	0.78084	0.95711	0.84734	0.93366
W_R		192	0.04880	0.5	0.35996	0.59384	0.38949	0.49266
				1.0	0.82397	0.94063	0.85733	0.92161

Table 6.15: (Continued)

Test	r	s	α	θ	Gamma(2)	Gamma(5)	Gamma(10)
$W_{\max,r}$	2	270	0.04741	0.5	0.56513	0.37409	0.30890
				1.0	0.96800	0.84783	0.77186
	3	250	0.04945	0.5	0.49010	0.37268	0.34484
				1.0	0.92751	0.83231	0.81347
	4	238	0.05094	0.5	0.47011	0.37558	0.36465
				1.0	0.90369	0.82992	0.83360
	5	227	0.04991	0.5	0.45016	0.37019	0.37839
				1.0	0.89642	0.82583	0.84798
	6	218	0.04912	0.5	0.44076	0.36872	0.38593
				1.0	0.89048	0.82710	0.85519
	7	211	0.04909	0.5	0.43359	0.36851	0.38762
				1.0	0.87472	0.82602	0.85636
$W_{E,r}$	2	205	0.04741	0.5	0.56513	0.37409	0.31930
				1.0	0.96800	0.84783	0.76648
	3	201	0.04891	0.5	0.51785	0.38108	0.33848
				1.0	0.94713	0.84800	0.78914
	4	199	0.05101	0.5	0.52614	0.40228	0.36567
				1.0	0.94489	0.86421	0.81757
	5	197	0.05179	0.5	0.50835	0.40157	0.36801
				1.0	0.93081	0.86332	0.82499
	6	195	0.04808	0.5	0.48849	0.39599	0.36644
				1.0	0.92083	0.85899	0.82760
	7	194	0.04902	0.5	0.47899	0.39464	0.36677
				1.0	0.91423	0.85858	0.83034
W_R		192	0.04880	0.5	0.47190	0.40033	0.38005
				1.0	0.90428	0.86491	0.84786

Table 6.16: Power of precedence tests, maximal precedence tests, minimal, maximal, and expected rank-sum precedence tests, and Wilcoxon's rank-sum test (based on complete samples) when $n_1 = 15, n_2 = 20$

Test	r	s	α	θ	Normal	Exp(1)	LN(0.25)	LN(0.5)
$P_{(r)}$	1	3	0.06952	0.5	0.26103	0.98034	0.39668	0.68983
				1.0	0.56610	0.99994	0.82145	0.99016
	2	5	0.04009	0.5	0.22799	0.87119	0.33392	0.56444
				1.0	0.58927	0.99777	0.80473	0.97101
	3	6	0.04615	0.5	0.26977	0.79414	0.36640	0.56539
				1.0	0.66905	0.99367	0.83116	0.96482
	4	7	0.04718	0.5	0.28676	0.70414	0.37028	0.53471
				1.0	0.70558	0.98285	0.83120	0.94992
	5	8	0.04492	0.5	0.28890	0.60639	0.35453	0.48667
				1.0	0.71823	0.96025	0.81410	0.92500
	6	9	0.04043	0.5	0.27957	0.50487	0.32527	0.42887
				1.0	0.71204	0.91834	0.78186	0.88644
	7	10	0.03446	0.5	0.25718	0.40124	0.28593	0.36215
				1.0	0.69120	0.84953	0.73249	0.82850
$M_{(r)}$	2	4	0.05187	0.5	0.23931	0.93260	0.33630	0.57112
				1.0	0.57697	0.99933	0.78992	0.97513
	3	5	0.02770	0.5	0.15938	0.82474	0.21598	0.38511
				1.0	0.46125	0.99674	0.65157	0.92585
	4	5	0.03690	0.5	0.18512	0.82560	0.23569	0.39570
				1.0	0.49491	0.99675	0.66553	0.92743
	5	5	0.04609	0.5	0.20491	0.82636	0.24992	0.40293
				1.0	0.51620	0.99675	0.67290	0.92805
	6	5	0.05526	0.5	0.22093	0.82712	0.26127	0.40826
				1.0	0.52976	0.99677	0.67774	0.92855
	7	5	0.06441	0.5	0.23365	0.82768	0.26984	0.41227
				1.0	0.53884	0.99678	0.68099	0.92895
$W_{min,r}$	2	143	0.05187	0.5	0.25985	0.95111	0.39406	0.67415
				1.0	0.62814	0.99960	0.85783	0.98926
	3	153	0.05302	0.5	0.28617	0.94575	0.42144	0.68614
				1.0	0.68770	0.99948	0.88455	0.99010
	4	162	0.04950	0.5	0.29737	0.90905	0.42392	0.67168
				1.0	0.72107	0.99860	0.89252	0.98788
	5	170	0.04684	0.5	0.30454	0.88945	0.42355	0.65678
				1.0	0.74464	0.99818	0.89601	0.98583
	6	178	0.05149	0.5	0.33433	0.86588	0.44830	0.66554
				1.0	0.78297	0.99670	0.90908	0.98538
	7	185	0.05333	0.5	0.35208	0.85075	0.45966	0.66309
				1.0	0.80616	0.99592	0.91387	0.98391

Table 6.16: (Continued)

Test	r	s	α	θ	Gamma(2)	Gamma(5)	Gamma(10)
$P_{(r)}$	1	3	0.06952	0.5	0.77808	0.47275	0.37926
				1.0	0.99538	0.90770	0.80025
	2	5	0.04009	0.5	0.58158	0.37354	0.31301
				1.0	0.97204	0.86035	0.77478
	3	6	0.04615	0.5	0.54880	0.39401	0.34462
				1.0	0.95792	0.86573	0.80654
	4	7	0.04718	0.5	0.50231	0.38713	0.34864
				1.0	0.93573	0.85376	0.80835
	5	8	0.04492	0.5	0.44743	0.36288	0.33594
				1.0	0.90163	0.82691	0.79161
	6	9	0.04043	0.5	0.38805	0.33032	0.30925
				1.0	0.85177	0.78776	0.76187
	7	10	0.03446	0.5	0.32210	0.28523	0.27515
				1.0	0.78421	0.73317	0.71587
$M_{(r)}$	2	4	0.05187	0.5	0.61612	0.37709	0.31680
				1.0	0.98278	0.85651	0.76077
	3	5	0.02770	0.5	0.41274	0.23401	0.20145
				1.0	0.94357	0.72230	0.61501
	4	5	0.03690	0.5	0.41999	0.25053	0.22094
				1.0	0.94424	0.73044	0.63081
	5	5	0.04609	0.5	0.42531	0.26210	0.23570
				1.0	0.94483	0.73513	0.63931
	6	5	0.05526	0.5	0.42947	0.27132	0.24728
				1.0	0.94521	0.73832	0.64463
	7	5	0.06441	0.5	0.43317	0.27822	0.25643
				1.0	0.94554	0.74029	0.64854
$W_{\min,r}$	2	143	0.05187	0.5	0.72250	0.45616	0.37273
				1.0	0.99115	0.91802	0.83418
	3	153	0.05302	0.5	0.71835	0.47530	0.39655
				1.0	0.99056	0.92903	0.86298
	4	162	0.04950	0.5	0.68584	0.47116	0.39971
				1.0	0.98636	0.92796	0.87272
	5	170	0.04684	0.5	0.65930	0.46511	0.40060
				1.0	0.98288	0.92547	0.87502
	6	178	0.05149	0.5	0.65908	0.48462	0.42481
				1.0	0.98084	0.93135	0.89051
	7	185	0.05333	0.5	0.65027	0.49153	0.43604
				1.0	0.97895	0.93276	0.89709

Table 6.16: (Continued)

Test	r	s	α	θ	Normal	Exp(1)	LN(0.25)	LN(0.5)
$W_{\max,r}$	2	340	0.05074	0.5	0.26086	0.93935	0.38197	0.63716
				1.0	0.62844	0.99940	0.84402	0.98431
	3	320	0.04874	0.5	0.27939	0.86398	0.38212	0.59491
				1.0	0.68127	0.99755	0.84752	0.97483
	4	302	0.04970	0.5	0.29838	0.80080	0.38983	0.57586
				1.0	0.72108	0.99403	0.85374	0.96904
	5	287	0.05002	0.5	0.31665	0.75027	0.39995	0.57100
				1.0	0.75442	0.98727	0.86472	0.96490
	6	274	0.04912	0.5	0.32750	0.67553	0.40172	0.55136
				1.0	0.77413	0.97046	0.86265	0.94976
	7	264	0.04971	0.5	0.33957	0.63792	0.39975	0.52510
				1.0	0.78969	0.96524	0.85464	0.93819
$W_{E,r}$	2	241	0.05074	0.5	0.26086	0.93935	0.38197	0.63716
				1.0	0.62844	0.99940	0.84402	0.98431
	3	236	0.05134	0.5	0.28841	0.88129	0.39637	0.61766
				1.0	0.69158	0.99800	0.85935	0.97942
	4	232	0.05161	0.5	0.31070	0.83120	0.41291	0.61575
				1.0	0.73815	0.99528	0.87482	0.97678
	5	229	0.05037	0.5	0.32774	0.76175	0.41912	0.59851
				1.0	0.76806	0.98759	0.87826	0.96846
	6	227	0.05094	0.5	0.33882	0.73913	0.42158	0.58270
				1.0	0.78779	0.98598	0.87873	0.96346
	7	225	0.04878	0.5	0.34994	0.71183	0.42436	0.57711
				1.0	0.80544	0.97755	0.88337	0.96087
W_R		220	0.0497	0.5	0.39901	0.64380	0.43600	0.54447
				1.0	0.87100	0.95604	0.89288	0.94262

Table 6.16: (Continued)

Test	r	s	α	θ	Gamma(2)	Gamma(5)	Gamma(10)
$W_{\max,r}$	2	340	0.05074	0.5	0.67390	0.43135	0.35907
				1.0	0.98716	0.89894	0.81719
	3	320	0.04874	0.5	0.59185	0.41456	0.36043
				1.0	0.97324	0.88498	0.82349
	4	302	0.04970	0.5	0.55660	0.41245	0.36698
				1.0	0.96218	0.88079	0.83173
	5	287	0.05002	0.5	0.54798	0.41820	0.37895
				1.0	0.95352	0.88593	0.84374
	6	274	0.04912	0.5	0.51837	0.41619	0.38002
				1.0	0.93297	0.87499	0.84373
	7	264	0.04971	0.5	0.48901	0.40905	0.38100
				1.0	0.91989	0.86293	0.83918
$W_{E,r}$	2	241	0.05074	0.5	0.67390	0.43135	0.35907
				1.0	0.98716	0.89894	0.81719
	3	236	0.05134	0.5	0.62077	0.43135	0.37338
				1.0	0.97798	0.89796	0.83638
	4	232	0.05161	0.5	0.60331	0.44114	0.38913
				1.0	0.97111	0.90309	0.85398
	5	229	0.05037	0.5	0.57470	0.44037	0.39778
				1.0	0.95769	0.89750	0.85834
	6	227	0.05094	0.5	0.55321	0.43890	0.39903
				1.0	0.95181	0.89346	0.86080
	7	225	0.04878	0.5	0.54746	0.44049	0.40365
				1.0	0.94720	0.89519	0.86713
W_R		220	0.0497	0.5	0.51830	0.44589	0.42322
				1.0	0.92949	0.89928	0.88529

Table 6.17: Power of precedence tests, maximal precedence tests, minimal, maximal, and expected rank-sum precedence tests, and Wilcoxon's rank-sum test (based on complete samples) when $n_1 = n_2 = 20$

Test	r	s	α	θ	Normal	Exp(1)	LN(0.25)	LN(0.5)
$P_{(r)}$	1	4	0.05301	0.5	0.23855	0.98866	0.37686	0.68551
				1.0	0.55525	1.00000	0.82751	0.99428
	2	6	0.04574	0.5	0.26616	0.93554	0.39035	0.64528
				1.0	0.65292	0.99981	0.86474	0.99011
	3	8	0.03242	0.5	0.24335	0.80663	0.34027	0.54800
				1.0	0.66299	0.99719	0.83996	0.97553
	4	9	0.04118	0.5	0.29470	0.74992	0.38078	0.56144
				1.0	0.73517	0.99322	0.86742	0.97162
	5	10	0.04792	0.5	0.33174	0.69035	0.40343	0.55552
				1.0	0.77868	0.98635	0.87638	0.96482
	6	11	0.05267	0.5	0.35648	0.62910	0.41012	0.53689
				1.0	0.80552	0.97332	0.87481	0.95402
	7	12	0.05548	0.5	0.36864	0.56701	0.40861	0.51070
				1.0	0.81861	0.95316	0.86600	0.93674
$M_{(r)}$	2	5	0.04691	0.5	0.24343	0.96341	0.34864	0.59833
				1.0	0.59939	0.99997	0.82297	0.98815
	3	6	0.03000	0.5	0.18471	0.90506	0.25067	0.44604
				1.0	0.52351	0.99969	0.72654	0.96538
	4	6	0.04026	0.5	0.21380	0.90533	0.27237	0.45715
				1.0	0.56051	0.99969	0.74016	0.96618
	5	6	0.05000	0.5	0.23598	0.90583	0.28786	0.46423
				1.0	0.58325	0.99969	0.74747	0.96660
	6	6	0.06025	0.5	0.25446	0.90621	0.29944	0.46884
				1.0	0.59768	0.99969	0.75171	0.96683
	7	6	0.07021	0.5	0.26872	0.90653	0.30885	0.47260
				1.0	0.60667	0.99970	0.75454	0.96700
$W_{\min,r}$	2	241	0.04691	0.5	0.26563	0.97529	0.41115	0.70839
				1.0	0.65211	0.99998	0.88839	0.99603
	3	255	0.05335	0.5	0.31306	0.97323	0.46196	0.74223
				1.0	0.73361	0.99998	0.92205	0.99728
	4	267	0.04671	0.5	0.31593	0.94911	0.45624	0.71938
				1.0	0.76021	0.99984	0.92596	0.99599
	5	279	0.05234	0.5	0.35496	0.93654	0.48976	0.73413
				1.0	0.80732	0.99972	0.94067	0.99618
	6	289	0.04987	0.5	0.36229	0.91749	0.48704	0.71781
				1.0	0.82412	0.99963	0.94163	0.99508
	7	298	0.04816	0.5	0.36966	0.89261	0.48555	0.70248
				1.0	0.83961	0.99913	0.94215	0.99393

Table 6.17: (Continued)

Test	r	s	α	θ	Gamma(2)	Gamma(5)	Gamma(10)
$P_{(r)}$	1	4	0.05301	0.5	0.78803	0.45093	0.35517
				1.0	0.99817	0.91502	0.80255
	2	6	0.04574	0.5	0.67111	0.43567	0.36379
				1.0	0.99140	0.91288	0.83709
	3	8	0.03242	0.5	0.53308	0.36703	0.31605
				1.0	0.96991	0.87542	0.81209
	4	9	0.04118	0.5	0.53174	0.40110	0.35826
				1.0	0.96147	0.88663	0.84176
	5	10	0.04792	0.5	0.51727	0.41772	0.38082
				1.0	0.94874	0.88877	0.85320
	6	11	0.05267	0.5	0.49662	0.41973	0.39200
				1.0	0.93225	0.88021	0.85581
	7	12	0.05548	0.5	0.46744	0.41259	0.39212
				1.0	0.90856	0.86660	0.84911
$M_{(r)}$	2	5	0.04691	0.5	0.65799	0.38935	0.32350
				1.0	0.99337	0.88558	0.78925
	3	6	0.03000	0.5	0.48784	0.27136	0.23271
				1.0	0.97866	0.79514	0.68755
	4	6	0.04026	0.5	0.49485	0.28888	0.25464
				1.0	0.97904	0.80249	0.70298
	5	6	0.05000	0.5	0.50036	0.30132	0.27088
				1.0	0.97931	0.80651	0.71154
	6	6	0.06025	0.5	0.50442	0.31157	0.28339
				1.0	0.97955	0.80929	0.71709
	7	6	0.07021	0.5	0.50806	0.31961	0.29313
				1.0	0.97970	0.81080	0.72051
$W_{\min,r}$	2	241	0.04691	0.5	0.76888	0.47784	0.38507
				1.0	0.99717	0.94085	0.86307
	3	255	0.05335	0.5	0.78152	0.52251	0.43477
				1.0	0.99743	0.95763	0.90239
	4	267	0.04671	0.5	0.73988	0.50644	0.42666
				1.0	0.99567	0.95465	0.90690
	5	279	0.05234	0.5	0.74228	0.53457	0.45978
				1.0	0.99501	0.96195	0.92479
	6	289	0.04987	0.5	0.71814	0.52715	0.45897
				1.0	0.99344	0.96028	0.92623
	7	298	0.04816	0.5	0.69538	0.51957	0.45696
				1.0	0.99143	0.95826	0.92733

Table 6.17: (Continued)

Test	r	s	α	θ	Normal	Exp(1)	LN(0.25)	LN(0.5)
$W_{\max,r}$	2	510	0.05326	0.5	0.28995	0.96892	0.42488	0.69261
				1.0	0.67845	0.99998	0.88741	0.99463
	3	477	0.04979	0.5	0.31303	0.90068	0.43704	0.67156
				1.0	0.73784	0.99940	0.90144	0.99091
	4	458	0.05023	0.5	0.33510	0.85888	0.44318	0.65618
				1.0	0.77776	0.99806	0.91038	0.98864
	5	439	0.04967	0.5	0.34947	0.80467	0.43694	0.62239
				1.0	0.80024	0.99561	0.90740	0.98374
	6	423	0.04952	0.5	0.36331	0.75011	0.43729	0.60253
				1.0	0.82318	0.99076	0.90781	0.97779
	7	410	0.05053	0.5	0.37765	0.70899	0.44447	0.59235
				1.0	0.84235	0.98622	0.90878	0.97186
$W_{E,r}$	2	375	0.05326	0.5	0.28995	0.96892	0.42488	0.69261
				1.0	0.67845	0.99998	0.88741	0.99463
	3	367	0.04979	0.5	0.32336	0.90076	0.44710	0.67790
				1.0	0.74637	0.99940	0.90416	0.99095
	4	363	0.05114	0.5	0.34072	0.86001	0.45265	0.66558
				1.0	0.78355	0.99806	0.91428	0.98906
	5	360	0.05071	0.5	0.36530	0.83452	0.46498	0.65793
				1.0	0.81679	0.99754	0.92136	0.98689
	6	357	0.05017	0.5	0.37277	0.80486	0.46004	0.63891
				1.0	0.83451	0.99571	0.92214	0.98437
	7	355	0.04941	0.5	0.38888	0.78140	0.46869	0.63603
				1.0	0.85360	0.99323	0.92578	0.98289
W_R		348	0.0482	0.5	0.44025	0.70110	0.47674	0.59412
				1.0	0.91358	0.98022	0.93438	0.97157

Table 6.17: (Continued)

Test	r	s	α	θ	Gamma(2)	Gamma(5)	Gamma(10)
$W_{\max,r}$	2	510	0.05326	0.5	0.73621	0.47815	0.39764
				1.0	0.99595	0.93288	0.86214
	3	477	0.04979	0.5	0.67241	0.47535	0.40846
				1.0	0.98864	0.93154	0.87985
	4	458	0.05023	0.5	0.64441	0.47383	0.41736
				1.0	0.98423	0.93206	0.89030
	5	439	0.04967	0.5	0.59906	0.45986	0.41287
				1.0	0.97622	0.92428	0.88658
	6	423	0.04952	0.5	0.57319	0.45612	0.41656
				1.0	0.96701	0.92037	0.88866
	7	410	0.05053	0.5	0.55775	0.45879	0.42413
				1.0	0.95727	0.91683	0.89187
$W_{E,r}$	2	375	0.05326	0.5	0.73621	0.47815	0.39764
				1.0	0.99595	0.93288	0.86214
	3	367	0.04979	0.5	0.67643	0.48555	0.41876
				1.0	0.98867	0.93312	0.88311
	4	363	0.05114	0.5	0.65310	0.48465	0.42571
				1.0	0.98455	0.93512	0.89458
	5	360	0.05071	0.5	0.63627	0.49216	0.44006
				1.0	0.98066	0.93694	0.90272
	6	357	0.05017	0.5	0.61393	0.48393	0.43713
				1.0	0.97775	0.93532	0.90421
	7	355	0.04941	0.5	0.60740	0.48754	0.44576
				1.0	0.97510	0.93648	0.91078
W_R		348	0.0482	0.5	0.56991	0.49170	0.46220
				1.0	0.96218	0.93918	0.92747

normal, standard exponential, standardized gamma, and standardized log-normal distributions, with location-shift being equal to 0.5 and 1.0 (for the case $n_1 = n_2 = 10$, we present the estimated power values up to $r = 10$). For comparison purposes, the corresponding critical values and the exact levels of significance are also included.

6.6 EVALUATION AND COMPARATIVE REMARKS

From Tables 6.13–6.17, we see that the power values of all tests increase with increasing sample sizes as well as with increasing location-shift. We can compare the power values of the precedence tests and the maximal precedence tests with the same value of r since they are both test procedures based on failures from the X-sample occurring before the rth failure from the Y-sample. When we compare the power values of the three Wilcoxon-type rank-sum precedence tests with those of precedence tests and the maximal precedence tests, we find that the Wilcoxon-type rank-sum precedence tests perform better than the precedence tests and maximal precedence tests for same value of r. For example, in Table 6.17, when $n_1 = n_2 = 20$, the underlying distribution is standard normal and the location-shift equals 0.5, the powers of the minimal, maximal, and expected rank-sum precedence tests with $r = 7$ are 0.370, 0.378, and 0.389 (exact levels of significance are 0.048, 0.051, and 0.049), respectively, while the power of the precedence test is 0.369 (exact level of significance is 0.055) and the power of the maximal precedence test is 0.269 (exact level of significance is 0.070).

From Tables 6.13–6.17, we can also see that when the underlying distributions are nearly symmetric, such as the normal distribution, gamma distribution with large values of shape parameter a, and lognormal distribution with small values of shape parameter σ, the power values of the Wilcoxon-type precedence tests increase with increasing values of r. However, under some right-skewed distributions such as the exponential distribution, gamma distribution with shape parameter $a = 2.0$, and lognormal distribution with shape parameter $\sigma = 0.5$, the power values of the Wilcoxon-type precedence tests may decrease with increasing values of r.

When we compare the power performance between the three Wilcoxon-type precedence tests, we find that the power values are almost the same for these three tests when the underlying distributions are nearly symmetric.

Yet, the minimal rank-sum precedence test compares quite favorably with the maximal and expected rank-sum precedence tests under right-skewed distributions. For example, in Table 6.16, when $n_1 = 15, n_2 = 20$, the underlying distribution is gamma with shape parameter 2.0, and the location-shift equals 0.5, the power of the minimal rank-sum precedence test with $r = 4$ is 0.686 (exact level of significance is 0.050) while the power values of maximal and expected rank-sum precedence tests are 0.557 and 0.603 (exact level of significance is 0.050 and 0.052), respectively.

So, our overall recommendation will be to use the minimal rank-sum precedence test in general for testing $H_0 : F_X = F_Y$ against $H_1 : F_X > F_Y$. This test has better performance, in terms of power, than the precedence and maximal precedence tests, and also in comparison with the maximal and expected rank-sum precedence tests.

6.7   ILLUSTRATIVE EXAMPLES

Let us use the data considered earlier in Examples 3.3 and 3.4 to illustrate the three Wilcoxon-type rank-sum precedence tests discussed in preceding sections.

Example 6.1. Let us consider the data of Example 3.3 once again. In this case, we have $n_1 = n_2 = 10$, $r = 7$, $m_1 = m_2 = 0$, $m_3 = m_4 = 1$, $m_5 = m_6 = 0$, and $m_7 = 3$. We then obtain the following results:

Test	Test statistic	p-value
Precedence test	$P_{(7)} = 5$	0.498
Maximal precedence test	$M_{(7)} = 3$	0.625
Minimal rank-sum precedence test	$W_{\min,7} = 110$	0.848
Maximal rank-sum precedence test	$W_{\max,7} = 125$	0.825
Expected rank-sum precedence test	$W_{E,7} = 117.5$	0.821
Wilcoxon rank-sum test (based on complete samples)	$W_R = 113$	0.711

Based on these p-values, we do not reject the null hypothesis that the two samples are identically distributed.

Example 6.2. Let us consider the data of Example 3.4 once again. In this case, we have $n_1 = n_2 = 10$, $r = 6$, $m_1 = 5$, $m_2 = m_3 = 0$, $m_4 = 2$, $m_5 = 0$, and $m_6 = 2$. We then obtain the following results:

Test	Test statistic	p-value
Precedence test	$P_{(6)} = 9$	0.065
Maximal precedence test	$M_{(6)} = 5$	0.097
Minimal rank-sum precedence test	$W_{\min,6} = 77$	0.029
Maximal rank-sum precedence test	$W_{\max,6} = 81$	0.023
Expected rank-sum precedence test	$W_{E,6} = 79$	0.025
Wilcoxon rank-sum test (based on complete samples)	$W_R = 80$	0.032

From these p-values, we observe that the data provide enough evidence to reject H_0 though the precedence and maximal precedence tests would lead to accepting H_0 at 5% and rejecting it at 10% level of significance.

All the testing procedures discussed so far are applicable even though only a few early failures from the two samples are available as data. Furthermore, the decisions are reached without making any assumption on the lifetime distributions underlying the two samples.

Chapter 7

Extension to Progressive Censoring

7.1 INTRODUCTION

In this chapter, we present generalizations of the weighted precedence test, the weighted maximal precedence test, and the maximal Wilcoxon rank-sum precedence test, as given by Ng and Balakrishnan (2005), for testing the hypotheses in (3.1), when the available sample is progressively censored.

As mentioned earlier, precedence-type tests are particularly useful (1) when life-tests involve expensive units since the units that had not failed could be used for some other testing purposes, and (2) to make quick and reliable decisions early in the life-testing experiment. We note here that in a precedence test, since the life-testing continues until the occurrence of the rth Y-failure, it may be viewed as a test based on a Type-II right censored Y-sample. Therefore, if we want to save the testing units at the early stage of the life-test, a Type-II progressive censoring scheme may instead be employed on the Y-sample, as explained earlier in Section 2.4; see Balakrishnan and Aggarwala (2000) for an elaborate treatment on progressive censoring and related developments. These, as given by Ng and Balakrishnan (2005), will be natural extensions of the weighted precedence, the weighted maximal precedence, and the maximal Wilcoxon rank-sum precedence tests.

7.2 TEST STATISTICS AND EXACT NULL DISTRIBUTIONS

Suppose a Type-II progressive censoring scheme is to be adopted on the Y-sample, which means that a number $r < n_2$ is prefixed for the number of complete failures to be observed and the progressive censoring scheme $(R_1, R_2, \cdots, R_r)$, with $R_j \geq 0$ and $\sum_{j=1}^{r} R_j + r = n_2$, is to be employed at the failure times; see Section 2.4 for more details. Let us denote such an observed ordered Y-sample by $Y_{1:r:n_2} \leq Y_{2:r:n_2} \leq \cdots \leq Y_{r:r:n_2}$. Moreover, we denote by M_1 the number of X-failures before $Y_{1:r:n_2}$, and by M_i the number of X-failures between $Y_{i-1:r:n_2}$ and $Y_{i:r:n_2}$, $i = 2, 3, \cdots, r$.

Then, based on $M_1, M_2, \cdots, M_r$, Ng and Balakrishnan (2005) proposed the weighted precedence test statistic $P_{(r)}^*$, analogous to Eq. (5.1), as

$$P_{(r)}^* = \sum_{i=1}^{r} \left[n_2 - \left(\sum_{j=1}^{i-1} R_j \right) - i + 1 \right] m_i, \tag{7.1}$$

and the weighted maximal precedence test statistic $M_{(r)}^*$, analogous to Eq. (5.2), as

$$M_{(r)}^* = \max_{1 \leq i \leq r} \left\{ \left[n_2 - \left(\sum_{j=1}^{i-1} R_j \right) - i + 1 \right] m_i \right\}. \tag{7.2}$$

Similarly, we can apply the idea of maximal Wilcoxon rank-sum precedence statistic for the progressively censored data. The Wilcoxon's rank-sum test statistic will be the largest when all the progressive censored Y-items in an interval fail before the smallest of X-failures in the corresponding interval. The maximal Wilcoxon rank-sum precedence statistic for progressively censored data then becomes

$$W_{\max,r}^* = \sum_{i=1}^{r+1} \left\{ m_i \left[\sum_{j=1}^{i-1} m_j + \sum_{j=1}^{i-1} R_j + (i-1) \right] + \frac{m_i(m_i + 1)}{2} \right\}, \tag{7.3}$$

where $m_{r+1} = n_1 - \sum_{i=1}^{r} m_i$.

For example, from Figure 7.1, with $n_1 = n_2 = 10$, $r = 4$, and the progressive censoring scheme $(2, 1, 1, 2)$, the weighted precedence test statistic becomes $P_{(4)}^* = (10 \times 0) + (7 \times 3) + (5 \times 4) + (3 \times 1) = 44$, while the weighted

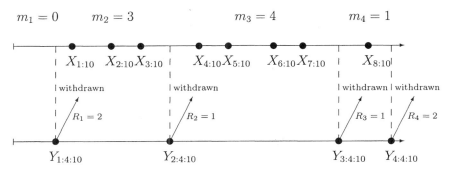

Figure 7.1: Schematic representation of a precedence life-test with progressive censoring

maximal precedence test statistic becomes $M^*_{(4)} = \max\{(10 \times 0), (7 \times 3), (5 \times 4), (3 \times 1)\} = 21$. It is clear that, in general, large values of $P^*_{(r)}$ or $M^*_{(r)}$ would lead to the rejection of H_0 and in favor of H_1 in (2.1). On the other hand, the maximal Wilcoxon rank-sum precedence test statistic becomes $W^*_{\max,r} = 4 + 5 + 6 + 9 + 10 + 11 + 12 + 15 + 19 + 20 = 111$ and small values of $W^*_{\max,r}$ would lead to the rejection of H_0 and in favor of H_1 in (2.1).

It is important to mention here that the weighted precedence and weighted maximal precedence test statistics defined earlier in (5.1) and (5.2) and the maximal Wilcoxon rank-sum precedence test statistic defined in Chapter 6 all become special cases when we set the progressive censoring scheme as $R_1 = R_2 = \cdots = R_{r-1} = 0$ and $R_r = n_2 - r$.

We can show that the maximal Wilcoxon rank-sum precedence test statistic in (7.3) is equivalent to the weighted precedence test statistic in (7.1) as follows:

$$W^*_{\max,r} = \sum_{i=1}^{r+1} \left\{ m_i \left[\sum_{j=1}^{i-1} m_j + \sum_{j=1}^{i-1} R_j + (i-1) \right] + \frac{m_i(m_i+1)}{2} \right\}$$

$$= \frac{n_1(n_1 + 2n_2 + 1)}{2} + \sum_{i=1}^{r} \left(m_i \sum_{j=1}^{i-1} R_j \right) - (n_2 - i + 1) \sum_{i=1}^{r} m_i$$

$$= \frac{n_1(n_1 + 2n_2 + 1)}{2} - P^*_{(r)}.$$

Since $\{n_1(n_1 + 2n_2 + 1)\}/2$ is just a constant, the above relationship readily reveals that the tests based on $P^*_{(r)}$ and $W^*_{\max,r}$ are equivalent, with large values of the former corresponding to small values of the latter.

The following theorem will enable the derivation of the null distribution of the weighted precedence and weighted maximal precedence tests under progressive censoring.

Theorem 7.1. With a progressive Type-II censoring on the Y-sample, under the null hypothesis $H_0 : F_X = F_Y$, the joint probability mass function of M_1, $M_2, \cdots, M_r$ is given by

$$\Pr(M_1 = m_1, M_2 = m_2, \cdots, M_r = m_r \mid H_0 : F_X = F_Y)$$

$$= C \sum_{\substack{j_k=0 \\ (k=1,2,\cdots,r-1)}}^{\sum_{i=1}^{k} R_i - \sum_{i=1}^{k-1} j_i} \left\{ \prod_{l=1}^{r-1} \binom{\sum_{i=1}^{l} R_i - \sum_{i=1}^{l} j_i}{j_l} \Gamma(m_l + j_{l+1} + 1) \right\}$$

$$\times \frac{\Gamma\left(n_1 + n_2 - r - \sum_{i=1}^{r} m_i - \sum_{i=1}^{r-1} j_i + 1 \right)}{\Gamma(n_1 + n_2 + 1)}, \tag{7.4}$$

where

$$C = \frac{n_1! n_2 (n_2 - R_1 - 1) \cdots (n_2 - R_1 - R_2 - \cdots - R_{r-1} - r + 1)}{m_1! m_2! \cdots m_r! \left(n_1 - \sum_{i=1}^{r} m_i \right)!}.$$

Proof. The derivation of the expression in (7.4) is presented in the following section.

$$\odot$$

7.3 EXACT POWER FUNCTION UNDER LEHMANN ALTERNATIVE

Theorem 7.2. With a progressive Type-II censoring on the Y-sample, under the Lehmann alternative $H_1 : [F_X]^\gamma = F_Y, \gamma > 1$, the joint probability mass function of $M_1, M_2, \cdots, M_r$ is given by

$$\Pr(M_1 = m_1, M_2 = m_2, \cdots, M_r = m_r \mid H_1 : [F_X]^\gamma = F_Y)$$

$$= C\gamma^r \sum_{\substack{j_i(i=1,2,\cdots,r)=0}}^{R_i} \left\{ (-1)^{\left(\sum_{i=1}^{r} j_i\right)} \binom{R_1}{j_1} \binom{R_2}{j_2} \cdots \binom{R_r}{j_r} \right.$$

$$\times \left[\prod_{k=1}^{r-1} B\left(\sum_{i=1}^{k} m_i + \gamma \left(\sum_{i=1}^{k} j_i \right) + k\gamma, m_{k+1} + 1 \right) \right]$$

$$\times B\left(\sum_{i=1}^{r} m_i + \gamma \left(\sum_{i=1}^{r} j_i \right) + r\gamma, n_1 - \sum_{i=1}^{r} m_i + 1 \right) \right\}. \quad (7.5)$$

Proof. First, conditional on the Y-failures, we have the probability that there are m_1 X-failures before $y_{1:r:n_2}$ and m_i X-failures between $y_{i-1:r:n_2}$ and $y_{i:r:n_2}$, $i = 2, 3, \cdots, r$, as the multinomial probability

$$\Pr\left\{ m_1 X's \le y_{1:r:n_2}, \cdots, m_r X's \in (y_{r-1:r:n_2}, y_{r:r:n_2}], \right.$$

$$\left(n_1 - \sum_{i=1}^{r} m_i \right) X's > y_{r:r:n_2} \middle| Y_{1:r:n_2} = y_{1:r:n_2}, \cdots, Y_{r:r:n_2} = y_{r:r:n_2} \right\}$$

$$= \frac{n_1!}{m_1! m_2! \cdots m_r! \left(n_1 - \sum_{i=1}^{r} m_i \right)!} [F_X(y_{1:r:n_2})]^{m_1}$$

$$\times \left\{ \prod_{i=2}^{r} [F_X(y_{i:r:n_2}) - F_X(y_{i-1:r:n_2})]^{m_i} \right\}$$

$$\times [1 - F_X(y_{r:r:n_2})]^{\left(n_1 - \sum_{i=1}^{r} m_i \right)}, \quad y_{1:r:n_2} \le \cdots \le y_{r:r:n_2}.$$

Next, we have the joint density of the progressively Type-II censored order statistics from the Y-sample as [see Eq. (2.33)]

$$f_{1,2,\cdots,r:r:n_2}(y_{1:r:n_2}, y_{2:r:n_2}, \cdots, y_{r:r:n_2})$$

$$= n_2(n_2 - R_1 - 1) \cdots (n_2 - R_1 - R_2 - \cdots - R_{r-1} - r + 1)$$

$$\times \prod_{j=1}^{r} f_Y(y_{j:r:n_2}) [1 - F_Y(y_{j:r:n_2})]^{R_j}, \quad y_{1:r:n_2} \le \cdots \le y_{r:r:n_2}.$$

Then, the unconditional probability of $\{M_1 = m_1, M_2 = m_2, \cdots, M_r = m_r\}$ is simply given by

$$\Pr\{M_1 = m_1, M_2 = m_2, \cdots, M_r = m_r\}$$

$$= C \int_{-\infty}^{\infty} \int_{-\infty}^{y_{r:r:n_2}} \cdots \int_{-\infty}^{y_{2:r:n_2}} [F_X(y_{1:r:n_2})]^{m_1}$$

$$\times \left\{ \prod_{i=2}^{r} [F_X(y_{i:r:n_2}) - F_X(y_{i-1:r:n_2})]^{m_i} \right\}$$

$$\times [1 - F_X(y_{r:r:n_2})]^{\left(n_1 - \sum\limits_{i=1}^{r} m_i\right)} \left\{ \prod_{i=1}^{r} f_Y(y_{i:r:n_2})[1 - F_Y(y_{i:r:n_2})]^{R_i} \right\}$$

$$\times dy_{1:r:n_2} dy_{2:r:n_2} \cdots dy_{r-1:r:n_2} dy_{r:r:n_2}, \tag{7.6}$$

where

$$C = \frac{n_1! n_2 (n_2 - R_1 - 1) \cdots (n_2 - R_1 - R_2 - \cdots - R_{r-1} - r + 1)}{m_1! m_2! \cdots m_r! \left(n_1 - \sum\limits_{i=1}^{r} m_i\right)!}.$$

Under Lehmann Alternative

Under the Lehmann alternative $H_1 : [F_X]^\gamma = F_Y, \gamma > 1$, the expression in (7.6) can be simplified as

$$\Pr\{M_1 = m_1, M_2 = m_2, \cdots, M_r = m_r \mid H_1 : [F_X]^\gamma = F_Y\}$$

$$= C \int_{-\infty}^{\infty} \int_{-\infty}^{y_{r:r:n_2}} \cdots \int_{-\infty}^{y_{2:r:n_2}} [F_X(y_{1:r:n_2})]^{m_1}$$

$$\times \left\{ \prod_{i=2}^{r} [F_X(y_{i:r:n_2}) - F_X(y_{i-1:r:n_2})]^{m_i} \right\} [1 - F_X(y_{r:r:n_2})]^{\left(n_1 - \sum\limits_{i=1}^{r} m_i\right)}$$

$$\times \left\{ \prod_{i=1}^{r} \gamma [F_X(y_{i:r:n_2})]^{\gamma-1} f_X(y_{i:r:n_2}) [1 - (F_X(y_{i:r:n_2}))^\gamma]^{R_i} \right\}$$

$$\times dy_{1:r:n_2} dy_{2:r:n_2} \cdots dy_{r-1:r:n_2} dy_{r:r:n_2}$$

$$= C\gamma^r \sum_{j_1=0}^{R_1} (-1)^{j_1} \binom{R_1}{j_1} \int_{-\infty}^{\infty} \int_{-\infty}^{y_{r:r:n_2}} \cdots \int_{-\infty}^{y_{2:r:n_2}} [F_X(y_{1:r:n_2})]^{m_1 + \gamma j_1 + \gamma - 1}$$

$$\times \left\{ \prod_{i=1}^{r} f_X(y_{i:r:n_2}) \right\} \left\{ \prod_{i=2}^{r} [F_X(y_{i:r:n_2}) - F_X(y_{i-1:r:n_2})]^{m_i} \right.$$

$$\times \left[\sum_{j_i=0}^{R_i} (-1)^{j_i} \binom{R_i}{j_i} [F_X(y_{i:r:n_2})]^{\gamma j_i + \gamma - 1} \right] \right\}$$

$$\times [1 - F_X(y_{r:r:n_2})]^{\left(n_1 - \sum\limits_{i=1}^{r} m_i\right)} dy_{1:r:n_2} \cdots dy_{r:r:n_2}.$$

Upon setting $u_i = F_X(y_{i:r:n_2})$, $i = 1, 2, \cdots, r$, we obtain

$$\Pr\{M_1 = m_1, M_2 = m_2, \cdots, M_r = m_r \mid [F_X]^\gamma = F_Y\}$$

$$= C\gamma^r \sum_{j_1=0}^{R_1} \binom{R_1}{j_1} (-1)^{j_1} \int_0^1 \int_0^{u_r} \cdots \int_0^{u_2} u_1^{m_1 + \gamma j_1 + \gamma - 1}$$

$$\times \left\{ \prod_{i=2}^{r} (u_i - u_{i-1})^{m_i} \left[\sum_{j_i=0}^{R_i} \binom{R_i}{j_i} (-1)^{j_i} u_i^{\gamma j_i + \gamma - 1} \right] \right\}$$

$$\times (1 - u_r)^{\left(n_1 - \sum_{i=1}^{r} m_i \right)} du_1 \cdots du_r.$$

Using the transformation $w_1 = u_1/u_2$, we readily obtain

$$\Pr\{M_1 = m_1, M_2 = m_2, \cdots, M_r = m_r \mid H_1 : [F_X]^\gamma = F_Y\}$$

$$= C\gamma^r \sum_{j_1=0}^{R_1} \sum_{j_2=0}^{R_2} (-1)^{j_1+j_2} \binom{R_1}{j_1} \binom{R_2}{j_2} B(m_1 + \gamma j_1 + \gamma, m_2 + 1)$$

$$\times \int_0^1 \int_0^{u_r} \cdots \int_0^{u_3} u_2^{m_1 + m_2 + \gamma(j_1+j_2) + 2\gamma - 1}$$

$$\times \left\{ \prod_{i=3}^{r} (u_i - u_{i-1})^{m_i} \left[\sum_{j_i=0}^{R_i} \binom{R_i}{j_i} (-1)^{j_i} u_i^{\gamma j_i + \gamma - 1} \right] \right\}$$

$$\times (1 - u_r)^{\left(n_1 - \sum_{i=1}^{r} m_i \right)} du_1 \cdots du_r,$$

where $B(a,b) = \int_0^1 x^{a-1}(1-x)^{b-1}dx$, as before, denotes the complete beta function.

Proceeding similarly by performing the transformations $w_l = u_l/u_{l+1}$ for $l = 2, 3, \cdots, r-1$, we arrive at the expression given in Eq. (7.5).

Under Null Hypothesis

Under the null hypothesis $H_0 : F_X = F_Y$, the expression in (7.5) can be simplified by employing the same technique as presented above by setting $\gamma = 1$, which results in the expression in Eq. (7.4).

$\odot$

For the weighted precedence test with specified values of n_1, n_2, s, r and the progressive censoring scheme $(R_1, R_2, \cdots, R_r)$, an expression for the level of significance α is obtained from (7.4) as

$$\alpha = \sum_{\substack{m_i(i=1,2,\cdots,r)=0 \\ s \le \sum_{i=1}^{r} \left[n_2 - \left(\sum_{j=1}^{i-1} R_j \right) - i + 1 \right] m_i \le n_1 n_2}}^{n_1} \Pr(M_1 = m_1, \cdots, M_r = m_r \mid F_X = F_Y).$$

(7.7)

Similarly, for the weighted maximal precedence test with specified values of n_1, n_2, t, r and the censoring scheme $(R_1, R_2, \cdots, R_r)$, an expression for the level of significance α is obtained from (7.4) as

$$\alpha = \sum_{\substack{m_i(i=1,2,\cdots,r)=0 \\ t \leq \max_{1 \leq i \leq r} \left[n_2 - \left(\sum_{j=1}^{i-1} R_j \right) - i + 1 \right] m_i \leq n_1 n_2}}^{n_1} \Pr(M_1 = m_1, \cdots, M_r = m_r \mid F_X = F_Y).$$

(7.8)

The critical values s, t and the exact level of significance α (as close to 5% as possible) for different choices of the sample sizes n_1 and n_2, $r = 2(1)5$, and different progressive censoring schemes are presented in Tables 7.1 and 7.2.

The power function of the weighted precedence and the weighted maximal precedence tests under the Lehmann alternative are similarly given by

$$\sum_{\substack{m_i(i=1,2,\cdots,r)=0 \\ s \leq \sum_{i=1}^{r} \left[n_2 - \left(\sum_{j=1}^{i-1} R_j \right) - i + 1 \right] m_i \leq n_1 n_2}}^{n_1} \Pr(M_1 = m_1, \cdots, M_r = m_r \mid [F_X]^\gamma = F_Y)$$

(7.9)

and

$$\sum_{\substack{m_i(i=1,2,\cdots,r)=0 \\ t \leq \max_{1 \leq i \leq r} \left[n_2 - \left(\sum_{j=1}^{i-1} R_j \right) - i + 1 \right] m_i \leq n_1 n_2}}^{n_1} \Pr(M_1 = m_1, \cdots, M_r = m_r \mid [F_X]^\gamma = F_Y),$$

(7.10)

respectively, where $\Pr(M_1 = m_1, \cdots, M_r = m_r \mid [F_X]^\gamma = F_Y)$ is as given in Eq. (7.5).

In Table 7.3, we have presented the exact power values computed from Eqs. (7.9) and (7.10) for $n_1 = n_2 = 10$, $r = 2$, $\gamma = 2(1)6$, and four different progressive censoring schemes.

Table 7.1: Near 5% critical values (s) and exact levels of significance (α) for the weighted precedence test statistic $P^*_{(r)}$ under progressive censoring

		$r = 2$		$r = 3$	
n_1	n_2	Censoring scheme	$s\ (\alpha)$	Censoring scheme	$s\ (\alpha)$
10	10	(8,0)	37(.05108)	(7,0,0)	42(.04829)
		(6,2)	41(.04782)	(5,1,1)	46(.04913)
		(4,4)	45(.05037)	(2,2,3)	51(.04956)
15	15	(13,0)	60(.04981)	(12,0,0)	66(.04949)
		(10,3)	65(.05164)	(10,1,1)	72(.04944)
		(6,7)	71(.04903)	(4,4,4)	85(.05235)
20	20	(18,0)	82(.04955)	(17,0,0)	90(.05052)
		(14,4)	95(.04691)	(14,2,1)	102(.04865)
		(9,9)	100(.05680)	(5,6,6)	119(.05061)
30	30	(28,0)	126(.04951)	(27,0,0)	139(.05008)
		(22,6)	146(.05004)	(22,3,2)	162(.05064)
		(14,14)	165(.04498)	(9,9,9)	190(.05528)
30	50	(48,0)	150(.04942)	(47,0,0)	153(.05032)
		(38,10)	165(.04862)	(38,5,4)	188(.04970)
		(24,24)	175(.06151)	(15,16,16)	219(.05007)

		$r = 4$		$r = 5$	
n_1	n_2	Censoring scheme	$s\ (\alpha)$	Censoring scheme	$s\ (\alpha)$
10	10	(6,0,0,0)	45(.05166)	(5,0,0,0,0)	49(.05151)
		(3,1,1,1)	52(.05349)	(3,0,0,0,2)	56(.05092)
		(1,1,2,2)	57(.04929)	(1,1,1,1,1)	60(.04923)
15	15	(11,0,0,0)	72(.05051)	(10,0,0,0,0)	79(.04875)
		(8,1,1,1)	82(.04943)	(6,1,1,1,1)	93(.04845)
		(2,3,3,3)	96(.05039)	(2,2,2,2,2)	105(.04884)
20	20	(16,0,0,0)	99(.05036)	(15,0,0,0,0)	108(.05045)
		(13,1,1,1)	112(.05105)	(11,1,1,1,1)	126(.05171)
		(4,4,4,4)	140(.04559)	(3,3,3,3,3)	152(.05002)
30	30	(26,0,0,0)	153(.05047)	(25,0,0,0,0)	168(.04947)
		(21,2,2,2)	180(.04980)	(20,2,1,1,1)	196(.05069)
		(6,6,7,7)	222(.05051)	(5,5,5,5,5)	252(.04813)
30	50	(46,0,0,0)	170(.04982)	(45,0,0,0,0)	185(.04990)
		(37,3,3,3)	212(.04965)	(36,3,2,2,2)	230(.04949)
		(11,11,12,12)	255(.04929)	(9,9,9,9,9)	290(.05121)

Table 7.2: Near 5% critical values (t) and exact levels of significance (α) for the weighted maximal precedence test statistic $M^*_{(r)}$ under progressive censoring

n_1	n_2	Censoring scheme ($r=2$)	$t\,(\alpha)$	Censoring scheme ($r=3$)	$t\,(\alpha)$
10	10	(8,0)	40(.04334)	(7,0,0)	40(.04334)
		(6,2)	40(.04334)	(5,1,1)	32(.05294)
		(4,4)	40(.04773)	(2,2,3)	40(.05567)
15	15	(13,0)	60(.04981)	(12,0,0)	60(.04981)
		(10,3)	60(.04994)	(10,1,1)	60(.04994)
		(6,7)	60(.05986)	(4,4,4)	65(.03202)
20	20	(18,0)	80(.05301)	(17,0,0)	80(.05301)
		(14,4)	80(.05463)	(14,2,1)	80(.05463)
		(9,9)	90(.03313)	(5,6,6)	84(.05222)
30	30	(28,0)	120(.05620)	(27,0,0)	120(.05620)
		(22,6)	120(.06012)	(22,3,2)	120(.06012)
		(14,14)	135(.04024)	(9,9,9)	130(.05092)
30	50	(48,0)	150(.04942)	(47,0,0)	150(.04942)
		(38,10)	150(.05302)	(38,5,4)	150(.05302)
		(24,24)	150(.06803)	(15,16,16)	153(.04330)

n_1	n_2	Censoring scheme ($r=4$)	$t\,(\alpha)$	Censoring scheme ($r=5$)	$t\,(\alpha)$
10	10	(6,0,0,0)	40(.04334)	(5,0,0,0,0)	32(.05294)
		(3,1,1,1)	40(.05059)	(3,0,0,0,2)	40(.05263)
		(1,1,2,2)	40(.06985)	(1,1,1,1,1)	40(.06994)
15	15	(11,0,0,0)	60(.04981)	(10,0,0,0,0)	60(.04994)
		(8,1,1,1)	60(.05600)	(6,1,1,1,1)	60(.06338)
		(2,3,3,3)	64(.04202)	(2,2,2,2,2)	63(.04669)
20	20	(16,0,0,0)	80(.05301)	(15,0,0,0,0)	80(.05306)
		(13,1,1,1)	80(.05595)	(11,1,1,1,1)	80(.06578)
		(4,4,4,4)	90(.05136)	(3,3,3,3,3)	96(.04975)
30	30	(26,0,0,0)	120(.05620)	(25,0,0,0,0)	120(.05621)
		(21,2,2,2)	120(.06491)	(20,2,1,1,1)	126(.03521)
		(6,6,7,7)	144(.04837)	(5,5,5,5,5)	150(.04488)
30	50	(46,0,0,0)	150(.04942)	(45,0,0,0,0)	108(.04999)
		(37,3,3,3)	150(.05430)	(36,3,2,2,2)	150(.05571)
		(11,11,12,12)	156(.04935)	(9,9,9,9,9)	170(.04295)

Table 7.3: Power values under Lehmann alternative for $n_1 = n_2 = 10, r = 2$, and $\gamma = 2(1)6$

(R_1, R_2)	Test	$\gamma = 1$	$\gamma = 2$	$\gamma = 3$	$\gamma = 4$	$\gamma = 5$	$\gamma = 6$
(0,8)	$P_{(2)}^*$	0.04954	0.35681	0.64389	0.80970	0.89619	0.94126
	$M_{(2)}^*$	0.05954	0.35732	0.62980	0.79435	0.88451	0.93342
(8,0)	$P_{(2)}^*$	0.05108	0.34707	0.62563	0.79312	0.88424	0.93340
	$M_{(2)}^*$	0.04334	0.31471	0.58850	0.76277	0.86219	0.91805
(6,2)	$P_{(2)}^*$	0.04782	0.35013	0.63832	0.80730	0.89596	0.94204
	$M_{(2)}^*$	0.04334	0.31471	0.58850	0.76277	0.86219	0.91805
(4,4)	$P_{(2)}^*$	0.05037	0.36112	0.65009	0.81566	0.90104	0.94495
	$M_{(2)}^*$	0.04773	0.32292	0.59495	0.76692	0.86471	0.91958

7.4 MONTE CARLO SIMULATION UNDER LOCATION-SHIFT ALTERNATIVE

Power values of the weighted precedence and maximal precedence tests with $r = 2(1)6$ were all estimated by Ng and Balakrishnan (2005) through Monte Carlo simulations for location-shift with $\theta = 0.5$ and 1.0. The lifetime distributions listed earlier in Section 4.4 were used in the Monte Carlo simulations. 10,000 sets of data were generated in order to determine the estimated rejection rates. In Tables 7.4–7.7, we have presented these estimated power values as well as the corresponding exact levels of significance.

From Tables 7.4–7.7, upon comparing the power values of the weighted precedence and maximal precedence tests for different censoring schemes, we find that the Type-II progressive censoring with $R_1 = n_2 - r$ gives better power performance when the underlying distributions are right-skewed. For example, in Table 7.4, under the gamma distribution with $a = 2.0$, the power of the weighted precedence test with $r = 3$, $\theta = 1.0$ and censoring scheme (7,0,0) is 0.8912 (with exact level of significance as 0.04829) while the corresponding power for the censoring scheme (0,0,7) is 0.7987 (with exact level of significance as 0.05096). Thus, in this case, we can withdraw 7 experimental units early in the life-testing process and still have a significant gain in power. Moreover, though the progressive censoring schemes may

Table 7.4: Power of the weighted precedence $(P^*_{(r)})$ and weighted maximal precedence tests $(M^*_{(r)})$ under progressive censoring for $n_1 = n_2 = 10$ with location-shift $= 0.5$

r	Censoring scheme	Test	Exact l.o.s.	N(0,1)	Exp(1)	Gamma(2)	Gamma(10)
2	(0,8)	$P^*_{(r)}$	0.04954	0.2275	0.6145	0.4080	0.2708
		$M^*_{(r)}$	0.05954	0.2309	0.7197	0.4344	0.2735
	(8,0)	$P^*_{(r)}$	0.05108	0.2059	0.7361	0.4524	0.2605
		$M^*_{(r)}$	0.04334	0.1790	0.7173	0.4167	0.2293
	(6,2)	$P^*_{(r)}$	0.04782	0.2056	0.6740	0.4379	0.2493
		$M^*_{(r)}$	0.04334	0.1726	0.7189	0.4222	0.2198
	(4,4)	$P^*_{(r)}$	0.05037	0.2191	0.6488	0.4322	0.2624
		$M^*_{(r)}$	0.04773	0.1832	0.7191	0.4243	0.2306
3	(0,0,7)	$P^*_{(r)}$	0.05096	0.2400	0.5124	0.3707	0.2724
		$M^*_{(r)}$	0.03792	0.1542	0.4886	0.2483	0.1739
	(7,0,0)	$P^*_{(r)}$	0.04829	0.1938	0.7194	0.4494	0.2585
		$M^*_{(r)}$	0.04334	0.1672	0.7168	0.4204	0.2291
	(5,1,1)	$P^*_{(r)}$	0.04913	0.2125	0.6738	0.4484	0.2725
		$M^*_{(r)}$	0.05294	0.1926	0.7174	0.4252	0.2465
	(2,3,3)	$P^*_{(r)}$	0.04956	0.2342	0.5605	0.4058	0.2756
		$M^*_{(r)}$	0.05567	0.2049	0.7179	0.4301	0.2589
4	(0,0,0,6)	$P^*_{(r)}$	0.05065	0.2492	0.4210	0.3360	0.2684
		$M^*_{(r)}$	0.04334	0.1632	0.4847	0.2573	0.1693
	(6,0,0,0)	$P^*_{(r)}$	0.05166	0.2159	0.7166	0.4691	0.2650
		$M^*_{(r)}$	0.04334	0.1763	0.7167	0.4211	0.2187
	(3,1,1,1)	$P^*_{(r)}$	0.05349	0.2546	0.6260	0.4452	0.2917
		$M^*_{(r)}$	0.05059	0.2014	0.7173	0.4264	0.2390
	(1,1,2,2)	$P^*_{(r)}$	0.04929	0.2490	0.5187	0.3849	0.2781
		$M^*_{(r)}$	0.06985	0.2464	0.7198	0.4446	0.2811
5	(0,0,0,0,5)	$P^*_{(r)}$	0.05108	0.2558	0.4246	0.3372	0.2669
		$M^*_{(r)}$	0.04489	0.1638	0.4857	0.2512	0.1739
	(5,0,0,0,0)	$P^*_{(r)}$	0.05151	0.2273	0.6963	0.4652	0.2800
		$M^*_{(r)}$	0.05294	0.2004	0.7198	0.4185	0.2456
	(3,0,0,0,2)	$P^*_{(r)}$	0.05092	0.2481	0.6031	0.4326	0.2873
		$M^*_{(r)}$	0.05263	0.2021	0.7200	0.4206	0.2487
	(1,1,1,1,1)	$P^*_{(r)}$	0.04923	0.2493	0.5207	0.3913	0.2765
		$M^*_{(r)}$	0.06994	0.2428	0.7224	0.4384	0.2865
6	(0,0,0,0,0,4)	$P^*_{(r)}$	0.04900	0.2589	0.4105	0.3255	0.2794
		$M^*_{(r)}$	0.04495	0.1715	0.4854	0.2522	0.1791
	(4,0,0,0,0,0)	$P^*_{(r)}$	0.05239	0.2417	0.6715	0.4606	0.2949
		$M^*_{(r)}$	0.04886	0.1870	0.7157	0.4168	0.2421
	(0,0,0,1,2,1)	$P^*_{(r)}$	0.05182	0.2668	0.4403	0.3473	0.2911
		$M^*_{(r)}$	0.04348	0.1700	0.4853	0.2517	0.1782
	(0,1,1,1,1,0)	$P^*_{(r)}$	0.05307	0.2673	0.4936	0.3783	0.3029
		$M^*_{(r)}$	0.04111	0.1653	0.4852	0.2504	0.1768

Table 7.4: (Continued)

r	Censoring scheme	Test	Exact l.o.s.	LN(0.1)	LN(0.5)	EV
2	(0,8)	$P_{(r)}^*$	0.04954	0.2379	0.4206	0.2002
		$M_{(r)}^*$	0.05954	0.2457	0.4301	0.2034
	(8,0)	$P_{(r)}^*$	0.05108	0.2218	0.4364	0.1742
		$M_{(r)}^*$	0.04334	0.1953	0.4031	0.1422
	(6,2)	$P_{(r)}^*$	0.04782	0.2219	0.4188	0.1836
		$M_{(r)}^*$	0.04334	0.1902	0.3858	0.1422
	(4,4)	$P_{(r)}^*$	0.05037	0.2370	0.4187	0.2055
		$M_{(r)}^*$	0.04773	0.2008	0.3897	0.1672
3	(0,0,7)	$P_{(r)}^*$	0.05096	0.2518	0.3839	0.2414
		$M_{(r)}^*$	0.03792	0.1562	0.2559	0.1584
	(7,0,0)	$P_{(r)}^*$	0.04829	0.2195	0.4215	0.1725
		$M_{(r)}^*$	0.04334	0.1915	0.3928	0.1427
	(5,1,1)	$P_{(r)}^*$	0.04913	0.2392	0.4283	0.2008
		$M_{(r)}^*$	0.05294	0.2133	0.3989	0.1854
	(2,3,3)	$P_{(r)}^*$	0.04956	0.2527	0.4061	0.2338
		$M_{(r)}^*$	0.05567	0.2261	0.4076	0.1949
4	(0,0,0,6)	$P_{(r)}^*$	0.05065	0.2560	0.3549	0.2720
		$M_{(r)}^*$	0.04334	0.1681	0.2618	0.1769
	(6,0,0,0)	$P_{(r)}^*$	0.05166	0.2373	0.4380	0.1893
		$M_{(r)}^*$	0.04334	0.1914	0.3907	0.1409
	(3,1,1,1)	$P_{(r)}^*$	0.05349	0.2704	0.4418	0.2509
		$M_{(r)}^*$	0.05059	0.2132	0.3996	0.1784
	(1,1,2,2)	$P_{(r)}^*$	0.04929	0.2589	0.3941	0.2611
		$M_{(r)}^*$	0.06985	0.2582	0.4301	0.2329
5	(0,0,0,0,5)	$P_{(r)}^*$	0.05108	0.2609	0.3541	0.2935
		$M_{(r)}^*$	0.04489	0.1684	0.2702	0.1699
	(5,0,0,0,0)	$P_{(r)}^*$	0.05151	0.2398	0.4547	0.2077
		$M_{(r)}^*$	0.05294	0.2138	0.4119	0.1836
	(3,0,0,0,2)	$P_{(r)}^*$	0.05092	0.2640	0.4384	0.2504
		$M_{(r)}^*$	0.05263	0.2162	0.4135	0.1842
	(1,1,1,1,1)	$P_{(r)}^*$	0.04923	0.2601	0.4068	0.2644
		$M_{(r)}^*$	0.06994	0.2593	0.4424	0.2309
6	(0,0,0,0,0,4)	$P_{(r)}^*$	0.04900	0.2597	0.3466	0.3021
		$M_{(r)}^*$	0.04495	0.1709	0.2671	0.1799
	(4,0,0,0,0,0)	$P_{(r)}^*$	0.05239	0.2533	0.4554	0.2249
		$M_{(r)}^*$	0.04886	0.2003	0.4052	0.1673
	(0,0,0,1,2,1)	$P_{(r)}^*$	0.05182	0.2722	0.3693	0.3002
		$M_{(r)}^*$	0.04348	0.1695	0.2671	0.1765
	(0,1,1,1,1,0)	$P_{(r)}^*$	0.05307	0.2777	0.4011	0.2840
		$M_{(r)}^*$	0.04111	0.1673	0.2660	0.1682

Table 7.5: Power of the weighted precedence $(P^*_{(r)})$ and weighted maximal precedence tests $(M^*_{(r)})$ under progressive censoring for $n_1 = n_2 = 10$ with location-shift $= 1.0$

r	Censoring scheme	Test	Exact l.o.s.	N(0,1)	Exp(1)	Gamma(2)	Gamma(10)
2	(0,8)	$P^*_{(r)}$	0.04954	0.5278	0.9468	0.8525	0.6487
		$M^*_{(r)}$	0.05954	0.5067	0.9784	0.8881	0.6313
	(8,0)	$P^*_{(r)}$	0.05108	0.4708	0.9807	0.8981	0.6212
		$M^*_{(r)}$	0.04334	0.4161	0.9784	0.8834	0.5768
	(6,2)	$P^*_{(r)}$	0.04782	0.5128	0.9614	0.8752	0.6543
		$M^*_{(r)}$	0.04334	0.4248	0.9774	0.8787	0.5857
	(4,4)	$P^*_{(r)}$	0.05037	0.5443	0.9551	0.8656	0.6656
		$M^*_{(r)}$	0.04773	0.4521	0.9774	0.8788	0.5939
3	(0,0,7)	$P^*_{(r)}$	0.05096	0.5839	0.8824	0.7987	0.6540
		$M^*_{(r)}$	0.03792	0.3942	0.9164	0.7338	0.4610
	(7,0,0)	$P^*_{(r)}$	0.04829	0.4817	0.9741	0.8912	0.6197
		$M^*_{(r)}$	0.04334	0.4233	0.9765	0.8777	0.5721
	(5,1,1)	$P^*_{(r)}$	0.04913	0.5372	0.9591	0.8842	0.6595
		$M^*_{(r)}$	0.05294	0.4618	0.9766	0.8779	0.5870
	(2,3,3)	$P^*_{(r)}$	0.04956	0.5851	0.9060	0.8318	0.6695
		$M^*_{(r)}$	0.05567	0.4930	0.9768	0.8792	0.6076
4	(0,0,0,6)	$P^*_{(r)}$	0.05065	0.6083	0.7921	0.7172	0.6435
		$M^*_{(r)}$	0.04334	0.4045	0.9173	0.7294	0.4611
	(6,0,0,0)	$P^*_{(r)}$	0.05166	0.5047	0.9718	0.9001	0.6580
		$M^*_{(r)}$	0.04334	0.4183	0.9765	0.8790	0.5770
	(3,1,1,1)	$P^*_{(r)}$	0.05349	0.6060	0.9410	0.8681	0.7031
		$M^*_{(r)}$	0.05059	0.4677	0.9765	0.8801	0.5970
	(1,1,2,2)	$P^*_{(r)}$	0.04929	0.6133	0.8910	0.8021	0.6804
		$M^*_{(r)}$	0.06985	0.5353	0.9768	0.8838	0.6404
5	(0,0,0,0,5)	$P^*_{(r)}$	0.05108	0.6332	0.8193	0.7373	0.6558
		$M^*_{(r)}$	0.04489	0.4047	0.9111	0.7409	0.4684
	(5,0,0,0,0)	$P^*_{(r)}$	0.05151	0.5440	0.9658	0.8955	0.6830
		$M^*_{(r)}$	0.05294	0.4626	0.9763	0.8829	0.5896
	(3,0,0,0,2)	$P^*_{(r)}$	0.05092	0.6140	0.9320	0.8576	0.7035
		$M^*_{(r)}$	0.05263	0.4744	0.9763	0.8837	0.5953
	(1,1,1,1,1)	$P^*_{(r)}$	0.04923	0.6288	0.8993	0.8121	0.6862
		$M^*_{(r)}$	0.06994	0.5356	0.9763	0.8881	0.6401
6	(0,0,0,0,0,4)	$P^*_{(r)}$	0.04900	0.6372	0.7965	0.7200	0.6643
		$M^*_{(r)}$	0.04495	0.3984	0.9155	0.7446	0.4711
	(4,0,0,0,0,0)	$P^*_{(r)}$	0.05239	0.5749	0.9586	0.8939	0.7038
		$M^*_{(r)}$	0.04886	0.4451	0.9781	0.8813	0.5935
	(0,0,0,1,2,1)	$P^*_{(r)}$	0.05182	0.6536	0.8254	0.7539	0.6836
		$M^*_{(r)}$	0.04348	0.3976	0.9155	0.7446	0.4709
	(0,1,1,1,1,0)	$P^*_{(r)}$	0.05307	0.6492	0.8693	0.8002	0.7030
		$M^*_{(r)}$	0.04111	0.3955	0.9153	0.7451	0.4698

Table 7.5: (Continued)

r	Censoring scheme	Test	Exact l.o.s.	LN(0.1)	LN(0.5)	EV
2	(0,8)	$P^*_{(r)}$	0.04954	0.5803	0.8727	0.4691
		$M^*_{(r)}$	0.05954	0.5535	0.8801	0.4504
	(8,0)	$P^*_{(r)}$	0.05108	0.5286	0.8859	0.3828
		$M^*_{(r)}$	0.04334	0.4771	0.8695	0.3177
	(6,2)	$P^*_{(r)}$	0.04782	0.5556	0.8775	0.4173
		$M^*_{(r)}$	0.04334	0.4776	0.8643	0.3177
	(4,4)	$P^*_{(r)}$	0.05037	0.5836	0.8714	0.4761
		$M^*_{(r)}$	0.04773	0.4982	0.8652	0.3869
3	(0,0,7)	$P^*_{(r)}$	0.05096	0.6087	0.8221	0.5769
		$M^*_{(r)}$	0.03792	0.4145	0.7289	0.4007
	(7,0,0)	$P^*_{(r)}$	0.04829	0.5304	0.8802	0.3836
		$M^*_{(r)}$	0.04334	0.4734	0.8622	0.3193
	(5,1,1)	$P^*_{(r)}$	0.04913	0.5826	0.8860	0.4622
		$M^*_{(r)}$	0.05294	0.5034	0.8632	0.4167
	(2,3,3)	$P^*_{(r)}$	0.04956	0.6156	0.8490	0.5593
		$M^*_{(r)}$	0.05567	0.5299	0.8676	0.4402
4	(0,0,0,6)	$P^*_{(r)}$	0.05065	0.6165	0.7765	0.6499
		$M^*_{(r)}$	0.04334	0.4263	0.7391	0.4219
	(6,0,0,0)	$P^*_{(r)}$	0.05166	0.5704	0.9016	0.4083
		$M^*_{(r)}$	0.04334	0.4847	0.8726	0.3171
	(3,1,1,1)	$P^*_{(r)}$	0.05349	0.6407	0.8917	0.5737
		$M^*_{(r)}$	0.05059	0.5192	0.8747	0.4104
	(1,1,2,2)	$P^*_{(r)}$	0.04929	0.6319	0.8419	0.6225
		$M^*_{(r)}$	0.06985	0.5798	0.8842	0.5039
5	(0,0,0,0,5)	$P^*_{(r)}$	0.05108	0.6334	0.7821	0.7073
		$M^*_{(r)}$	0.04489	0.4262	0.7464	0.4223
	(5,0,0,0,0)	$P^*_{(r)}$	0.05151	0.5959	0.9016	0.4575
		$M^*_{(r)}$	0.05294	0.5102	0.8740	0.4172
	(3,0,0,0,2)	$P^*_{(r)}$	0.05092	0.6453	0.8747	0.5984
		$M^*_{(r)}$	0.05263	0.5170	0.8746	0.4247
	(1,1,1,1,1)	$P^*_{(r)}$	0.04923	0.6472	0.8434	0.6385
		$M^*_{(r)}$	0.06994	0.5735	0.8852	0.5066
6	(0,0,0,0,0,4)	$P^*_{(r)}$	0.04900	0.6380	0.7620	0.7407
		$M^*_{(r)}$	0.04495	0.4170	0.7381	0.4330
	(4,0,0,0,0,0)	$P^*_{(r)}$	0.05239	0.6194	0.8971	0.5133
		$M^*_{(r)}$	0.04886	0.4947	0.8721	0.3994
	(0,0,0,1,2,1)	$P^*_{(r)}$	0.05182	0.6530	0.7909	0.7316
		$M^*_{(r)}$	0.04348	0.4166	0.7381	0.4291
	(0,1,1,1,1,0)	$P^*_{(r)}$	0.05307	0.6597	0.8310	0.6831
		$M^*_{(r)}$	0.04111	0.4160	0.7381	0.4188

Table 7.6: Power of the weighted precedence $(P_{(r)}^*)$ and weighted maximal precedence tests $(M_{(r)}^*)$ under progressive censoring for $n_1 = n_2 = 20$ with location-shift $= 0.5$

r	Censoring scheme	Test	Exact l.o.s.	N(0,1)	Exp(1)	Gamma(2)	Gamma(10)
2	(0,18)	$P_{(r)}^*$	0.05326	0.2906	0.9700	0.7352	0.3929
		$M_{(r)}^*$	0.04691	0.2401	0.9642	0.6563	0.3240
	(18,0)	$P_{(r)}^*$	0.04955	0.2401	0.9859	0.7805	0.3401
		$M_{(r)}^*$	0.05301	0.2452	0.9891	0.7919	0.3490
	(14,4)	$P_{(r)}^*$	0.04691	0.2631	0.9764	0.7557	0.3677
		$M_{(r)}^*$	0.05463	0.2526	0.9891	0.7924	0.3547
	(9,9)	$P_{(r)}^*$	0.05680	0.3037	0.9746	0.7657	0.4113
		$M_{(r)}^*$	0.03313	0.1979	0.9637	0.6375	0.2715
3	(0,0,17)	$P_{(r)}^*$	0.04979	0.3096	0.9015	0.6741	0.4102
		$M_{(r)}^*$	0.04359	0.2328	0.9626	0.6488	0.3169
	(17,0,0)	$P_{(r)}^*$	0.05052	0.2413	0.9841	0.7780	0.3661
		$M_{(r)}^*$	0.05301	0.2309	0.9880	0.7854	0.3562
	(14,2,1)	$P_{(r)}^*$	0.04865	0.2630	0.9741	0.7659	0.3842
		$M_{(r)}^*$	0.05463	0.2397	0.9880	0.7860	0.3611
	(5,6,6)	$P_{(r)}^*$	0.05061	0.3126	0.9461	0.7240	0.4175
		$M_{(r)}^*$	0.05222	0.2504	0.9631	0.6525	0.3333
4	(0,0,0,16)	$P_{(r)}^*$	0.05023	0.3301	0.8582	0.6453	0.4231
		$M_{(r)}^*$	0.05357	0.2576	0.9617	0.6603	0.3442
	(16,0,0,0)	$P_{(r)}^*$	0.05036	0.2582	0.9821	0.7812	0.3799
		$M_{(r)}^*$	0.05301	0.2348	0.9878	0.7850	0.3625
	(13,1,1,1)	$P_{(r)}^*$	0.05105	0.2921	0.9706	0.7773	0.4129
		$M_{(r)}^*$	0.05595	0.2499	0.9878	0.7863	0.3722
	(4,4,4,4)	$P_{(r)}^*$	0.04559	0.3201	0.9111	0.6855	0.4199
		$M_{(r)}^*$	0.05136	0.2518	0.9616	0.6578	0.3424
5	(0,0,0,0,15)	$P_{(r)}^*$	0.04967	0.3501	0.8043	0.5978	0.4189
		$M_{(r)}^*$	0.04439	0.2251	0.9038	0.4968	0.2631
	(15,0,0,0,0)	$P_{(r)}^*$	0.05045	0.2641	0.9809	0.7822	0.3922
		$M_{(r)}^*$	0.05306	0.2317	0.9885	0.7839	0.3588
	(11,1,1,1,1)	$P_{(r)}^*$	0.05171	0.3157	0.9617	0.7687	0.4387
		$M_{(r)}^*$	0.06578	0.2810	0.9886	0.7905	0.4010
	(3,3,3,3,3)	$P_{(r)}^*$	0.05002	0.3550	0.8921	0.6814	0.4441
		$M_{(r)}^*$	0.04975	0.2535	0.9615	0.6544	0.3336
6	(0,0,0,0,0,14)	$P_{(r)}^*$	0.04952	0.3598	0.7498	0.5678	0.4210
		$M_{(r)}^*$	0.04852	0.2252	0.9023	0.4969	0.2667
	(14,0,0,0,0,0)	$P_{(r)}^*$	0.05087	0.2750	0.9772	0.7853	0.3890
		$M_{(r)}^*$	0.05464	0.2419	0.9887	0.7871	0.3537
	(7,2,2,1,1,1)	$P_{(r)}^*$	0.05077	0.3440	0.9326	0.7315	0.4509
		$M_{(r)}^*$	0.04917	0.2477	0.9636	0.6506	0.3221
	(2,2,3,3,2,2)	$P_{(r)}^*$	0.04918	0.3634	0.8660	0.6659	0.4464
		$M_{(r)}^*$	0.05024	0.2531	0.9636	0.6533	0.3298

Table 7.6: (Continued)

r	Censoring scheme	Test	Exact l.o.s.	LN(0.1)	LN(0.5)	EV
2	(0,18)	$P_{(r)}^*$	0.05326	0.3201	0.6917	0.2154
		$M_{(r)}^*$	0.04691	0.2664	0.6048	0.1845
	(18,0)	$P_{(r)}^*$	0.04955	0.2675	0.6743	0.1555
		$M_{(r)}^*$	0.05301	0.2733	0.6834	0.1601
	(14,4)	$P_{(r)}^*$	0.04691	0.2928	0.6818	0.1977
		$M_{(r)}^*$	0.05463	0.2803	0.6844	0.1769
	(9,9)	$P_{(r)}^*$	0.05680	0.3336	0.7137	0.2299
		$M_{(r)}^*$	0.03313	0.2209	0.5607	0.1503
3	(0,0,17)	$P_{(r)}^*$	0.04979	0.3486	0.6769	0.2493
		$M_{(r)}^*$	0.04359	0.2699	0.5879	0.1933
	(17,0,0)	$P_{(r)}^*$	0.05052	0.2900	0.6939	0.1782
		$M_{(r)}^*$	0.05301	0.2800	0.6917	0.1629
	(14,2,1)	$P_{(r)}^*$	0.04865	0.3083	0.6943	0.2036
		$M_{(r)}^*$	0.05463	0.2885	0.6929	0.1819
	(5,6,6)	$P_{(r)}^*$	0.05061	0.3549	0.6989	0.2650
		$M_{(r)}^*$	0.05222	0.2870	0.5988	0.2216
4	(0,0,0,16)	$P_{(r)}^*$	0.05023	0.3695	0.6640	0.2864
		$M_{(r)}^*$	0.05357	0.2964	0.5993	0.2304
	(16,0,0,0)	$P_{(r)}^*$	0.05036	0.3028	0.6975	0.1901
		$M_{(r)}^*$	0.05301	0.2834	0.6882	0.1626
	(13,1,1,1)	$P_{(r)}^*$	0.05105	0.3354	0.7135	0.2388
		$M_{(r)}^*$	0.05595	0.2995	0.6919	0.1892
	(4,4,4,4)	$P_{(r)}^*$	0.04559	0.3582	0.6768	0.2763
		$M_{(r)}^*$	0.05136	0.2892	0.5992	0.2197
5	(0,0,0,0,15)	$P_{(r)}^*$	0.04967	0.3816	0.6250	0.3128
		$M_{(r)}^*$	0.04439	0.2471	0.4574	0.2010
	(15,0,0,0,0)	$P_{(r)}^*$	0.05045	0.3149	0.7017	0.2019
		$M_{(r)}^*$	0.05306	0.2810	0.6855	0.1625
	(11,1,1,1,1)	$P_{(r)}^*$	0.05171	0.3670	0.7193	0.2732
		$M_{(r)}^*$	0.06578	0.3288	0.7036	0.2338
	(3,3,3,3,3)	$P_{(r)}^*$	0.05002	0.3999	0.6860	0.3173
		$M_{(r)}^*$	0.04975	0.2859	0.5943	0.2167
6	(0,0,0,0,0,14)	$P_{(r)}^*$	0.04952	0.3777	0.5982	0.3500
		$M_{(r)}^*$	0.04852	0.2428	0.4581	0.2254
	(14,0,0,0,0,0)	$P_{(r)}^*$	0.05087	0.3219	0.7044	0.2183
		$M_{(r)}^*$	0.05464	0.2853	0.6847	0.1813
	(7,2,2,1,1,1)	$P_{(r)}^*$	0.05077	0.3881	0.7026	0.3180
		$M_{(r)}^*$	0.04917	0.2749	0.5743	0.2170
	(2,2,3,3,2,2)	$P_{(r)}^*$	0.04918	0.3931	0.6627	0.3373
		$M_{(r)}^*$	0.05024	0.2862	0.5817	0.2210

Table 7.7: Power of the weighted precedence $(P_{(r)}^*)$ and weighted maximal precedence tests $(M_{(r)}^*)$ under progressive censoring for $n_1 = n_2 = 20$ with location-shift $= 1.0$

r	Censoring scheme	Test	Exact l.o.s.	N(0,1)	Exp(1)	Gamma(2)	Gamma(10)
2	(0,18)	$P_{(r)}^*$	0.05326	0.6780	0.9998	0.9958	0.8599
		$M_{(r)}^*$	0.04691	0.5985	0.9998	0.9932	0.7905
	(18,0)	$P_{(r)}^*$	0.04955	0.5512	0.9999	0.9977	0.7991
		$M_{(r)}^*$	0.05301	0.5550	0.9999	0.9978	0.8026
	(14,4)	$P_{(r)}^*$	0.04691	0.6338	0.9999	0.9969	0.8430
		$M_{(r)}^*$	0.05463	0.5754	0.9999	0.9978	0.8078
	(9,9)	$P_{(r)}^*$	0.05680	0.6938	0.9999	0.9968	0.8731
		$M_{(r)}^*$	0.03313	0.5376	0.9998	0.9926	0.7468
3	(0,0,17)	$P_{(r)}^*$	0.04979	0.7359	0.9994	0.9886	0.8816
		$M_{(r)}^*$	0.04359	0.6088	1.0000	0.9913	0.7941
	(17,0,0)	$P_{(r)}^*$	0.05052	0.5907	1.0000	0.9975	0.8161
		$M_{(r)}^*$	0.05301	0.5598	1.0000	0.9975	0.7962
	(14,2,1)	$P_{(r)}^*$	0.04865	0.6476	1.0000	0.9966	0.8523
		$M_{(r)}^*$	0.05463	0.5813	1.0000	0.9975	0.8025
	(5,6,6)	$P_{(r)}^*$	0.05061	0.7409	0.9999	0.9941	0.8882
		$M_{(r)}^*$	0.05222	0.6216	1.0000	0.9915	0.7998
4	(0,0,0,16)	$P_{(r)}^*$	0.05023	0.7800	0.9986	0.9829	0.8900
		$M_{(r)}^*$	0.05357	0.6381	1.0000	0.9919	0.7923
	(16,0,0,0)	$P_{(r)}^*$	0.05036	0.6049	1.0000	0.9972	0.8366
		$M_{(r)}^*$	0.05301	0.5506	1.0000	0.9971	0.8022
	(13,1,1,1)	$P_{(r)}^*$	0.05105	0.6900	0.9999	0.9969	0.8817
		$M_{(r)}^*$	0.05595	0.5858	1.0000	0.9971	0.8121
	(4,4,4,4)	$P_{(r)}^*$	0.04559	0.7700	0.9993	0.9907	0.8953
		$M_{(r)}^*$	0.05136	0.6284	1.0000	0.9919	0.7857
5	(0,0,0,0,15)	$P_{(r)}^*$	0.04967	0.7999	0.9964	0.9749	0.8852
		$M_{(r)}^*$	0.04439	0.5641	0.9993	0.9770	0.7071
	(15,0,0,0,0)	$P_{(r)}^*$	0.05045	0.6414	1.0000	0.9989	0.8499
		$M_{(r)}^*$	0.05306	0.5584	1.0000	0.9983	0.8071
	(11,1,1,1,1)	$P_{(r)}^*$	0.05171	0.7564	0.9998	0.9976	0.8987
		$M_{(r)}^*$	0.06578	0.6504	1.0000	0.9984	0.8421
	(3,3,3,3,3)	$P_{(r)}^*$	0.05002	0.8118	0.9991	0.9899	0.9101
		$M_{(r)}^*$	0.04975	0.6293	1.0000	0.9933	0.7952
6	(0,0,0,0,0,14)	$P_{(r)}^*$	0.04952	0.8208	0.9912	0.9642	0.8813
		$M_{(r)}^*$	0.04852	0.5729	0.9997	0.9751	0.7015
	(14,0,0,0,0,0)	$P_{(r)}^*$	0.05087	0.6552	1.0000	0.9974	0.8502
		$M_{(r)}^*$	0.05464	0.5707	1.0000	0.9985	0.7962
	(7,2,2,1,1,1)	$P_{(r)}^*$	0.05077	0.8007	0.9997	0.9939	0.9143
		$M_{(r)}^*$	0.04917	0.6103	0.9999	0.9938	0.7729
	(2,2,3,3,2,2)	$P_{(r)}^*$	0.04918	0.8281	0.9976	0.9868	0.9101
		$M_{(r)}^*$	0.05024	0.6287	0.9999	0.9938	0.7869

Table 7.7: (Continued)

r	Censoring scheme	Test	Exact l.o.s.	LN(0.1)	LN(0.5)	EV
2	(0,18)	$P^*_{(r)}$	0.05326	0.7575	0.9940	0.5001
		$M^*_{(r)}$	0.04691	0.6743	0.9885	0.4481
	(18,0)	$P^*_{(r)}$	0.04955	0.6508	0.9936	0.3517
		$M^*_{(r)}$	0.05301	0.6553	0.9942	0.3549
	(14,4)	$P^*_{(r)}$	0.04691	0.7237	0.9936	0.4687
		$M^*_{(r)}$	0.05463	0.6678	0.9942	0.4235
	(9,9)	$P^*_{(r)}$	0.05680	0.7682	0.9947	0.5275
		$M^*_{(r)}$	0.03313	0.6137	0.9843	0.4106
3	(0,0,17)	$P^*_{(r)}$	0.04979	0.8047	0.9908	0.5919
		$M^*_{(r)}$	0.04359	0.6874	0.9857	0.4823
	(17,0,0)	$P^*_{(r)}$	0.05052	0.6770	0.9941	0.3953
		$M^*_{(r)}$	0.05301	0.6504	0.9934	0.3403
	(14,2,1)	$P^*_{(r)}$	0.04865	0.7297	0.9949	0.4738
		$M^*_{(r)}$	0.05463	0.6652	0.9935	0.4103
	(5,6,6)	$P^*_{(r)}$	0.05061	0.8073	0.9938	0.6144
		$M^*_{(r)}$	0.05222	0.6942	0.9857	0.5157
4	(0,0,0,16)	$P^*_{(r)}$	0.05023	0.8298	0.9881	0.6751
		$M^*_{(r)}$	0.05357	0.7160	0.9872	0.5459
	(16,0,0,0)	$P^*_{(r)}$	0.05036	0.7021	0.9946	0.4098
		$M^*_{(r)}$	0.05301	0.6545	0.9937	0.3389
	(13,1,1,1)	$P^*_{(r)}$	0.05105	0.7760	0.9955	0.5481
		$M^*_{(r)}$	0.05595	0.6805	0.9938	0.4273
	(4,4,4,4)	$P^*_{(r)}$	0.04559	0.8271	0.9928	0.6696
		$M^*_{(r)}$	0.05136	0.7022	0.9873	0.5158
5	(0,0,0,0,15)	$P^*_{(r)}$	0.04967	0.8401	0.9840	0.7448
		$M^*_{(r)}$	0.04439	0.6320	0.9641	0.5358
	(15,0,0,0,0)	$P^*_{(r)}$	0.05045	0.7192	0.9955	0.4551
		$M^*_{(r)}$	0.05306	0.6510	0.9942	0.3617
	(11,1,1,1,1)	$P^*_{(r)}$	0.05171	0.8098	0.9959	0.6459
		$M^*_{(r)}$	0.06578	0.7218	0.9948	0.5233
	(3,3,3,3,3)	$P^*_{(r)}$	0.05002	0.8556	0.9919	0.7414
		$M^*_{(r)}$	0.04975	0.6926	0.9862	0.5356
6	(0,0,0,0,0,14)	$P^*_{(r)}$	0.04952	0.8604	0.9770	0.8039
		$M^*_{(r)}$	0.04852	0.6303	0.9668	0.5609
	(14,0,0,0,0,0)	$P^*_{(r)}$	0.05087	0.7476	0.9961	0.4873
		$M^*_{(r)}$	0.05464	0.6735	0.9939	0.4174
	(7,2,2,1,1,1)	$P^*_{(r)}$	0.05077	0.8602	0.9957	0.7336
		$M^*_{(r)}$	0.04917	0.6857	0.9873	0.5167
	(2,2,3,3,2,2)	$P^*_{(r)}$	0.04918	0.8773	0.9908	0.7850
		$M^*_{(r)}$	0.05024	0.7039	0.9882	0.5408

not give a better power performance in all the cases when the underlying distribution is nearly symmetric or left-skewed, they can still retain good power to be useful, in addition to being cost-effective.

7.5   ILLUSTRATIVE EXAMPLE

Example 7.1. We shall now illustrate the weighted precedence and the weighted maximal precedence tests under progressive Type-II censoring for the Y-sample. Referring to Example 5.2, let us consider X- and Y-samples to be Samples 3 and 6 in Nelson (1982), respectively. If we had observed only up to the 5th breakdown from the Y-sample with progressive censoring scheme (3,0,0,0,2), three items from the Y-sample had to be randomly removed from the test at the time of the first Y-failure resulting in the removal of breakdown times 2.10, 3.83, and 3.97, thus resulting in all 10 observations from the X-sample being observed. The observations are presented in Table 7.8.

Nonparametric Procedure. In this case, we have $n_1 = n_2 = 10$, $r = 5$, $m_1 = 5$, $m_2 = m_3 = 0$, $m_4 = 2$, and $m_5 = 3$. The following are the values of the test statistics and the corresponding p-values:

Test	Test statistic	p-value
Weighted precedence test	$P^*_{(5)} = 67$	0.009
Weighted maximal precedence test	$M^*_{(5)} = 50$	0.006

From the small p-values of the weighted precedence and weighted maximal precedence tests, we observe that the data provide strong evidence against H_0 and in favor of H_1. In this example, since five X-failures occurred before the first Y-failure, it appears that this decision may indeed be a good one.

It is of interest to note here that the proposed tests are applicable even though only a few early failures from the two samples are observed as data and the Y-sample is under progressive censoring. In addition, the decisions are reached without making any assumption on the underlying lifetime distributions.

Parametric Procedure. If we assume exponential distributions for the two populations and consider the X-sample to be a complete sample of size $n_1 = 10$ and the Y-sample to be a Type-II progressively censored sample

Table 7.8: Times to insulating fluid breakdown data from Nelson (1982) for Samples 3 and 6

X-sample (Sample 3)	0.49	0.64	0.82	0.93	1.08	1.99	2.06	2.15	2.57	4.75
Y-sample (Sample 6)	1.34	*1	*1	*1	1.49	1.56	2.12	5.13	*2	*2

[1] Three surviving fluids removed randomly at time 1.34.
[2] Two remaining surviving fluids removed randomly at time 5.13.

with $n_2 = 10$, $r = 5$ and progressive censoring scheme (3,0,0,0,2), we find the MLEs of θ_X and θ_Y to be (see Section 2.4.3)

$$\hat{\theta}_X = \frac{1}{n_1} \sum_{i=1}^{n_1} X_{i:10} = 1.748$$

$$\text{and} \quad \hat{\theta}_Y = \frac{1}{r} \left[\sum_{i=1}^{r} (R_i + 1) Y_{i:r:10} \right] = 5.184,$$

respectively. Further, it is also known that $2n_1 \hat{\theta}_X / \theta_X$ and $2r\hat{\theta}_Y / \theta_Y$ have independent central chi-square distributions with $2n_1$ and $2r$ degrees of freedom (see Section 2.4.3), respectively. So, using a central F-distribution with (20, 10) degrees of freedom, we find the p-value for testing hypotheses in (3.6) to be 0.0185. Once again, there is a strong evidence against H_0 and in favor of H_1.

This is the same conclusion that was reached earlier with the use of the weighted precedence and the weighted maximal precedence tests.

Chapter 8

Generalization to k-Sample Situation

8.1 INTRODUCTION

So far, the reliability testing problem we have discussed is a two-sample problem and our primary interest was in testing whether a new manufacturing process (Y) produces units just as reliable as the standard manufacturing process (X) does or it produces more reliable units. Since the Y-variable is stochastically larger than the X-variable under the alternative hypothesis, we proposed and discussed various precedence-type test procedures. Note that for all the procedures we have considered, the termination of the life-testing experiment occurs when the rth Y-failure occurs and the data observed are then ($M_1, M_2, \cdots, M_r$), where M_1 is the number of X-failures that preceded the first Y-failure, M_2 is the number of X-failures that occurred between the first and second Y-failures, and so on.

In this chapter, we discuss a generalization of this problem when there are k processes being compared instead of just two. In Section 8.2, we outline some problems that will be of interest in this k-sample situation. In Section 8.3, we consider the problem of comparing $k-1$ treatments with a control and propose precedence-type and maximal precedence-type statistics based on a fixed sampling scheme. In Section 8.4, we consider the problem of comparing k treatments and propose precedence-type and maximal precedence-type statistics based on an inverse sampling scheme.

8.2   SOME PERTINENT PROBLEMS

While in the two-sample situation the problem of comparing the new process with the standard process is quite a natural one and one that is of great practical interest, one could consider several different problems in the k-processes case. For example, a manufacturer may wish to compare $k - 1$ new processes with the standard process and determine, for example, whether any of these new processes would produce more reliable units than the standard process. As a second example, we may consider the scenario wherein there are k different treatments for an illness and one is interested in deciding whether these k treatments have the same cure rate.

Different test procedures can be proposed for these problems by means of precedence-type statistics. These may be based on either fixed sampling or inverse sampling rules, as explained in the following sections. Of course, in the problems outlined above, it will often be not enough just to reject the null hypothesis and conclude that the k populations (or processes or treatments) are not all equal. The experimenter, in such a case, would naturally be interested in selecting the best population. This problem is addressed later in Chapters 9 and 10.

8.3   COMPARING TREATMENTS WITH CONTROL

Let us consider here the first problem described in the preceding section. In this case, we have $k - 1$ treatments that we wish to compare to a control, or $k - 1$ new processes that we wish to compare to a standard process. With $F_1(x)$ denoting the lifetime distribution associated with the control (or the standard process) and $F_{i+1}(x)$ denoting the lifetime distribution associated with the ith treatment (or the ith new process) for $i = 1, 2, \cdots, k - 1$, our null hypothesis is simply

$$H_0 : F_1(x) = F_2(x) = \cdots = F_k(x) \text{ for all } x. \tag{8.1}$$

As before, we will consider a stochastically ordered alternative of the form

$$F_2(x) \leq F_1(x) \ \cdots \ \text{or } F_k(x) \leq F_1(x) \text{ with at least one strict inequality.} \tag{8.2}$$

Suppose k independent random samples of sizes $n_1, n_2, \cdots, n_k$ from $F_1(x)$, $F_2(x)$, $\cdots$, $F_k(x)$, respectively, are placed simultaneously on a life-testing experiment. When the sample sizes are all equal, we have a balanced case that usually provides a favorable setting for carrying out a precedence-type procedure for testing H_0 in (8.1) against the alternative in (8.2); however, the test can be carried out even in the unbalanced case although the power of the test may be adversely affected in this case.

A precedence-type test procedure for this specific testing problem may be constructed as follows. After pre-fixing an r ($\leq n_1$), the life-test continues until the rth failure in the sample from the standard process or the control group. We then observe $(M_{12}, M_{22}, \cdots, M_{r2}), \cdots, (M_{1k}, M_{2k}, \cdots, M_{rk})$ from the $k - 1$ new processes or the $k - 1$ treatments, where $M_{1i}, M_{2i}, \cdots, M_{ri}$ are the number of failures in the sample from the $(i - 1)$th new process (for $i = 2, 3, \cdots, k$) before the first failure, between the first and second failures, $\cdots$, and between the $(r - 1)$th and rth failures from the standard process, respectively. Now, let us set

$$P_{(r)i} = \sum_{j=1}^{r} M_{ji} \quad \text{for} \quad i = 2, 3, \cdots, k \tag{8.3}$$

for the precedence statistic corresponding to the sample from the $(i-1)$th new process. We may then propose the following precedence-type test statistics:

$$P_1 = \sum_{i=2}^{k} P_{(r)i} = \sum_{i=2}^{k} \sum_{j=1}^{r} M_{ji} \tag{8.4}$$

and

$$P_2 = \min_{2 \leq i \leq k} P_{(r)i} = \min_{2 \leq i \leq k} \left\{ \sum_{j=1}^{r} M_{ji} \right\}. \tag{8.5}$$

The rationale for the use of the statistics in (8.4) and (8.5) is that, under the stochastically ordered alternative H_1 in (8.2), we would expect at least one of the precedence statistics $P_{(r)i}$ in (8.3) to be too small. Consequently, we will tend to reject H_0 in (8.1) in favor of H_1 in (8.2) for small values of P_1 and P_2. Specifically, $\{0 \leq P_1 \leq s_1\}$ and $\{0 \leq P_2 \leq s_2\}$ will serve as critical regions, where s_1 and s_2 are determined such that

$$\Pr(P_1 \leq s_1 | H_0) = \alpha \quad \text{and} \quad \Pr(P_2 \leq s_2 | H_0) = \alpha, \tag{8.6}$$

with α being the pre-fixed level of significance.

The null distributions of the test statistics P_1 and P_2 can be derived along the same lines as in Section 3.3.

Remark 8.1. Observe that the life-testing experiment here gets terminated when the rth failure occurs from the standard process or the control group, while in Chapter 3 the termination occurred when the rth failure occurred from the new process or the treatment group. The basic nature of the problem of comparing treatments with control necessitated this switch around in the termination process of the experiment. Furthermore, because of this change, small values of the precedence-type test statistics in (8.4) and (8.5) result in the rejection of H_0 in favor of H_1, while in Chapter 3 large values of precedence test statistic led to the rejection of H_0.

We can proceed similarly and propose maximal precedence-type statistics for the testing problem discussed here. Once again, we terminate the lifetest when the rth failure occurs in the sample from the standard process. Then, with $(M_{1i}, M_{2i}, \cdots, M_{ri})$, for $i = 2, \cdots, k$, being observed from the $k-1$ new processes, where M_{ji} are the number of failures in the sample from the $(i-1)$th new process between the $(j-1)$th and jth failures from the standard process, we may set

$$M_{(r)i} = \max_{1 \le j \le r} M_{ji} \quad \text{for} \quad i = 2, 3, \cdots, k \tag{8.7}$$

for the maximal precedence statistic corresponding to the sample from the $(i-1)$th new process. We may then propose the maximal precedence-type test statistics as

$$M_1 = \sum_{i=2}^{k} M_{(r)i} = \sum_{i=2}^{k} \max_{1 \le j \le r} M_{ji} \tag{8.8}$$

and

$$M_2 = \min_{2 \le i \le k} M_{(r)i} = \min_{2 \le i \le k} \left\{ \max_{1 \le j \le r} M_{ji} \right\}. \tag{8.9}$$

Here again, the rationale for the use of the statistics in (8.8) and (8.9) is that, under the stochastically ordered alternative H_1 in (8.2), we would expect at least one of the maximal precedence statistics $M_{(r)i}$ in (8.7) to be too small. Therefore, we would reject H_0 in (8.1) in favor of H_1 in (8.2) for small values of M_1 and M_2. Specifically, $\{0 \le M_1 \le s_1\}$ and $\{0 \le M_2 \le s_2\}$ will serve as critical regions, where s_1 and s_2 are determined such that

$$\Pr(M_1 \le s_1 | H_0) = \alpha \quad \text{and} \quad \Pr(M_2 \le s_2 | H_0) = \alpha, \tag{8.10}$$

with α being the pre-fixed level of significance.

The null distributions of the test statistics M_1 and M_2 can be derived along lines similar to those Section 4.2. For the same reason as described in Remark 8.1, small values of the maximal precedence-type test statistics in (8.8) and (8.9) result in the rejection of H_0 in favor of H_1, while in Chapter 4 large values of maximal precedence test statistic led to the rejection of H_0.

8.4 COMPARISON OF TREATMENTS

Let us now consider the second problem described earlier in Section 8.2. In this case, we have k different treatments for an illness and we are interested in deciding whether these k treatments have the same cure rate. With $F_i(x)$ denoting the lifetime distribution associated with the ith treatment for $i = 1, 2, \cdots, k$, our null hypothesis is once again (8.1) while the alternative we are interested in now is

$$H_1 : \text{At least one } F_i(x) \text{ is less than all others.} \tag{8.11}$$

Suppose k independent random samples of size n each from $F_1(x)$, $\cdots$, $F_k(x)$ are placed simultaneously on a life-testing experiment. Note that, for convenience in presentation, we are considering here the balanced case; in the unbalanced case, the sampling procedure needs to be changed as done in Chapter 9, for example.

Let us now adopt an inverse sampling procedure as follows. After pre-fixing an r ($\le n$), the life-test continues until the occurrence of the rth failure in any one sample. Suppose this occurs in the lth sample ($l = 1, 2, \cdots, k$). Then, with M_{1i}, M_{2i}, $\cdots$, M_{ri} denoting the number of failures in the ith sample ($1 \le i \ne l \le k$) before the first failure, between the first and second failures, $\cdots$, and between the $(r-1)$th and rth failures from the lth sample, we may find

$$P^*_{(r)i} = \sum_{j=1}^{r} M_{ji} \quad \text{for } 1 \le i \le k, \ i \ne l \tag{8.12}$$

for the precedence statistic corresponding to the ith treatment. Then, analogous to (8.4) and (8.5), we may propose the following precedence-type test statistics:

$$P_1^* = \sum_{\substack{i=1 \\ i\neq l}}^{k} P_{(r)i}^* = \sum_{\substack{i=1 \\ i\neq l}}^{k} \sum_{j=1}^{r} M_{ji} \qquad (8.13)$$

and

$$P_2^* = \min_{\substack{1\leq i\leq k \\ i\neq l}} P_{(r)i}^* = \min_{\substack{1\leq i\leq k \\ i\neq l}} \left\{ \sum_{j=1}^{r} M_{ji} \right\}. \qquad (8.14)$$

In a similar manner, we may find

$$M_{(r)i}^* = \max_{1\leq j\leq r} M_{ji} \quad \text{for } 1\leq i\leq k,\ i\neq l \qquad (8.15)$$

for the maximal precedence statistic corresponding to the sample from the ith treatment. Then, analogous to (8.8) and (8.9), we may propose the following maximal precedence-type test statistics:

$$M_1^* = \sum_{\substack{i=1 \\ i\neq l}}^{k} M_{(r)i}^* = \sum_{\substack{i=1 \\ i\neq l}}^{k} \max_{1\leq j\leq r} M_{ji} \qquad (8.16)$$

and

$$M_2^* = \min_{\substack{1\leq i\leq k \\ i\neq l}} M_{(r)i}^* = \min_{\substack{1\leq i\leq k \\ i\neq l}} \left\{ \max_{1\leq j\leq r} M_{ji} \right\}. \qquad (8.17)$$

As in the preceding section, small values of all these test statistics would lead to the rejection of H_0 in (8.1) in favor of H_1 in (8.11). The null distributions of all these test statistics can be derived along the same lines as in Chapters 3 and 4.

Chapter 9

Selecting the Best Population Using a Test for Equality Based on Precedence Statistic

9.1 INTRODUCTION

When we have two or more competing populations, a problem of interest often is to make a decision as to which is the best if we reject the null hypothesis of homogeneity of populations. To this end, we formulate the problem of selecting the best population as a multiple-decision problem. We still control the probability of rejecting a true null hypothesis. However, we use the probability of selecting the best population as a performance characteristic. It is important to note that this is not the usual ranking and selection formulation that does not have a null hypothesis and typically controls the probability of selecting the best population. For a general introduction to ranking and selection problems, see Bechhofer, Santner, and Goldsman (1995), Gibbons, Olkin, and Sobel (1999), and Gupta and Panchapakesan (2002). In this chapter, we first introduce a nonparametric test based on early failures for the equality of two lifetime distributions against two alternatives regarding the best population, as proposed by Balakrishnan, Ng, and Panchapakesan (2006). This procedure makes use of the precedence statistic, which can determine the difference between populations based on early ($100q\%$) failures as already mentioned in Chapter 3. Then, we present the generalization of this test procedure to the k-sample case, as given by Balakrishnan, Ng, and Panchapakesan (2006).

In Section 9.2, we describe the nonparametric test procedure to test the equality of two distributions against two alternative hypotheses that terminates the life-testing experiment with no more than $100q\%$ of the two samples having failed. The test statistic is introduced and its null distribution is derived. Both equal and unequal sample size cases are considered and some critical values close to 5% level of significance are presented. Exact probability of correct selection of these tests under the Lehmann alternative is also discussed. In Section 9.3, extension of this test procedure to the k-sample case is discussed. We present the test statistic and derive its null distribution. We examine the properties of the test procedure under a location-shift between the populations by means of Monte Carlo simulations for $k = 2$ and 3. Next, in Section 9.4, we discuss the simulation results and also the choice of the value of q. Finally, we present an example in Section 9.5 to illustrate the precedence-type selection procedures.

9.2   TWO-SAMPLE SELECTION PROBLEM

Suppose there are independent samples coming from two different populations. Let $X_1, X_2, \cdots, X_{n_1}$ denote the sample from a continuous population π_1 with distribution F_X, and $Y_1, Y_2, \cdots, Y_{n_2}$ denote an independent sample from another continuous population π_2 with distribution F_Y. The number of units under the life-testing experiment is thus $N = n_1 + n_2$. These two groups of units are placed simultaneously on a life-testing experiment. We are concerned with the equality of the two populations and the problem of selecting the best population among π_1 and π_2 if they are not equal. This problem is, of course, related to testing the hypotheses

$$H_0 : F_X = F_Y \quad \text{against} \quad \begin{cases} H_{A1} : F_X < F_Y \\ H_{A2} : F_X > F_Y. \end{cases} \tag{9.1}$$

We will first describe the selection procedure for equal sample size situation (i.e., $n_1 = n_2 = n$) and then generalize this selection procedure to the unequal sample size situation appropriately.

9.2.1   Equal Sample Size Situation

We assume that the sample sizes are equal, i.e., $n_1 = n_2 = n$. As we mentioned earlier, there are many situations where one may not want to wait for all failures to occur in a life-testing experiment. Therefore, we suggest that the life-testing experiment be terminated as soon as the rth failure from either one of the two samples is observed. That is, we will terminate the experiment as soon as the rth X-failure or the rth Y-failure occurs. In order words, the termination here is a *sooner waiting time*; see, for example, Balakrishnan and Koutras (2002). Then, we count the number of failures from the other sample and denote this value by $P_{(r)}$ $(0 \leq P_{(r)} < r)$. $P_{(r)}$ is a *precedence statistic*, as in Chapter 3.

Let us further denote

$$T_r(n_1, n_2) = \min(X_{r:n_1}, Y_{r:n_2}). \tag{9.2}$$

For convenience, in the sequel we will suppress n_1, n_2, and r in $T_r(n_1, n_2)$. Then, we can write

$$P_{(r)} = \begin{cases} \#Y's < X_{r:n_1} & \text{if } T = X_{r:n_1} \\ \#X's < Y_{r:n_2} & \text{if } T = Y_{r:n_2} \end{cases}. \tag{9.3}$$

The test procedure we describe here is based on the precedence statistic $P_{(r)}$ in (9.3). Since we are interested in selecting the best population among π_1 and π_2 (longer life is assumed to be better, of course), it is clear that smaller values of $P_{(r)}$ will lead to the rejection of the null hypothesis H_0 in (9.1). Then, the critical region for the test procedure based on $P_{(r)}$ will be $\{0, 1, \ldots, s\}$ $(s < r)$. For this reason, a decision rule can be proposed as

Do not reject H_{A1} (select π_1 as the best)

if and only if $T_r = Y_{r:n_2}$ and $P_{(r)} \leq s$;

Do not reject H_{A2} (select π_2 as the best)

if and only if $T_r = X_{r:n_1}$ and $P_{(r)} \leq s$. \tag{9.4}

Under the null hypothesis that two populations are same, i.e., $H_0 : F_X = F_Y$, the probabilities that the experiment terminates at the X-sample and Y-sample are

$$\Pr(T = X_{r:n_1} \mid H_0 : F_X = F_Y) = \frac{\sum_{l=0}^{r-1} \binom{l+r-1}{l} \binom{N-l-r}{n_2-l}}{\binom{N}{n_1}}, \tag{9.5}$$

$$\Pr(T = Y_{r:n_1} \mid H_0 : F_X = F_Y) = \frac{\sum_{l=0}^{r-1} \binom{l+r-1}{l} \binom{N-l-r}{n_1-l}}{\binom{N}{n_2}}, \tag{9.6}$$

respectively.

In the case of equal sample size (viz., $n_1 = n_2 = n$), we have

$$\Pr(T = X_{r:n} \mid H_0 : F_X = F_Y) = \Pr(T = Y_{r:n} \mid H_0 : F_X = F_Y) = \frac{1}{2}. \quad (9.7)$$

Under the null hypothesis, the distribution of $P_{(r)}$ is given by

$$\begin{aligned}
\Pr(P_{(r)} &= l, T = X_{r:n} \mid H_0 : F_X = F_Y) \\
&= \Pr(P_{(r)} = l, T = Y_{r:n} \mid H_0 : F_X = F_Y) \\
&= \frac{\binom{l+r-1}{l}\binom{N-l-r}{n-l}}{\binom{N}{n}}
\end{aligned} \quad (9.8)$$

for $l = 0, 1, \cdots, r-1$. For a fixed level of significance α, therefore, the critical region for this test procedure will be $\{0, 1, \cdots, s\}$, where

$$\begin{aligned}
\alpha &= \Pr(P_{(r)} \le s, T = X_{r:n} \mid H_0 : F_X = F_Y) \\
&\quad + \Pr(P_{(r)} \le s, T = Y_{r:n} \mid H_0 : F_X = F_Y).
\end{aligned} \quad (9.9)$$

From (9.7)–(9.9), for specified values of n, s, and r, an expression for α is given by

$$\alpha = \frac{\sum_{j=0}^{s} \binom{j+r-1}{l}\binom{N-j-r}{n-l}}{\sum_{j=0}^{r-1} \binom{j+r-1}{j}\binom{N-j-r}{n-j}}.$$

The critical value s and the exact level of significance α as close as possible to 5% for different choices of the sample size $n_1 = n_2 = n$ and $r = 5(1)n$ are presented in Table 9.1.

9.2.2   Unequal Sample Size Situation

Now, let us consider the situation in which the sizes of the two samples under the life-test are unequal. In this case, the above test procedure needs to be modified as one can readily see that the termination rule used in Section 9.2.1 may not be suitable because the probability that the experiment would terminate at the sample with the larger size will be higher. Consequently, if

Table 9.1: Near 5% critical values and exact levels of significance (in parentheses) for the procedure in (9.4)

r	$n = 10$	$n = 15$	$n = 20$	$n = 30$
5	0(0.0325)	0(0.0421)	0(0.0471)	0(0.0522)
6	1(0.0573)	1(0.0801)	0(0.0202)	0(0.0237)
7	2(0.0698)	1(0.0352)	1(0.0436)	1(0.0523)
8	3(0.0698)	2(0.0502)	2(0.0648)	1(0.0257)
9	4(0.0573)	3(0.0604)	2(0.0310)	2(0.0419)
10	5(0.0325)	4(0.0656)	3(0.0407)	3(0.0575)
11		5(0.0656)	4(0.0484)	3(0.0303)
12		6(0.0604)	5(0.0536)	4(0.0391)
13		7(0.0502)	6(0.0562)	5(0.0470)
14		8(0.0352)	7(0.0562)	6(0.0539)
15		10(0.0421)	8(0.0536)	7(0.0596)
16			9(0.0484)	8(0.0641)
17			10(0.0407)	8(0.0352)
18			12(0.0648)	9(0.0370)
19			13(0.0436)	10(0.0379)
20			15(0.0471)	11(0.0379)
21				12(0.0370)
22				14(0.0641)
23				15(0.0596)
24				16(0.0539)
25				17(0.0470)
26				18(0.0391)
27				20(0.0575)
28				21(0.0419)
29				23(0.0523)
30				25(0.0522)

the sample with larger sample size is coming from a better population, then the probability of correct selection will decrease dramatically. For this reason, it is suggested that the life-testing experiment be terminated as soon as the r_1th X-failure or the r_2th Y-failure occurs, where $r_1 = \lfloor n_1 q \rfloor$, $r_2 = \lfloor n_2 q \rfloor$, and $\lfloor h \rfloor$ is the integer part of h. When the experiment is terminated, we count the number of failures from the other sample and denoted it by P_q, i.e.,

$$P_q = \begin{cases} \#Y's < X_{r_1:n_1} & \text{if } T = X_{r_1:n_1} \\ \#X's < Y_{r_2:n_2} & \text{if } T = Y_{r_2:n_2} \end{cases}, \tag{9.10}$$

where $T = T_{r_1,r_2}(n_1, n_2) = \min(X_{r_1:n_1}, Y_{r_2:n_2})$.

The test procedure is then based on the statistic P_q. The critical region for this test procedure will be $\{0, 1, \cdots, s_1\}$ if $T = X_{r_1:n_1}$ or $\{0, 1, \cdots, s_2\}$ if $T = Y_{r_2:n_2}$, where $s_1 < r_2$ and $s_2 < r_1$. More specifically, the decision rule can be written as

Do not reject H_{A1} (select π_1 as the best)

if and only if $T = Y_{r_2:n_2}$ and $P_q \leq s_2$;

Do not reject H_{A2} (select π_2 as the best)

if and only if $T = X_{r_1:n_1}$ and $P_q \leq s_1$.   (9.11)

Under the null hypothesis $H_0 : F_X = F_Y$, the probabilities that the experiment would terminate at X-sample and Y-sample are given by (9.5) and (9.6), respectively, and the null distribution of P_q is

$$\Pr(P_q = l, T = X_{r_1:n_1} \mid H_0 : F_X = F_Y)$$
$$= \frac{\binom{l + r_1 - 1}{l} \binom{N - l - r_1}{n_2 - l}}{\binom{N}{n_1}} \quad \text{for } l = 0, 1, \cdots, r_2 - 1; \tag{9.12}$$

$$\Pr(P_q = l, T = Y_{r_2:n_2} \mid F_X = F_Y)$$
$$= \frac{\binom{l + r_2 - 1}{l} \binom{N - l - r_2}{n_1 - l}}{\binom{N}{n_2}} \quad \text{for } l = 0, 1, \cdots, r_1 - 1. \tag{9.13}$$

For a fixed level of significance α, the critical region for this test procedure will be $\{0, 1, \cdots, s_1\}$ if the experiment terminated at X-sample, and

$\{0, 1, \cdots, s_2\}$ if the experiment terminated at Y-sample, where

$$\begin{aligned}
\alpha &= \Pr(P_q \leq s_1, T = X_{r_1:n_1} \mid H_0 : F_X = F_Y) \\
&\quad + \Pr(P_q \leq s_2, T = Y_{r_2:n_2} \mid H_0 : F_X = F_Y).
\end{aligned} \tag{9.14}$$

In order to obtain the critical values s_1 and s_2 for a specified level of significance α, we may solve (9.14) by using (9.5), (9.6), (9.12), and (9.13). However, if the probabilities of the experiment terminating at X-sample and Y-sample are close to 0.5, in order to obtain the critical values s_1 and s_2 at α level of significance (or close to α level of significance), we can find s_1 and s_2 such that

$$\begin{aligned}
\alpha &= \alpha_1 \Pr(T = X_{r_1:n_1} \mid H_0 : F_X = F_Y) \\
&\quad + \alpha_2 \Pr(T = Y_{r_2:n_2} \mid H_0 : F_X = F_Y),
\end{aligned} \tag{9.15}$$

where

$$\begin{aligned}
\alpha_1 &= \Pr(P_q \leq s_1 \mid T = X_{r_1:n_1}, H_0 : F_X = F_Y) \\
&= \frac{\sum_{l=0}^{s_1} \binom{l + r_2 - 1}{l} \binom{N - l - r_2}{n_1 - l}}{\sum_{j=0}^{r_1 - 1} \binom{j + r_2 - 1}{j} \binom{N - j - r_2}{n_1 - j}}
\end{aligned} \tag{9.16}$$

and

$$\begin{aligned}
\alpha_2 &= \Pr(P_q \leq s_2 \mid T = Y_{r_2:n_2}, H_0 : F_X = F_Y) \\
&= \frac{\sum_{l=0}^{s_2} \binom{l + r_1 - 1}{l} \binom{N - l - r_1}{n_2 - l}}{\sum_{j=0}^{r_2 - 1} \binom{j + r_1 - 1}{j} \binom{N - j - r_1}{n_2 - j}},
\end{aligned} \tag{9.17}$$

with α_1 and α_2 close or equal to α. This is the case when the difference between n_1 and n_2 is not too large (see Table 9.2).

Note that when $n_1 = n_2 = n$, the test procedure and all the computational formulas will reduce to the one presented earlier in Section 9.2.1. The probabilities of terminating at X-sample and Y-sample, critical values s_1 and s_2, α_1 and α_2, and the exact level of significance α as close as possible to 5%, are presented in Table 9.2 for different choices of the sample sizes n_1, n_2, and q.

Table 9.2: Near 5% critical values and exact levels of significance for the procedure in (9.4) when $n_1 \neq n_2$

q	r_1	r_2	$\Pr(T = X_{r_1:n_1})$	$\Pr(T = Y_{r_2:n_2})$	$s_1(\alpha_1)$	$s_2\ (\alpha_2)$	Exact l.o.s.
				$n_1 = 10, n_2 = 15$			
0.50	5	7	0.4660	0.5340	1(0.0483)	0(0.0251)	0.0359
0.60	6	9	0.5340	0.4660	2(0.0414)	1(0.0348)	0.0383
0.70	7	10	0.4697	0.5303	3(0.0387)	2(0.0541)	0.0469
0.80	8	12	0.5447	0.4553	5(0.0527)	3(0.0400)	0.0469
0.90	9	13	0.4676	0.5324	6(0.0347)	4(0.0415)	0.0383
1.00	10	15	0.6000	0.4000	9(0.0471)	6(0.0415)	0.0449
				$n_1 = 15, n_2 = 20$			
0.30	4	6	0.6058	0.3942	0(0.0430)	0(0.0606)	0.0499
0.35	5	7	0.5597	0.4403	1(0.0716)	0(0.0262)	0.0516
0.40	6	8	0.5183	0.4817	1(0.0306)	1(0.0605)	0.0450
0.45	6	9	0.6347	0.3653	2(0.0727)	1(0.0403)	0.0609
0.50	7	10	0.5959	0.4041	2(0.0322)	2(0.0654)	0.0456
0.55	8	11	0.5582	0.4418	3(0.0358)	2(0.0301)	0.0333
0.60	9	12	0.5206	0.4794	4(0.0364)	3(0.0424)	0.0393
0.65	9	13	0.6370	0.3630	5(0.0635)	3(0.0269)	0.0502
0.70	10	14	0.6030	0.3970	6(0.0571)	4(0.0336)	0.0478
0.75	11	15	0.5671	0.4329	7(0.0486)	5(0.0382)	0.0441
0.80	12	16	0.5274	0.4726	8(0.0386)	6(0.0400)	0.0393
0.85	12	17	0.6679	0.3321	9(0.0586)	7(0.0603)	0.0592
0.90	13	18	0.6404	0.3596	10(0.0413)	8(0.0534)	0.0456
0.95	14	19	0.6096	0.3904	12(0.0478)	9(0.0406)	0.0450
1.00	15	20	0.5714	0.4286	14(0.0418)	11(0.0608)	0.0499

Table 9.2: (Continued)

q	r_1	r_2	$\Pr(T = X_{r_1:n_1})$	$\Pr(T = Y_{r_2:n_2})$	$s_1(\alpha_1)$	$s_2\ (\alpha_2)$	Exact l.o.s.
\multicolumn{8}{c}{$n_1 = 20, n_2 = 30$}							
0.20	4	6	0.5230	0.4770	0(0.0402)	0(0.0783)	0.0584
0.30	6	9	0.5216	0.4784	2(0.0690)	1(0.0643)	0.0668
0.40	8	12	0.5217	0.4783	4(0.0666)	2(0.0423)	0.0550
0.50	10	15	0.5227	0.4773	6(0.0534)	4(0.0649)	0.0589
0.60	12	18	0.5245	0.4755	8(0.0372)	5(0.0320)	0.0347
0.70	14	21	0.5276	0.4724	11(0.0399)	7(0.0327)	0.0365
0.80	16	24	0.5330	0.4670	15(0.0581)	10(0.0598)	0.0589
0.90	18	27	0.5444	0.4556	18(0.0372)	12(0.0347)	0.0360
1.00	20	30	0.6000	0.4000	24(0.0623)	16(0.0526)	0.0584
\multicolumn{8}{c}{$n_1 = 30, n_2 = 50$}							
0.20	6	10	0.5230	0.4770	2(0.0551)	1(0.0691)	0.0618
0.25	7	12	0.5497	0.4503	3(0.0528)	1(0.0282)	0.0417
0.30	9	15	0.5217	0.4783	5(0.0488)	3(0.0682)	0.0581
0.35	10	17	0.5461	0.4539	6(0.0418)	3(0.0304)	0.0366
0.40	12	20	0.5218	0.4782	9(0.0566)	5(0.0516)	0.0542
0.45	13	22	0.5452	0.4548	10(0.0455)	6(0.0544)	0.0495
0.50	15	25	0.5228	0.4772	13(0.0514)	7(0.0332)	0.0427
0.55	16	27	0.5463	0.4537	14(0.0396)	8(0.0328)	0.0365
0.60	18	30	0.5247	0.4753	18(0.0602)	10(0.0388)	0.0500
0.65	19	32	0.5494	0.4506	19(0.0445)	11(0.0353)	0.0403
0.70	21	35	0.5279	0.4721	23(0.0585)	13(0.0362)	0.0480
0.75	22	37	0.5552	0.4448	24(0.0407)	15(0.0604)	0.0495
0.80	24	40	0.5335	0.4665	28(0.0464)	17(0.0532)	0.0495
0.85	25	42	0.5667	0.4333	30(0.0436)	18(0.0399)	0.0420
0.90	27	45	0.5458	0.4542	35(0.0598)	21(0.0561)	0.0581
0.95	28	47	0.5978	0.4022	37(0.0466)	22(0.0317)	0.0406
1.00	30	50	0.6250	0.3750	43(0.0503)	26(0.0462)	0.0488

9.2.3   Performance Under Lehmann Alternative

To compare the performance of this test procedure with selection as an objective for different choices of q, we will be interested in the probability of "Correct Selection" (CS). The probability of correctly choosing the best population under the alternative hypothesis is given by

$$
\begin{aligned}
\Pr[CS \mid F_X &\neq F_Y] \\
&= \Pr[\text{do not reject } H_{A1} \mid H_{A1} : F_X > F_Y] \\
&\quad + \Pr[\text{do not reject } H_{A2} \mid H_{A2} : F_X < F_Y] \\
&= \Pr[P_{(r)} \leq s_1, T_{r_1,r_2} = X_{r_1:n_1} \mid H_{A1} : F_X > F_Y] \\
&\quad + \Pr[P_{(r)} \leq s_2, T_{r_1,r_2} = Y_{r_2:n_2} \mid H_{A2} : F_X < F_Y] \\
&= \sum_{j=0}^{s_1} \binom{n_2}{j} \int_{-\infty}^{\infty} [F_Y(x)]^j [1 - F_Y(x)]^{n_2-j} f_{X_{r_1:n_1}}(x)\,dx \\
&\quad + \sum_{j=0}^{s_2} \binom{n_1}{j} \int_{-\infty}^{\infty} [F_X(y)]^j [1 - F_X(y)]^{n_1-j} f_{Y_{r_2:n_2}}(y)\,dy \\
&= \sum_{j=0}^{s_1} r_1 \binom{n_2}{j} \binom{n_1}{r_1} \int_{-\infty}^{\infty} \Big\{ [F_Y(x)]^j [1 - F_Y(x)]^{n_2-j} \\
&\qquad\qquad\qquad\qquad\qquad \times [F_X(x)]^{r_1-1} [1 - F_X(x)]^{n_1-r_1} f_X(x) \Big\}\,dx \\
&\quad + \sum_{j=0}^{s_2} r_2 \binom{n_1}{j} \binom{n_2}{r_2} \int_{-\infty}^{\infty} \Big\{ [F_X(y)]^j [1 - F_X(y)]^{n_1-j} \\
&\qquad\qquad\qquad\qquad\qquad \times [F_Y(y)]^{r_2-1} [1 - F_Y(y)]^{n_2-r_2} f_Y(y) \Big\}\,dy.
\end{aligned}
$$

$$(9.18)$$

For specific distributions F_X and F_Y, the probability of CS can be computed from (9.18). We can, however, derive an explicit expression for $\Pr[CS]$ under the Lehmann alternative $H_1 : [F_X]^\gamma = F_Y$ for some γ.

Under the Lehmann alternative $H_1 : F_X^\gamma = F_Y$, $\gamma > 1$, the probability of CS can be computed as

$$
\begin{aligned}
\Pr[CS \mid H_1 &: F_X^\gamma = F_Y] \\
&= \Pr[T_{r_1,r_2} = X_{r_1:n_1}, P_{(r)} \leq s_1 \mid H_1 : F_X^\gamma = F_Y] \\
&= \sum_{j=0}^{s_1} r_1 \binom{n_2}{j} \binom{n_1}{r_1} \int_{-\infty}^{\infty} \Big\{ [F_Y(x)]^j [1 - F_Y(x)]^{n_2-j} \\
&\qquad\qquad\qquad\qquad \times [F_X(x)]^{r_1-1} [1 - F_X(x)]^{n_1-r_1} f_X(x) \Big\}\,dx
\end{aligned}
$$

$$= \sum_{j=0}^{s_1} r_1 \binom{n_2}{j} \binom{n_1}{r_1} \int_{-\infty}^{\infty} \left\{ [F_X(x)]^{\gamma j} [1 - F_X^{\gamma}(x)]^{n_2-j} \right.$$
$$\left. \times [F_X(x)]^{r_1-1} [1 - F_X(x)]^{n_1-r_1} f_X(x) \right\} dx$$

$$= \sum_{j=0}^{s_1} r_1 \binom{n_2}{j} \binom{n_1}{r_1} \int_{-\infty}^{\infty} \left\{ [F_X(x)]^{\gamma j + r_1 - 1} \right.$$
$$\left. \times \left[\sum_{l=0}^{n_2-j} \binom{n_2 - j}{l} (-1)^l [F_X(x)]^{\gamma l} \right] [1 - F_X(x)]^{n_1 - r_1} f_X(x) \right\} dx$$

$$= r_1 \binom{n_1}{r_1} \sum_{j=0}^{s_1} \sum_{l=0}^{n_2-j} (-1)^l \binom{n_2}{j} \binom{n_2 - j}{l}$$
$$\times \int_{-\infty}^{\infty} [F_X(x)]^{\gamma(j+l)+r_1-1} [1 - F_X(x)]^{n_1-r_1} f_X(x) dx$$

$$= r_1 \binom{n_1}{r_1} \sum_{j=0}^{s_1} \sum_{l=0}^{n_2-j} (-1)^l \binom{n_2}{j} \binom{n_2 - j}{l}$$
$$\times \int_0^1 u^{\gamma(j+l)+r_1-1} (1 - u)^{n_1-r_1} du$$

$$= r_1 \binom{n_1}{r} \sum_{j=0}^{s_1} \sum_{l=0}^{n_2-j} (-1)^l \binom{n_2}{j} \binom{n_2 - j}{l}$$
$$\times B(\gamma(j+l) + r_1, n_1 - r_1 + 1). \tag{9.19}$$

Now, we demonstrate the above explicit expression as well as the use of Monte Carlo simulation method for the computation of $\Pr[CS]$ under the Lehmann alternative. For this purpose, we generated 100,000 sets of data (from F_X and F_X^{γ}) and carried out the selection procedure. The $\Pr[CS]$ were estimated by the selection rates of π_2 for different values of γ. For $n_1 = n_2 = n = 10, 20$, $r = 5(1)n$, and $\gamma = 2(1)7$, the $\Pr[CS]$ computed from the above expression and those estimated through the above-described Monte Carlo simulation method are presented in Table 9.3. We observe that the estimated values of the $\Pr[CS]$ determined from Monte Carlo simulations are quite close to the exact values. In addition to revealing the correctness of the expression of the $\Pr[CS]$ presented above, these results also suggest that the Monte Carlo simulation method provides a feasible and accurate way to estimate the $\Pr[CS]$ of the proposed test procedure with selection as an objective.

Table 9.3: $\Pr[CS]$ under the Lehmann alternative for $n_1 = n_2 = n = 10, 20, r = 5(1)n$, and $\gamma = 2(1)7$

		$\gamma = 2$		$\gamma = 3$		$\gamma = 4$	
n	r	Exact	Sim.	Exact	Sim.	Exact	Sim.
10	5	0.1772	0.1783	0.4087	0.4065	0.5997	0.6003
	6	0.2397	0.2396	0.4905	0.4899	0.6730	0.6718
	7	0.2458	0.2453	0.4809	0.4797	0.6521	0.6518
	8	0.2168	0.2156	0.4200	0.4187	0.5771	0.5768
	9	0.1611	0.1583	0.3153	0.3143	0.4460	0.4438
	10	0.0839	0.0819	0.1682	0.1692	0.2481	0.2464
20	5	0.3676	0.3681	0.7231	0.7235	0.8917	0.8906
	6	0.2523	0.2539	0.6056	0.6067	0.8196	0.8185
	7	0.3868	0.3899	0.7591	0.7605	0.9180	0.9172
	8	0.4580	0.4584	0.8140	0.8138	0.9435	0.9413
	9	0.3321	0.3324	0.7101	0.7097	0.8922	0.8897
	10	0.3646	0.3677	0.7344	0.7340	0.9028	0.9004
	11	0.3769	0.3782	0.7367	0.7354	0.9006	0.8988
	12	0.3741	0.3746	0.7236	0.7233	0.8891	0.8870
	13	0.3593	0.3604	0.6976	0.6989	0.8683	0.8666
	14	0.3342	0.3345	0.6585	0.6603	0.8365	0.8358
	15	0.3000	0.3017	0.6053	0.6056	0.7900	0.7896
	16	0.2575	0.2594	0.5362	0.5348	0.7242	0.7242
	17	0.2075	0.2076	0.4492	0.4477	0.6325	0.6311
	18	0.2225	0.2225	0.4356	0.4353	0.5966	0.5962
	19	0.1452	0.1460	0.2990	0.2996	0.4323	0.4313
	20	0.1088	0.1086	0.2054	0.2032	0.2918	0.2891

Table 9.3: (Continued)

n	r	$\gamma = 5$		$\gamma = 6$		$\gamma = 7$	
		Exact	Sim.	Exact	Sim.	Exact	Sim.
10	5	0.7325	0.7315	0.8202	0.8196	0.8775	0.8776
	6	0.7896	0.7889	0.8622	0.8616	0.9076	0.9084
	7	0.7646	0.7674	0.8373	0.8360	0.8849	0.8855
	8	0.6883	0.6906	0.7661	0.7645	0.8210	0.8210
	9	0.5483	0.5493	0.6272	0.6262	0.6882	0.6871
	10	0.3182	0.3168	0.3783	0.3786	0.4298	0.4267
20	5	0.9578	0.9583	0.9831	0.9824	0.9929	0.9928
	6	0.9201	0.9196	0.9642	0.9635	0.9835	0.9838
	7	0.9723	0.9716	0.9903	0.9901	0.9964	0.9964
	8	0.9826	0.9831	0.9943	0.9944	0.9980	0.9980
	9	0.9606	0.9615	0.9851	0.9856	0.9941	0.9940
	10	0.9645	0.9640	0.9865	0.9866	0.9946	0.9945
	11	0.9622	0.9614	0.9849	0.9849	0.9936	0.9936
	12	0.9549	0.9549	0.9808	0.9812	0.9913	0.9914
	13	0.9418	0.9412	0.9731	0.9734	0.9869	0.9869
	14	0.9205	0.9201	0.9597	0.9608	0.9787	0.9792
	15	0.8869	0.8868	0.9369	0.9364	0.9634	0.9635
	16	0.8344	0.8333	0.8979	0.8986	0.9350	0.9344
	17	0.7535	0.7522	0.8314	0.8315	0.8819	0.8810
	18	0.7078	0.7076	0.7839	0.7861	0.8366	0.8368
	19	0.5373	0.5371	0.6184	0.6208	0.6812	0.6792
	20	0.3648	0.3653	0.4257	0.4266	0.4768	0.4750

9.3 k-SAMPLE SELECTION PROBLEM

Suppose there are independent random samples coming from k different populations. Let $X_{11}, X_{12}, \cdots, X_{1n_1}$ denote the sample from population π_1 with distribution F_1, $X_{21}, X_{22}, \cdots, X_{2n_2}$ denote the sample from population π_2 with distribution F_2, $\cdots$, and $X_{k1}, X_{k2}, \cdots, X_{kn_k}$ denote the sample from population π_k with distribution F_k. The total number of experimental units involved in the experiment is, thus, $N = \sum_{i=1}^{k} n_i$. These k samples are placed simultaneously on a life-testing experiment. In this case, the null hypothesis of interest is the homogeneity of the distributions, viz.,

$$H_0 : F_1 = F_2 = \cdots = F_k. \tag{9.20}$$

Since we are interested in selecting the best population among $\pi_1, \pi_2, \cdots, \pi_k$ if H_0 is rejected, we can consider the alternative to homogeneity in (9.20) as

$$\begin{cases} H_{A1} : F_1 < F_j & \text{for all } 2 \le j \le k \\ H_{A2} : F_2 < F_j & \text{for all } 1 \le j \le k, j \ne 2 \\ \quad \vdots & \quad \vdots \\ H_{Ai} : F_i < F_j & \text{for all } 1 \le j \le k, j \ne i \\ \quad \vdots & \quad \vdots \\ H_{Ak} : F_k < F_j & \text{for all } 1 \le j \le k-1. \end{cases} \tag{9.21}$$

If H_{Ai} is not rejected, then the best population is that corresponding to F_i.

   For the reason described earlier in Section 9.2, we wish to have a balanced setting (that is, $n_1 = n_2 = \cdots = n_k = n$) for our test procedure. However, this may not always be the case. So, we describe the test procedure for the case of unequal sample sizes.

9.3.1   Selection Procedure and Null Distribution

Suppose k samples are placed simultaneously on a life-testing experiment and a value q, $0 < q < 1$, is fixed prior to the experiment, where q is the upper limit for the proportion of units or subjects allowed to fail from any one sample. Then, the experiment is terminated as soon as the r_ith failure from X_i-sample is observed, where $r_i = \lfloor n_i q \rfloor$ for $i = 1, 2, \ldots, k$. When the experiment is terminated, we count the number of failures corresponding to the other $k - 1$ samples, and denote them by $P_{q,j}^{(i)}$, $j = 1, 2, \ldots, k, j \ne i$; i.e.,

$$P_{q,j}^{(i)} = \{\#X_j\text{'s} < X_{r_i:n_i} \text{ if } T = X_{r_i:n_i}\}, \qquad j = 1, 2, \cdots, k, j \ne i, \tag{9.22}$$

where $T = T_{r_1, \cdots, r_k}(n_1, \cdots, n_k) = \min_{1 \leq i \leq k} X_{r_i:n_i}$ (once again, we use the simpler notation, for convenience). Then the selection procedure proposed by Balakrishnan, Ng, and Panchapakesan (2006) is based on the test statistic

$$P_q^{*(i)} = \min_{\substack{1 \leq j \leq k \\ j \neq i}} \left(\frac{P_{q,j}^{(i)}}{n_j} \right),$$

which has a discrete support

$$\Omega_i = \left\{ 0, \left(\frac{1}{n_1}, \cdots, \frac{r_1 - 1}{n_1} \right), \cdots, \left(\frac{1}{n_{i-1}}, \cdots, \frac{r_{i-1} - 1}{n_{i-1}} \right), \right.$$
$$\left. \left(\frac{1}{n_{i+1}}, \cdots, \frac{r_{i+1} - 1}{n_{i+1}} \right), \cdots, \left(\frac{1}{n_k}, \cdots, \frac{r_k - 1}{n_k} \right) \right\}.$$

It is clear that small values of $P_q^{*(i)}$ will lead to the rejection of H_0 in (9.20) and will not reject at least one alternative hypothesis in (9.21), i.e., we will select at least one of the populations as the best. The critical region for this procedure will be the interval $[0, s_i]$ if $T = X_{r_i:n_i}$. Therefore, the decision rule can be written as

Do not reject H_{Aj} (select π_j as the best)

if and only if $T = X_{r_i:n_i}, i \neq j$, and $\dfrac{P_{q,j}^{(i)}}{n_j} = P_q^{*(i)} \leq s_i$ (9.23)

for $j = 1, 2, \cdots, k$.

Under the null hypothesis $H_0 : F_1 = F_2 = \cdots = F_k$, the probability that the experiment would terminate at the ith sample is

$$\Pr(T = X_{r_i:n_i} \mid H_0 : F_1 = F_2 = \cdots = F_k)$$
$$= \frac{1}{\binom{N}{n_1, n_2, \cdots, n_k}}$$
$$\times \sum_{\substack{l_m (m=1,\cdots,k)=0 \\ m \neq i}}^{r_m - 1} \left\{ \binom{r_i - 1 + l_1 + \cdots + l_{i-1} + l_{i+1} + \cdots + l_k}{r_i - 1, l_1, \cdots, l_{i-1}, l_{i+1}, \cdots, l_k} \right.$$
$$\left. \times \binom{N - l_1 - \cdots - l_{i-1} - l_{i+1} - \cdots - l_k - r_i}{n_1 - l_1, \cdots, n_{i-1} - l_{i-1}, n_{i+1} - l_{i+1}, \cdots, n_k - l_k, n_i - r_i} \right\}$$
$$(9.24)$$

for $i = 1, 2, \cdots, k$.

The null distribution of $P_q^{*(i)}$ is then given by

$$\Pr(P_q^{*(i)} = a, T = X_{r_i:n_i} \mid H_0 : F_1 = F_2 = \cdots = F_k)$$

$$= \frac{1}{\binom{N}{n_1, n_2, \cdots, n_k}}$$

$$\times \sum_{\substack{l_m (m=1,\cdots,k)=0, m \neq i \\ \min_{m \neq i}\left(\frac{l_m}{n_m}\right)=a}}^{r_m-1} \left\{ \binom{r_i - 1 + l_1 + \cdots + l_{i-1} + l_{i+1} + \cdots + l_k}{r_i - 1, l_1, \cdots, l_{i-1}, l_{i+1}, \cdots, l_k} \right.$$

$$\left. \times \binom{N - l_1 - \cdots - l_{i-1} - l_{i+1} - \cdots - l_k - r_i}{n_1 - l_1, \cdots, n_{i-1} - l_{i-1}, n_{i+1} - l_{i+1}, \cdots, n_k - l_k, n_i - r_i} \right\}$$

$$(9.25)$$

for $a \in \Omega_i$.

Let us denote

$$\alpha_i = \sum_{a \leq s_i} \Pr(P_q^{*(i)} = a \mid T = X_{r_i:n_i}, H_0 : F_1 = F_2 = \cdots = F_k); \qquad (9.26)$$

then, for specified values of n, r, and s_i, $i = 1, \cdots, k$, an expression for the level of significance α is given by

$$\alpha = \sum_{i=1}^{k} \alpha_i \Pr(T = X_{r_i:n_i} \mid H_0 : F_1 = F_2 = \cdots = F_k). \qquad (9.27)$$

For a fixed level of significance α, the critical region for this procedure $[0, s_i], i = 1, 2, \cdots, k$, can be determined from (9.25)–(9.27). For equal sample sizes $n_1 = \cdots = n_k = n$, we have $r_1 = r_2 = \cdots = r_k = r$, $\alpha_1 = \alpha_2 = \cdots = \alpha_k$, and

$$\Pr(T = X_{r_i:n_i} \mid H_0 : F_1 = F_2 = \cdots = F_k) = \frac{1}{k}, \qquad i = 1, 2, \cdots, k. \qquad (9.28)$$

As a result, we can determine the critical value $s_1 = s_2 = \cdots = s_k = s$ by solving

$$\sum_{a \leq s} \Pr(P_q^{*(i)} = a \mid T = X_{r_i:n_i}, H_0 : F_1 = F_2 = \cdots = F_k) = \alpha. \qquad (9.29)$$

For $k = 3$ and 4, the critical value s and the exact level of significance α as close as possible to 5% for the sample sizes $n_1 = \cdots = n_k = n = 10, 15, 20, 30$ and $r = 6(1)n$ are presented in Tables 9.4 and 9.5, respectively. Note that the test statistic in this special case reduces to

$$P^{*(i)}_{(r)} = \min_{\substack{1 \le j \le k \\ j \ne i}} \left(P^{(i)}_{q,j} \right),$$

and the null hypothesis is rejected when $P^{*(i)}_{(r)} \le s$.

For unequal sample sizes, one can find the critical values $s_1, s_2, \cdots, s_k$ for a fixed level of significance α by solving Eqs. (9.25)–(9.27). If $\Pr(T = X_{r_i:n_i} \mid H_0 : F_1 = F_2 = \cdots = F_k)$, $i = 1, 2, \cdots, k$, are close to $1/k$, the formula in (9.27) can still be used to obtain the critical values. This is the case when the difference among $n_1, n_2, \cdots, n_k$ is not too large (see Table 9.6).

For $k = 3$, the critical values $s_1, s_2, s_3, \alpha_1, \alpha_2, \alpha_3$ and the exact level of significance α as close as possible to 5%, for different choices of the sample sizes n_1, n_2, n_3, and q, are presented in Table 9.6.

9.3.2 Handling Ties

In practice, it is possible that two or more samples have the same value $P^{(i)}_{q,j}/n_j$ equaling $P^{*(i)}_q$ and that H_0 is rejected. When such ties occur, randomization may be used, in which case we will randomly pick one of the corresponding populations as the best. For example, when $k = 3$, suppose π_1 is the best population among the three populations; then the $\Pr[CS]$ can be computed in this case as

$$\Pr[CS \mid F_1 < F_2, F_1 < F_3]$$

$$= \Pr\left[T_{r_1, r_2, r_3} = X_{r_2:n_2}, \frac{P^{(2)}_{q,1}}{n_1} \le s_2, \frac{P^{(2)}_{q,1}}{n_1} < \frac{P^{(2)}_{q,3}}{n_3} \,\middle|\, F_1 < F_2, F_1 < F_3 \right]$$

$$+ \Pr\left[T_{r_1, r_2, r_3} = X_{r_3:n_3}, \frac{P^{(3)}_{q,1}}{n_1} \le s_2, \frac{P^{(3)}_{q,1}}{n_1} < \frac{P^{(3)}_{q,2}}{n_2} \,\middle|\, F_1 < F_2, F_1 < F_3 \right]$$

$$+ \frac{1}{2} \Pr\left[T_{r_1, r_2, r_3} = X_{r_2:n_2}, \frac{P^{(2)}_{q,1}}{n_1} = \frac{P^{(2)}_{q,3}}{n_3} \le s_2 \,\middle|\, F_1 < F_2, F_1 < F_3 \right]$$

$$+ \frac{1}{2} \Pr\left[T_{r_1, r_2, r_3} = X_{r_3:n_3}, \frac{P^{(3)}_{q,1}}{n_1} = \frac{P^{(3)}_{q,2}}{n_2} \le s_3 \,\middle|\, F_1 < F_2, F_1 < F_3 \right].$$

Table 9.4: Near 5% critical values and exact levels of significance (in parentheses) for the procedure in (9.23) with $k = 3$ and equal sample sizes $n_1 = \cdots = n_k = n$

r	$n = 10$	$n = 15$	$n = 20$	$n = 30$
6	0(0.0291)	0(0.0442)	0(0.0523)	0(0.0607)
7	1(0.0518)	0(0.0173)	0(0.0224)	0(0.0281)
8	2(0.0598)	1(0.0377)	1(0.0512)	1(0.0658)
9	3(0.0518)	2(0.0546)	1(0.0227)	1(0.0325)
10	4(0.0291)	3(0.0654)	2(0.0368)	2(0.0549)
11		4(0.0690)	3(0.0491)	2(0.0276)
12		5(0.0654)	4(0.0584)	3(0.0406)
13		6(0.0546)	5(0.0642)	4(0.0531)
14		7(0.0377)	6(0.0661)	5(0.0646)
15		9(0.0442)	7(0.0642)	5(0.0340)
16			8(0.0584)	6(0.0398)
17			9(0.0491)	7(0.0445)
18			10(0.0368)	8(0.0480)
19			12(0.0512)	9(0.0501)
20			14(0.0523)	10(0.0508)
21				11(0.0501)
22				12(0.0480)
23				13(0.0445)
24				14(0.0398)
25				16(0.0646)
26				17(0.0531)
27				18(0.0406)
28				20(0.0549)
29				22(0.0658)
30				24(0.0607)

Table 9.5: Near 5% critical values and exact levels of significance (in parentheses) for the procedure in (9.23) with $k = 4$ and equal sample sizes $n_1 = \cdots = n_k = n$

r	$n = 10$	$n = 15$	$n = 20$	$n = 30$
6	0(0.0527)	0(0.0782)	0(0.0916)	0(0.1054)
7	0(0.0164)	0(0.0317)	0(0.0408)	0(0.0505)
8	1(0.0282)	1(0.0676)	0(0.0170)	0(0.0232)
9	2(0.0282)	1(0.0267)	1(0.0414)	1(0.0583)
10	4(0.0527)	2(0.0390)	2(0.0660)	1(0.0283)
11		3(0.0458)	2(0.0293)	2(0.0500)
12		4(0.0458)	3(0.0396)	3(0.0724)
13		5(0.0390)	4(0.0471)	3(0.0371)
14		7(0.0676)	5(0.0510)	4(0.0496)
15		8(0.0317)	6(0.0510)	5(0.0612)
16			7(0.0471)	5(0.0311)
17			8(0.0396)	6(0.0366)
18			10(0.0660)	7(0.0410)
19			11(0.0414)	8(0.0441)
20			13(0.0408)	9(0.0456)
21				10(0.0456)
22				11(0.0441)
23				12(0.0410)
24				13(0.0366)
25				15(0.0612)
26				16(0.0496)
27				17(0.0371)
28				19(0.0500)
29				21(0.0583)
30				23(0.0505)

Table 9.6: Near 5% critical values and exact levels of significance (in parentheses) for the procedure in (9.23) with $k = 3$ and unequal sample sizes

q	r_1	r_2	r_3	$\Pr(T = X_{r_1:n_1})$ $s_1\ (\alpha_1)$	$\Pr(T = X_{r_2:n_2})$ $s_2\ (\alpha_2)$	$\Pr(T = X_{r_3:n_3})$ $s_3\ (\alpha_3)$	Exact l.o.s.
				$n_1 = n_2 = 10, n_3 = 15$			
0.4	4	4	6	0.3569	0.3569	0.2863	
				0.000(0.0502)	0.000(0.0502)	0.000(0.0415)	0.0477
0.5	5	5	7	0.3245	0.3245	0.3510	
				0.067(0.0327)	0.067(0.0327)	0.000(0.0221)	0.0290
0.6	6	6	9	0.3585	0.3585	0.2831	
				0.134*(0.0425)	0.134*(0.0425)	0.100(0.0254)	0.0377
0.7	7	7	10	0.3265	0.3265	0.3469	
				0.200(0.0436)	0.200(0.0436)	0.200(0.0452)	0.0442
0.8	8	8	12	0.3632	0.3632	0.2737	
				0.334*(0.0521)	0.334*(0.0521)	0.300(0.0283)	0.0456
0.9	9	9	13	0.3256	0.3256	0.3488	
				0.467(0.0502)	0.467(0.0502)	0.400(0.0359)	0.0452
1.0	10	10	15	0.3857	0.3857	0.2286	
				0.600(0.0582)	0.600(0.0582)	0.700(0.0690)	0.0606
				$n_1 = n_2 = 15, n_3 = 20$			
0.3	4	4	6	0.3896	0.3896	0.2208	
				0.000(0.0632)	0.000(0.0632)	0.000(0.0332)	0.0566
0.4	6	6	8	0.3486	0.3486	0.3027	
				0.067(0.0466)	0.067(0.0466)	0.067(0.0443)	0.0459
0.5	7	7	10	0.3847	0.3847	0.2306	
				0.134*(0.0621)	0.134*(0.0621)	0.134*(0.0376)	0.0565
0.6	9	9	12	0.3497	0.3497	0.3006	
				0.250(0.0568)	0.250(0.0568)	0.200(0.0322)	0.0494
0.7	10	10	14	0.3873	0.3873	0.2253	
				0.300(0.0552)	0.300(0.0552)	0.334*(0.0478)	0.0535
0.8	12	12	16	0.3528	0.3528	0.2944	
				0.450(0.0558)	0.450(0.0558)	0.467(0.0653)	0.0586
0.9	13	13	18	0.4019	0.4019	0.1962	
				0.500(0.0431)	0.500(0.0431)	0.600(0.0592)	0.0463
1.0	15	15	20	0.3714	0.3714	0.2571	
				0.734*(0.0600)	0.734*(0.0600)	0.734*(0.0397)	0.0548

* The critical values are rounded up.

Table 9.6: (Continued)

q	r_1	r_2	r_3	$\Pr(T = X_{r_1:n_1})$ $s_1\ (\alpha_1)$	$\Pr(T = X_{r_2:n_2})$ $s_2\ (\alpha_2)$	$\Pr(T = X_{r_3:n_3})$ $s_3\ (\alpha_3)$	Exact l.o.s.
				$n_1 = n_2 = 20, n_3 = 30$			
0.2	4	4	6	0.3535 0.000(0.0605)	0.3535 0.000(0.0605)	0.2929 0.000(0.0533)	0.0584
0.3	6	6	9	0.3528 0.050(0.0491)	0.3528 0.050(0.0491)	0.2943 0.050(0.0453)	0.0480
0.4	8	8	12	0.3528 0.100(0.0407)	0.3528 0.100(0.0407)	0.2944 0.100(0.0312)	0.0379
0.5	10	10	15	0.3533 0.200(0.0613)	0.3533 0.200(0.0613)	0.2935 0.200(0.0460)	0.0568
0.6	12	12	18	0.3541 0.267(0.0389)	0.3541 0.267(0.0389)	0.2918 0.300(0.0525)	0.0429
0.7	14	14	21	0.3555 0.367(0.0410)	0.3555 0.367(0.0410)	0.2889 0.400(0.0512)	0.0440
0.8	16	16	24	0.3580 0.500(0.0633)	0.3580 0.500(0.0633)	0.2841 0.500(0.0420)	0.0572
0.9	18	18	27	0.3630 0.634*(0.0540)	0.3630 0.634*(0.0540)	0.2739 0.650(0.0530)	0.0537
1.0	20	20	30	0.3857 0.767(0.0365)	0.3857 0.767(0.0365)	0.2286 0.800(0.0317)	0.0354
				$n_1 = 15, n_2 = n_3 = 20$			
0.2	3	4	4	0.3663 0.000(0.1051)	0.3168 0.000(0.1016)	0.3168 0.000(0.1016)	0.1029
0.3	4	6	6	0.4595 0.000(0.0456)	0.2702 0.050(0.0501)	0.2702 0.050(0.0501)	0.0481
0.4	6	8	8	0.3649 0.050(0.0276)	0.3175 0.100(0.0479)	0.3175 0.100(0.0479)	0.0405
0.5	7	10	10	0.4474 0.100(0.0339)	0.2763 0.150(0.0365)	0.2763 0.150(0.0365)	0.0353
0.6	9	12	12	0.3671 0.250(0.0650)	0.3164 0.267(0.0606)	0.3164 0.267(0.0606)	0.0622
0.7	10	14	14	0.4542 0.300(0.0579)	0.2729 0.350(0.0475)	0.2729 0.350(0.0475)	0.0522
0.8	12	16	16	0.3737 0.450(0.0629)	0.3132 0.467(0.0540)	0.3132 0.467(0.0540)	0.0573
0.9	13	18	18	0.4936 0.500(0.0450)	0.2532 0.600(0.0578)	0.2532 0.600(0.0578)	0.0515
1.0	15	20	20	0.4156 0.700(0.0396)	0.2922 0.750(0.0393)	0.2922 0.750(0.0393)	0.0394

* The critical values are rounded up.

Table 9.6: (Continued)

q	r_1	r_2	r_3	$\Pr(T = X_{r_1:n_1})$ $s_1\ (\alpha_1)$	$\Pr(T = X_{r_2:n_2})$ $s_2\ (\alpha_2)$	$\Pr(T = X_{r_3:n_3})$ $s_3\ (\alpha_3)$	Exact l.o.s.
					$n_1 = 10, n_2 = 20, n_3 = 30$		
0.2	2	4	6	0.4377 0.000(0.1282)	0.3077 0.000(0.1422)	0.2545 0.000(0.1150)	0.1291
0.3	3	6	9	0.4335 0.034*(0.0605)	0.3090 0.000(0.0549)	0.2575 0.050(0.0541)	0.0571
0.4	4	8	12	0.4332 0.067(0.0486)	0.3091 0.067(0.0229)	0.2576 0.100(0.0672)	0.0455
0.5	5	10	15	0.4355 0.134*(0.0458)	0.3086 0.167(0.0409)	0.2559 0.150(0.0275)	0.0396
0.6	6	12	18	0.4401 0.200(0.0535)	0.3074 0.267(0.0505)	0.2525 0.250(0.0336)	0.0475
0.7	7	14	21	0.4478 0.300(0.0626)	0.3053 0.367(0.0519)	0.2469 0.350(0.0334)	0.0521
0.8	8	16	24	0.4609 0.367(0.0384)	0.3017 0.467(0.0455)	0.2375 0.500(0.0694)	0.0479
0.9	9	18	27	0.4872 0.500(0.0517)	0.2942 0.567(0.0309)	0.2185 0.600(0.0465)	0.0445
1.0	10	20	30	0.5833 0.650(0.0497)	0.2667 0.767(0.0395)	0.1500 0.800(0.0498)	0.0470
					$n_1 = 10, n_2 = 15, n_3 = 20$		
0.2	2	3	4	0.4012 0.000(0.1808)	0.3218 0.000(0.1751)	0.2770 0.000(0.1535)	0.1714
0.3	3	4	6	0.3550 0.000(0.0675)	0.4077 0.000(0.1078)	0.2373 0.000(0.0566)	0.0813
0.4	4	6	8	0.3980 0.050(0.0417)	0.3225 0.050(0.0360)	0.2794 0.067(0.0369)	0.0385
0.5	5	7	10	0.3624 0.100(0.0409)	0.3956 0.100(0.0702)	0.2421 0.134*(0.0433)	0.0531
0.6	6	9	12	0.4025 0.150(0.0350)	0.3217 0.200(0.0608)	0.2758 0.200(0.0473)	0.0467
0.7	7	10	14	0.3687 0.267(0.0547)	0.3962 0.250(0.0387)	0.2351 0.334*(0.0530)	0.0480
0.8	8	12	16	0.4160 0.350(0.0451)	0.3191 0.400(0.0588)	0.2649 0.400(0.0426)	0.0488
0.9	9	13	18	0.3793 0.467(0.0457)	0.4138 0.450(0.0308)	0.2069 0.534*(0.0316)	0.0366
1.0	10	15	20	0.4889 0.650(0.0540)	0.3048 0.700(0.0605)	0.2063 0.734*(0.0393)	0.0529

* The critical values are rounded up.

9.4 MONTE CARLO SIMULATION UNDER LOCATION-SHIFT

In order to further examine the performance of the test procedure with selection as an objective in the alternative hypotheses proposed here, we consider the location-shift alternative. For the two-sample problem, we consider H_1 : $F_X(x) = F_Y(x+\theta)$ for some $\theta > 0$, where θ is the shift in location. For the k-sample problem with $k = 3$, we consider $H_1 : F_1(x) = F_2(x+\theta_1) = F_3(x+\theta_2)$ for some $\theta_1, \theta_2 > 0$, where θ_1 and θ_2 are shift in location parameters of F_2 and F_3, respectively.

The $\Pr[CS]$ for the proposed procedure was estimated by Balakrishnan, Ng, and Panchapakesan (2006) through Monte Carlo simulations when $\theta = 0.5$ and 1.0 for the two-sample problem and $(\theta_1 = 0.5, \theta_2 = 1.0)$, $(\theta_1 = 0.0, \theta_2 = 1.0)$, $(\theta_1 = 1.0, \theta_2 = 2.0)$, $(\theta_1 = 0.5, \theta_2 = 1.5)$ for the k-sample problem with $k = 3$. The following lifetime distributions were used in this study:

1. Standard normal distribution

2. Standard exponential distribution

3. Gamma distribution with shape parameter 3.0 and standardized by mean 3.0 and standard deviation $\sqrt{3.0}$

4. Lognormal distribution with shape parameter 0.1 and standardized by mean $e^{0.005}$ and standard deviation $\sqrt{e^{0.01}(e^{0.01} - 1)}$

A brief description of these distributions and their properties has been provided in Section 2.5. For different choices of sample sizes, 100,000 sets of data were generated in order to obtain the estimated probabilities of selection of each population.

In Tables 9.7–9.10, we have presented the estimated probabilities of selection of π_1 and π_2 for the two-sample case for different choices of n_1, n_2, and q for the distributions listed above with location-shift θ being equal to 0.5 and 1.0. In Tables 9.11–9.18, we have presented the estimated probabilities of selection of π_1, π_2, and π_3 for the three-sample case for different choices of n_1, n_2, n_3, and q for the distributions listed above with the location-shift (θ_1, θ_2) being equal to $(0.5,1.0)$, $(0.0, 1.0)$, $(1.0, 2.0)$, and $(0.5, 1.5)$. For comparison purposes, the corresponding exact levels of significance are also included in these tables.

Table 9.7: Estimated probabilities of selection for $n_1 = n_2 = n$, with location-shift $\theta = 0.5$

n	r	α	Normal		Exp(1)		Gamma(3)		LN(0.1)	
			(a)	(b)	(a)	(b)	(a)	(b)	(a)	(b)
10	5	0.033	0.001	0.092	0.000	0.490	0.000	0.174	0.001	0.102
	6	0.057	0.003	0.150	0.001	0.392	0.001	0.216	0.002	0.159
	7	0.070	0.003	0.178	0.002	0.307	0.002	0.213	0.003	0.183
	8	0.070	0.003	0.177	0.003	0.225	0.003	0.185	0.003	0.176
	9	0.057	0.003	0.147	0.003	0.145	0.003	0.135	0.003	0.141
	10	0.033	0.002	0.088	0.003	0.069	0.003	0.070	0.002	0.081
15	5	0.042	0.002	0.125	0.000	0.839	0.000	0.311	0.001	0.146
	6	0.080	0.003	0.221	0.000	0.768	0.001	0.392	0.002	0.246
	7	0.035	0.001	0.134	0.000	0.596	0.000	0.252	0.001	0.151
	8	0.050	0.001	0.183	0.000	0.517	0.000	0.275	0.001	0.197
	9	0.060	0.001	0.211	0.000	0.440	0.001	0.273	0.001	0.222
	10	0.066	0.002	0.226	0.000	0.369	0.001	0.259	0.001	0.232
	11	0.066	0.002	0.227	0.001	0.301	0.002	0.235	0.001	0.227
	12	0.060	0.002	0.213	0.001	0.233	0.002	0.202	0.002	0.207
	13	0.050	0.001	0.181	0.002	0.167	0.002	0.154	0.001	0.172
	14	0.035	0.001	0.136	0.002	0.104	0.002	0.102	0.001	0.125
	15	0.042	0.002	0.125	0.004	0.079	0.003	0.084	0.002	0.110
20	5	0.047	0.002	0.152	0.000	0.963	0.000	0.441	0.001	0.181
	6	0.020	0.000	0.092	0.000	0.904	0.000	0.296	0.000	0.112
	7	0.044	0.001	0.176	0.000	0.857	0.000	0.393	0.001	0.205
	8	0.065	0.001	0.243	0.000	0.807	0.000	0.440	0.001	0.273
	9	0.031	0.001	0.162	0.000	0.672	0.000	0.308	0.000	0.183
	10	0.041	0.001	0.200	0.000	0.604	0.000	0.322	0.000	0.220
	11	0.048	0.001	0.230	0.000	0.538	0.000	0.331	0.001	0.246
	12	0.054	0.001	0.249	0.000	0.474	0.000	0.321	0.001	0.260
	13	0.056	0.001	0.259	0.000	0.412	0.000	0.308	0.001	0.265
	14	0.056	0.001	0.258	0.000	0.350	0.001	0.282	0.001	0.259
	15	0.054	0.001	0.249	0.001	0.291	0.001	0.250	0.001	0.244
	16	0.048	0.000	0.231	0.001	0.232	0.001	0.215	0.001	0.221
	17	0.041	0.000	0.201	0.001	0.174	0.001	0.172	0.000	0.188
	18	0.065	0.001	0.243	0.003	0.178	0.002	0.190	0.002	0.223
	19	0.044	0.001	0.176	0.002	0.111	0.001	0.126	0.001	0.156
	20	0.047	0.002	0.149	0.005	0.084	0.003	0.100	0.002	0.128

Remarks: (a) $\Pr(\pi_1$ is selected); (b) $\Pr(\pi_2$ is selected) $= \Pr[CS]$.

Table 9.8: Estimated probabilities of selection for $n_1 = n_2 = n$, with location-shift $\theta = 1.0$

n	r	α	Normal		Exp(1)		Gamma(3)		LN(0.1)	
			(a)	(b)	(a)	(b)	(a)	(b)	(a)	(b)
10	5	0.033	0.000	0.283	0.000	0.917	0.000	0.602	0.000	0.327
	6	0.057	0.000	0.424	0.000	0.838	0.000	0.621	0.000	0.457
	7	0.070	0.000	0.481	0.000	0.724	0.000	0.578	0.000	0.496
	8	0.070	0.000	0.480	0.000	0.577	0.000	0.495	0.000	0.476
	9	0.057	0.000	0.424	0.000	0.396	0.000	0.374	0.000	0.403
	10	0.033	0.000	0.285	0.000	0.190	0.000	0.205	0.000	0.254
15	5	0.042	0.000	0.378	0.000	0.997	0.000	0.855	0.000	0.454
	6	0.080	0.000	0.572	0.000	0.993	0.000	0.882	0.000	0.637
	7	0.035	0.000	0.447	0.000	0.973	0.000	0.779	0.000	0.506
	8	0.050	0.000	0.547	0.000	0.944	0.000	0.779	0.000	0.592
	9	0.060	0.000	0.601	0.000	0.900	0.000	0.755	0.000	0.630
	10	0.066	0.000	0.625	0.000	0.832	0.000	0.716	0.000	0.638
	11	0.066	0.000	0.625	0.000	0.739	0.000	0.655	0.000	0.624
	12	0.060	0.000	0.601	0.000	0.621	0.000	0.578	0.000	0.585
	13	0.050	0.000	0.546	0.000	0.476	0.000	0.471	0.000	0.514
	14	0.035	0.000	0.445	0.000	0.310	0.000	0.335	0.000	0.401
	15	0.042	0.000	0.376	0.001	0.198	0.000	0.242	0.000	0.321
20	5	0.047	0.000	0.439	0.000	1.000	0.000	0.955	0.000	0.537
	6	0.020	0.000	0.335	0.000	1.000	0.000	0.896	0.000	0.423
	7	0.044	0.000	0.543	0.000	0.999	0.000	0.927	0.000	0.627
	8	0.065	0.000	0.663	0.000	0.997	0.000	0.930	0.000	0.728
	9	0.031	0.000	0.556	0.000	0.990	0.000	0.865	0.000	0.621
	10	0.041	0.000	0.633	0.000	0.980	0.000	0.861	0.000	0.683
	11	0.048	0.000	0.681	0.000	0.962	0.000	0.849	0.000	0.717
	12	0.054	0.000	0.711	0.000	0.934	0.000	0.827	0.000	0.733
	13	0.056	0.000	0.724	0.000	0.893	0.000	0.793	0.000	0.735
	14	0.056	0.000	0.724	0.000	0.831	0.000	0.759	0.000	0.725
	15	0.054	0.000	0.712	0.000	0.751	0.000	0.702	0.000	0.700
	16	0.048	0.000	0.683	0.000	0.652	0.000	0.629	0.000	0.659
	17	0.041	0.000	0.632	0.000	0.529	0.000	0.539	0.000	0.594
	18	0.065	0.000	0.664	0.000	0.472	0.000	0.508	0.000	0.610
	19	0.044	0.000	0.546	0.000	0.311	0.000	0.359	0.000	0.478
	20	0.047	0.000	0.439	0.001	0.198	0.001	0.248	0.000	0.366

Remarks: (a) $\Pr(\pi_1$ is selected); (b) $\Pr(\pi_2$ is selected) $= \Pr[CS]$.

Table 9.9: Estimated probabilities of selection for n_1, n_2, with location-shift $\theta = 0.5$

n_1	n_2	q	r_1	r_2	α	Normal (a)	Normal (b)	Exp(1) (a)	Exp(1) (b)
10	15	0.5	5	7	0.036	0.001	0.138	0.000	0.541
		0.6	6	9	0.038	0.001	0.145	0.000	0.399
		0.7	7	10	0.047	0.002	0.131	0.001	0.272
		0.8	8	12	0.047	0.001	0.177	0.001	0.230
		0.9	9	13	0.038	0.001	0.116	0.002	0.123
		1.0	10	15	0.045	0.002	0.145	0.003	0.107
15	20	0.3	4	6	0.050	0.002	0.145	0.000	0.931
		0.4	6	8	0.045	0.001	0.131	0.000	0.734
		0.5	7	10	0.046	0.001	0.163	0.000	0.631
		0.6	9	12	0.039	0.001	0.175	0.000	0.420
		0.7	10	14	0.048	0.000	0.255	0.000	0.404
		0.8	12	16	0.039	0.001	0.184	0.001	0.213
		0.9	13	18	0.046	0.001	0.202	0.002	0.183
		1.0	15	20	0.050	0.002	0.145	0.006	0.089
20	30	0.2	4	6	0.058	0.003	0.142	0.000	0.987
		0.3	6	9	0.067	0.001	0.269	0.000	0.939
		0.4	8	12	0.055	0.000	0.303	0.000	0.834
		0.5	10	15	0.059	0.001	0.288	0.000	0.675
		0.6	12	18	0.035	0.000	0.247	0.000	0.485
		0.7	14	21	0.036	0.000	0.259	0.000	0.357
		0.8	16	24	0.059	0.001	0.299	0.001	0.286
		0.9	18	27	0.036	0.000	0.213	0.001	0.152
		1.0	20	30	0.058	0.001	0.211	0.005	0.119
30	50	0.2	6	10	0.062	0.001	0.276	0.000	0.997
		0.3	9	15	0.058	0.000	0.325	0.000	0.974
		0.4	12	20	0.054	0.000	0.400	0.000	0.910
		0.5	15	25	0.043	0.000	0.408	0.000	0.785
		0.6	18	30	0.050	0.000	0.445	0.000	0.649
		0.7	21	35	0.048	0.000	0.436	0.000	0.495
		0.8	24	40	0.050	0.000	0.376	0.000	0.323
		0.9	27	45	0.058	0.000	0.360	0.002	0.229
		1.0	30	50	0.049	0.001	0.221	0.004	0.107

Remarks: (a) $\Pr(\pi_1$ is selected); (b) $\Pr(\pi_2$ is selected) $= \Pr[CS]$.

Table 9.9: (Continued)

						Gamma(3)		LN(0.1)	
n_1	n_2	q	r_1	r_2	α	(a)	(b)	(a)	(b)
10	15	0.5	5	7	0.036	0.000	0.251	0.001	0.154
		0.6	6	9	0.038	0.001	0.220	0.001	0.156
		0.7	7	10	0.047	0.001	0.173	0.002	0.137
		0.8	8	12	0.047	0.001	0.187	0.001	0.177
		0.9	9	13	0.038	0.002	0.110	0.002	0.113
		1.0	10	15	0.045	0.003	0.114	0.002	0.133
15	20	0.3	4	6	0.050	0.001	0.414	0.001	0.173
		0.4	6	8	0.045	0.001	0.302	0.001	0.153
		0.5	7	10	0.046	0.001	0.299	0.001	0.183
		0.6	9	12	0.039	0.001	0.253	0.001	0.186
		0.7	10	14	0.048	0.001	0.302	0.000	0.261
		0.8	12	16	0.039	0.001	0.182	0.001	0.179
		0.9	13	18	0.046	0.001	0.177	0.001	0.191
		1.0	15	20	0.050	0.005	0.103	0.003	0.128
20	30	0.2	4	6	0.058	0.000	0.519	0.002	0.178
		0.3	6	9	0.067	0.000	0.567	0.001	0.313
		0.4	8	12	0.055	0.000	0.511	0.000	0.338
		0.5	10	15	0.059	0.000	0.424	0.001	0.311
		0.6	12	18	0.035	0.000	0.321	0.000	0.260
		0.7	14	21	0.036	0.001	0.276	0.000	0.260
		0.8	16	24	0.059	0.001	0.268	0.001	0.287
		0.9	18	27	0.036	0.001	0.157	0.000	0.195
		1.0	20	30	0.058	0.003	0.137	0.002	0.183
30	50	0.2	6	10	0.062	0.000	0.735	0.000	0.339
		0.3	9	15	0.058	0.000	0.676	0.000	0.383
		0.4	12	20	0.054	0.000	0.640	0.000	0.443
		0.5	15	25	0.043	0.000	0.563	0.000	0.436
		0.6	18	30	0.050	0.000	0.514	0.000	0.456
		0.7	21	35	0.048	0.000	0.432	0.000	0.428
		0.8	24	40	0.050	0.000	0.317	0.000	0.353
		0.9	27	45	0.058	0.001	0.255	0.000	0.322
		1.0	30	50	0.049	0.003	0.129	0.001	0.183

Remarks: (a) $\Pr(\pi_1$ is selected); (b) $\Pr(\pi_2$ is selected) $= \Pr[CS]$.

Table 9.10: Estimated probabilities of selection for n_1, n_2, with location-shift $\theta = 1.0$

n_1	n_2	q	r_1	r_2	α	Normal (a)	Normal (b)	Exp(1) (a)	Exp(1) (b)
10	15	0.5	5	7	0.036	0.000	0.421	0.000	0.930
		0.6	6	9	0.038	0.000	0.454	0.000	0.845
		0.7	7	10	0.047	0.000	0.437	0.000	0.709
		0.8	8	12	0.047	0.000	0.514	0.000	0.587
		0.9	9	13	0.038	0.000	0.391	0.000	0.369
		1.0	10	15	0.045	0.000	0.396	0.001	0.252
15	20	0.3	4	6	0.050	0.000	0.419	0.000	0.999
		0.4	6	8	0.045	0.000	0.453	0.000	0.991
		0.5	7	10	0.046	0.000	0.535	0.000	0.976
		0.6	9	12	0.039	0.000	0.584	0.000	0.895
		0.7	10	14	0.048	0.000	0.694	0.000	0.850
		0.8	12	16	0.039	0.000	0.590	0.000	0.610
		0.9	13	18	0.046	0.000	0.600	0.000	0.505
		1.0	15	20	0.050	0.000	0.424	0.001	0.219
20	30	0.2	4	6	0.058	0.000	0.435	0.000	1.000
		0.3	6	9	0.067	0.000	0.701	0.000	1.000
		0.4	8	12	0.055	0.000	0.770	0.000	0.998
		0.5	10	15	0.059	0.000	0.775	0.000	0.987
		0.6	12	18	0.035	0.000	0.747	0.000	0.941
		0.7	14	21	0.036	0.000	0.755	0.000	0.844
		0.8	16	24	0.059	0.000	0.781	0.000	0.715
		0.9	18	27	0.036	0.000	0.658	0.000	0.456
		1.0	20	30	0.058	0.000	0.543	0.001	0.260
30	50	0.2	6	10	0.062	0.000	0.754	0.000	1.000
		0.3	9	15	0.058	0.000	0.853	0.000	1.000
		0.4	12	20	0.054	0.000	0.915	0.000	1.000
		0.5	15	25	0.043	0.000	0.927	0.000	0.997
		0.6	18	30	0.050	0.000	0.942	0.000	0.986
		0.7	21	35	0.048	0.000	0.936	0.000	0.938
		0.8	24	40	0.050	0.000	0.900	0.000	0.799
		0.9	27	45	0.058	0.000	0.858	0.000	0.593
		1.0	30	50	0.049	0.000	0.582	0.001	0.240

Remarks: (a) $\Pr(\pi_1$ is selected); (b) $\Pr(\pi_2$ is selected) $= \Pr[CS]$.

Table 9.10: (Continued)

n_1	n_2	q	r_1	r_2	α	Gamma(3) (a)	(b)	LN(0.1) (a)	(b)
10	15	0.5	5	7	0.036	0.000	0.717	0.000	0.475
		0.6	6	9	0.038	0.000	0.650	0.000	0.490
		0.7	7	10	0.047	0.000	0.550	0.000	0.456
		0.8	8	12	0.047	0.000	0.521	0.000	0.508
		0.9	9	13	0.038	0.000	0.346	0.000	0.370
		1.0	10	15	0.045	0.000	0.279	0.000	0.352
15	20	0.3	4	6	0.050	0.000	0.927	0.000	0.511
		0.4	6	8	0.045	0.000	0.852	0.000	0.529
		0.5	7	10	0.046	0.000	0.835	0.000	0.596
		0.6	9	12	0.039	0.000	0.745	0.000	0.616
		0.7	10	14	0.048	0.000	0.756	0.000	0.703
		0.8	12	16	0.039	0.000	0.561	0.000	0.573
		0.9	13	18	0.046	0.000	0.507	0.000	0.564
		1.0	15	20	0.050	0.001	0.263	0.000	0.363
20	30	0.2	4	6	0.058	0.000	0.975	0.000	0.548
		0.3	6	9	0.067	0.000	0.976	0.000	0.783
		0.4	8	12	0.055	0.000	0.959	0.000	0.823
		0.5	10	15	0.059	0.000	0.925	0.000	0.812
		0.6	12	18	0.035	0.000	0.849	0.000	0.767
		0.7	14	21	0.036	0.000	0.775	0.000	0.753
		0.8	16	24	0.059	0.000	0.707	0.000	0.753
		0.9	18	27	0.036	0.000	0.490	0.000	0.602
		1.0	20	30	0.058	0.001	0.324	0.000	0.460
30	50	0.2	6	10	0.062	0.000	0.998	0.000	0.854
		0.3	9	15	0.058	0.000	0.996	0.000	0.911
		0.4	12	20	0.054	0.000	0.992	0.000	0.942
		0.5	15	25	0.043	0.000	0.980	0.000	0.942
		0.6	18	30	0.050	0.000	0.961	0.000	0.944
		0.7	21	35	0.048	0.000	0.918	0.000	0.928
		0.8	24	40	0.050	0.811	0.000	0.000	0.871
		0.9	27	45	0.058	0.000	0.666	0.000	0.800
		1.0	30	50	0.049	0.000	0.320	0.000	0.482

Remarks: (a) $\Pr(\pi_1$ is selected); (b) $\Pr(\pi_2$ is selected) $= \Pr[CS]$.

Table 9.11: Estimated probabilities of selection for $n_1 = n_2 = n_3 = n$, with location-shift $\theta_1 = 0.5$, $\theta_2 = 1.0$

n	r	α	Normal (a)	Normal (b)	Normal (c)	Exp(1) (a)	Exp(1) (b)	Exp(1) (c)
10	5	0.082	0.001	0.066	0.283	0.000	0.246	0.709
	6	0.029	0.000	0.033	0.179	0.000	0.132	0.702
	7	0.052	0.000	0.047	0.278	0.000	0.058	0.644
	8	0.060	0.000	0.047	0.312	0.000	0.024	0.514
	9	0.052	0.000	0.040	0.284	0.000	0.016	0.338
	10	0.029	0.000	0.022	0.187	0.000	0.010	0.151
15	6	0.044	0.000	0.051	0.267	0.000	0.343	0.653
	7	0.017	0.000	0.028	0.183	0.000	0.249	0.728
	8	0.038	0.000	0.046	0.320	0.000	0.150	0.798
	9	0.055	0.000	0.054	0.406	0.000	0.078	0.820
	10	0.065	0.000	0.057	0.454	0.000	0.034	0.790
	11	0.069	0.000	0.054	0.473	0.000	0.016	0.709
	12	0.065	0.000	0.049	0.463	0.000	0.012	0.584
	13	0.055	0.000	0.041	0.422	0.000	0.013	0.434
	14	0.038	0.000	0.030	0.339	0.000	0.013	0.268
	15	0.044	0.000	0.026	0.293	0.001	0.016	0.164
20	6	0.052	0.000	0.065	0.323	0.000	0.455	0.545
	7	0.022	0.000	0.039	0.243	0.000	0.401	0.599
	8	0.051	0.000	0.064	0.418	0.000	0.326	0.673
	9	0.023	0.000	0.041	0.326	0.000	0.245	0.749
	10	0.037	0.000	0.051	0.434	0.000	0.158	0.826
	11	0.049	0.000	0.055	0.509	0.000	0.090	0.876
	12	0.058	0.000	0.056	0.555	0.000	0.045	0.891
	13	0.064	0.000	0.056	0.582	0.000	0.021	0.870
	14	0.066	0.000	0.053	0.593	0.000	0.010	0.818
	15	0.064	0.000	0.049	0.590	0.000	0.007	0.735
	16	0.058	0.000	0.044	0.568	0.000	0.009	0.627
	17	0.049	0.000	0.038	0.525	0.000	0.011	0.498
	18	0.037	0.000	0.031	0.455	0.000	0.013	0.351
	19	0.051	0.000	0.032	0.450	0.000	0.018	0.274
	20	0.052	0.000	0.026	0.359	0.001	0.019	0.168

Remarks: (a) $\Pr(\pi_1$ is selected); (b) $\Pr(\pi_2$ is selected); (c) $\Pr(\pi_3$ is selected) $= \Pr[CS]$.

Table 9.11: (Continued)

n	r	α	Gamma(3)			LN(0.1)		
			(a)	(b)	(c)	(a)	(b)	(c)
10	5	0.082	0.000	0.093	0.553	0.001	0.070	0.321
	6	0.029	0.000	0.046	0.384	0.000	0.035	0.204
	7	0.052	0.000	0.039	0.418	0.000	0.046	0.298
	8	0.060	0.000	0.034	0.374	0.000	0.046	0.320
	9	0.052	0.000	0.028	0.280	0.000	0.036	0.277
	10	0.029	0.000	0.016	0.145	0.000	0.020	0.168
15	6	0.044	0.000	0.094	0.660	0.000	0.058	0.325
	7	0.017	0.000	0.051	0.543	0.000	0.032	0.228
	8	0.038	0.000	0.041	0.624	0.000	0.048	0.367
	9	0.055	0.000	0.035	0.634	0.000	0.055	0.445
	10	0.065	0.000	0.035	0.610	0.000	0.055	0.482
	11	0.069	0.000	0.032	0.555	0.000	0.051	0.487
	12	0.065	0.000	0.030	0.484	0.000	0.046	0.461
	13	0.055	0.000	0.029	0.387	0.000	0.037	0.405
	14	0.038	0.000	0.021	0.265	0.000	0.027	0.312
	15	0.044	0.000	0.018	0.190	0.000	0.024	0.253
20	6	0.052	0.000	0.149	0.759	0.000	0.074	0.400
	7	0.022	0.000	0.095	0.722	0.000	0.045	0.311
	8	0.051	0.000	0.064	0.809	0.000	0.065	0.491
	9	0.023	0.000	0.039	0.737	0.000	0.042	0.392
	10	0.037	0.000	0.031	0.765	0.000	0.050	0.491
	11	0.049	0.000	0.025	0.771	0.000	0.053	0.554
	12	0.058	0.000	0.026	0.753	0.000	0.052	0.590
	13	0.064	0.000	0.026	0.726	0.000	0.052	0.605
	14	0.066	0.000	0.026	0.687	0.000	0.049	0.603
	15	0.064	0.000	0.025	0.637	0.000	0.045	0.588
	16	0.058	0.000	0.026	0.553	0.000	0.041	0.553
	17	0.049	0.000	0.026	0.471	0.000	0.034	0.499
	18	0.037	0.000	0.019	0.361	0.000	0.028	0.416
	19	0.051	0.000	0.022	0.311	0.000	0.029	0.397
	20	0.052	0.001	0.022	0.208	0.000	0.025	0.302

Remarks: (a) $\Pr(\pi_1$ is selected); (b) $\Pr(\pi_2$ is selected); (c) $\Pr(\pi_3$ is selected) $= \Pr[CS]$.

Table 9.12: Estimated probabilities of selection for $n_1 = n_2 = n_3 = n$, with location-shift $\theta_1 = 0.0$, $\theta_2 = 1.0$

n	r	α	Normal			Exp(1)		
			(a)	(b)	(c)	(a)	(b)	(c)
10	5	0.082	0.011	0.011	0.387	0.008	0.008	0.977
	6	0.029	0.004	0.004	0.257	0.003	0.003	0.943
	7	0.052	0.005	0.005	0.394	0.001	0.001	0.860
	8	0.060	0.005	0.005	0.440	0.000	0.000	0.710
	9	0.052	0.003	0.003	0.412	0.001	0.001	0.495
	10	0.029	0.002	0.001	0.286	0.001	0.001	0.233
15	6	0.044	0.006	0.006	0.373	0.004	0.004	0.991
	7	0.017	0.003	0.002	0.267	0.002	0.002	0.995
	8	0.038	0.003	0.003	0.451	0.001	0.001	0.991
	9	0.055	0.004	0.003	0.560	0.000	0.000	0.978
	10	0.065	0.003	0.003	0.620	0.000	0.000	0.943
	11	0.069	0.003	0.003	0.643	0.000	0.000	0.875
	12	0.065	0.002	0.002	0.634	0.000	0.000	0.766
	13	0.055	0.002	0.002	0.586	0.000	0.000	0.603
	14	0.038	0.001	0.001	0.491	0.001	0.001	0.397
	15	0.044	0.001	0.001	0.434	0.001	0.002	0.245
20	6	0.052	0.007	0.007	0.438	0.005	0.005	0.990
	7	0.022	0.003	0.003	0.340	0.002	0.002	0.996
	8	0.051	0.005	0.004	0.564	0.001	0.001	0.999
	9	0.023	0.002	0.002	0.453	0.000	0.000	0.999
	10	0.037	0.002	0.002	0.586	0.000	0.000	0.999
	11	0.049	0.003	0.002	0.670	0.000	0.000	0.996
	12	0.058	0.002	0.002	0.721	0.000	0.000	0.989
	13	0.064	0.002	0.002	0.748	0.000	0.000	0.973
	14	0.066	0.002	0.001	0.762	0.000	0.000	0.942
	15	0.064	0.001	0.001	0.759	0.000	0.000	0.888
	16	0.058	0.001	0.001	0.740	0.000	0.000	0.802
	17	0.049	0.001	0.001	0.698	0.000	0.000	0.676
	18	0.037	0.001	0.001	0.627	0.000	0.000	0.510
	19	0.051	0.001	0.001	0.624	0.001	0.001	0.400
	20	0.052	0.001	0.001	0.517	0.002	0.002	0.249

Remarks: (a) $\Pr(\pi_1$ is selected); (b) $\Pr(\pi_2$ is selected); (c) $\Pr(\pi_3$ is selected) $= \Pr[CS]$.

Table 9.12: (Continued)

n	r	α	Gamma(3)			LN(0.1)		
			(a)	(b)	(c)	(a)	(b)	(c)
10	5	0.082	0.008	0.009	0.776	0.010	0.010	0.447
	6	0.029	0.003	0.003	0.581	0.004	0.004	0.299
	7	0.052	0.002	0.002	0.600	0.005	0.004	0.429
	8	0.060	0.002	0.002	0.539	0.004	0.004	0.456
	9	0.052	0.002	0.002	0.418	0.003	0.003	0.405
	10	0.029	0.001	0.001	0.226	0.002	0.001	0.262
15	6	0.044	0.006	0.005	0.884	0.006	0.005	0.450
	7	0.017	0.002	0.001	0.762	0.003	0.002	0.333
	8	0.038	0.001	0.001	0.818	0.003	0.003	0.514
	9	0.055	0.001	0.001	0.813	0.003	0.003	0.608
	10	0.065	0.000	0.001	0.788	0.002	0.002	0.648
	11	0.069	0.001	0.000	0.738	0.002	0.002	0.656
	12	0.065	0.001	0.001	0.662	0.002	0.002	0.630
	13	0.055	0.000	0.001	0.548	0.002	0.001	0.568
	14	0.038	0.001	0.001	0.391	0.001	0.001	0.453
	15	0.044	0.002	0.002	0.278	0.001	0.002	0.376
20	6	0.052	0.008	0.005	0.966	0.006	0.006	0.541
	7	0.022	0.003	0.003	0.934	0.003	0.003	0.433
	8	0.051	0.001	0.001	0.960	0.004	0.003	0.652
	9	0.023	0.001	0.000	0.908	0.002	0.002	0.539
	10	0.037	0.000	0.000	0.917	0.002	0.002	0.657
	11	0.049	0.000	0.000	0.913	0.002	0.002	0.724
	12	0.058	0.000	0.000	0.901	0.002	0.001	0.759
	13	0.064	0.000	0.000	0.879	0.001	0.001	0.775
	14	0.066	0.000	0.000	0.846	0.001	0.001	0.776
	15	0.064	0.001	0.001	0.799	0.001	0.001	0.760
	16	0.058	0.001	0.000	0.736	0.001	0.001	0.728
	17	0.049	0.001	0.000	0.642	0.001	0.001	0.673
	18	0.037	0.001	0.000	0.511	0.001	0.001	0.588
	19	0.051	0.001	0.001	0.442	0.001	0.001	0.565
	20	0.052	0.002	0.002	0.305	0.001	0.001	0.440

Remarks: (a) $\Pr(\pi_1$ is selected); (b) $\Pr(\pi_2$ is selected); (c) $\Pr(\pi_3$ is selected) $= \Pr[CS]$.

Table 9.13: Estimated probabilities of selection for $n_1 = n_2 = n_3 = n$, with location-shift $\theta_1 = 1.0$, $\theta_2 = 2.0$

n	r	α	Normal			Exp(1)		
			(a)	(b)	(c)	(a)	(b)	(c)
10	5	0.082	0.000	0.156	0.681	0.000	0.459	0.541
	6	0.029	0.000	0.099	0.627	0.000	0.391	0.608
	7	0.052	0.000	0.086	0.777	0.000	0.278	0.714
	8	0.060	0.000	0.065	0.828	0.000	0.151	0.809
	9	0.052	0.000	0.043	0.822	0.000	0.055	0.788
	10	0.029	0.000	0.023	0.705	0.000	0.011	0.523
15	6	0.044	0.000	0.149	0.709	0.000	0.492	0.508
	7	0.017	0.000	0.103	0.683	0.000	0.478	0.522
	8	0.038	0.000	0.096	0.835	0.000	0.445	0.555
	9	0.055	0.000	0.076	0.891	0.000	0.383	0.617
	10	0.065	0.000	0.059	0.920	0.000	0.288	0.711
	11	0.069	0.000	0.044	0.938	0.000	0.185	0.811
	12	0.065	0.000	0.032	0.948	0.000	0.094	0.886
	13	0.055	0.000	0.024	0.944	0.000	0.036	0.891
	14	0.038	0.000	0.015	0.915	0.000	0.010	0.772
	15	0.044	0.000	0.010	0.848	0.000	0.004	0.531
20	6	0.052	0.000	0.180	0.727	0.000	0.501	0.499
	7	0.022	0.000	0.136	0.726	0.000	0.502	0.498
	8	0.051	0.000	0.124	0.848	0.000	0.495	0.505
	9	0.023	0.000	0.094	0.855	0.000	0.489	0.511
	10	0.037	0.000	0.077	0.905	0.000	0.472	0.528
	11	0.049	0.000	0.060	0.931	0.000	0.439	0.561
	12	0.058	0.000	0.046	0.949	0.000	0.381	0.619
	13	0.064	0.000	0.035	0.961	0.000	0.300	0.700
	14	0.066	0.000	0.026	0.970	0.000	0.209	0.790
	15	0.064	0.000	0.020	0.976	0.000	0.124	0.874
	16	0.058	0.000	0.015	0.979	0.000	0.061	0.929
	17	0.049	0.000	0.011	0.980	0.000	0.023	0.942
	18	0.037	0.000	0.008	0.973	0.000	0.006	0.889
	19	0.051	0.000	0.006	0.964	0.000	0.002	0.774
	20	0.052	0.000	0.005	0.902	0.000	0.005	0.530

Remarks: (a) $\Pr(\pi_1$ is selected); (b) $\Pr(\pi_2$ is selected); (c) $\Pr(\pi_3$ is selected) $= \Pr[CS]$.

Table 9.13: (Continued)

n	r	α	Gamma(3)			LN(0.1)		
			(a)	(b)	(c)	(a)	(b)	(c)
10	5	0.082	0.000	0.301	0.694	0.000	0.174	0.718
	6	0.029	0.000	0.199	0.774	0.000	0.109	0.687
	7	0.052	0.000	0.110	0.852	0.000	0.081	0.808
	8	0.060	0.000	0.048	0.875	0.000	0.055	0.841
	9	0.052	0.000	0.018	0.805	0.000	0.034	0.813
	10	0.029	0.000	0.008	0.558	0.000	0.019	0.651
15	6	0.044	0.000	0.367	0.633	0.000	0.173	0.752
	7	0.017	0.000	0.282	0.717	0.000	0.121	0.751
	8	0.038	0.000	0.202	0.797	0.000	0.097	0.864
	9	0.055	0.000	0.129	0.868	0.000	0.070	0.910
	10	0.065	0.000	0.075	0.920	0.000	0.050	0.934
	11	0.069	0.000	0.036	0.952	0.000	0.035	0.948
	12	0.065	0.000	0.017	0.956	0.000	0.024	0.952
	13	0.055	0.000	0.007	0.921	0.000	0.016	0.940
	14	0.038	0.000	0.004	0.809	0.000	0.012	0.887
	15	0.044	0.000	0.005	0.616	0.000	0.009	0.775
20	6	0.052	0.000	0.447	0.553	0.000	0.216	0.748
	7	0.022	0.000	0.397	0.603	0.000	0.167	0.772
	8	0.051	0.000	0.332	0.668	0.000	0.134	0.856
	9	0.023	0.000	0.272	0.728	0.000	0.101	0.878
	10	0.037	0.000	0.198	0.802	0.000	0.074	0.918
	11	0.049	0.000	0.127	0.873	0.000	0.055	0.942
	12	0.058	0.000	0.078	0.922	0.000	0.039	0.959
	13	0.064	0.000	0.045	0.955	0.000	0.029	0.969
	14	0.066	0.000	0.023	0.976	0.000	0.020	0.977
	15	0.064	0.000	0.011	0.984	0.000	0.015	0.981
	16	0.058	0.000	0.004	0.983	0.000	0.010	0.982
	17	0.049	0.000	0.002	0.964	0.000	0.008	0.979
	18	0.037	0.000	0.001	0.915	0.000	0.006	0.963
	19	0.051	0.000	0.003	0.844	0.000	0.005	0.939
	20	0.052	0.000	0.005	0.649	0.000	0.005	0.833

Remarks: (a) $\Pr(\pi_1$ is selected); (b) $\Pr(\pi_2$ is selected); (c) $\Pr(\pi_3$ is selected) $= \Pr[CS]$.

Table 9.14: Estimated probabilities of selection for $n_1 = n_2 = n_3 = n$, with location-shift $\theta_1 = 0.5$, $\theta_2 = 1.5$

n	r	α	Normal			Exp(1)		
			(a)	(b)	(c)	(a)	(b)	(c)
10	5	0.082	0.001	0.055	0.556	0.000	0.245	0.754
	6	0.029	0.000	0.028	0.432	0.000	0.131	0.861
	7	0.052	0.000	0.028	0.604	0.000	0.055	0.907
	8	0.060	0.000	0.022	0.662	0.000	0.017	0.859
	9	0.052	0.000	0.015	0.635	0.000	0.004	0.684
	10	0.029	0.000	0.008	0.482	0.000	0.002	0.370
15	6	0.044	0.000	0.042	0.575	0.000	0.339	0.661
	7	0.017	0.000	0.024	0.479	0.000	0.242	0.758
	8	0.038	0.000	0.027	0.692	0.000	0.151	0.849
	9	0.055	0.000	0.023	0.791	0.000	0.077	0.921
	10	0.065	0.000	0.018	0.837	0.000	0.033	0.958
	11	0.069	0.000	0.014	0.856	0.000	0.011	0.960
	12	0.065	0.000	0.012	0.852	0.000	0.003	0.916
	13	0.055	0.000	0.008	0.821	0.000	0.001	0.803
	14	0.038	0.000	0.006	0.741	0.000	0.001	0.599
	15	0.044	0.000	0.005	0.655	0.000	0.003	0.375
20	6	0.052	0.000	0.052	0.642	0.000	0.448	0.552
	7	0.022	0.000	0.032	0.570	0.000	0.402	0.598
	8	0.051	0.000	0.034	0.790	0.000	0.327	0.673
	9	0.023	0.000	0.023	0.723	0.000	0.241	0.759
	10	0.037	0.000	0.021	0.833	0.000	0.155	0.845
	11	0.049	0.000	0.017	0.887	0.000	0.090	0.910
	12	0.058	0.000	0.014	0.916	0.000	0.045	0.955
	13	0.064	0.000	0.011	0.931	0.000	0.020	0.978
	14	0.066	0.000	0.008	0.938	0.000	0.007	0.985
	15	0.064	0.000	0.007	0.937	0.000	0.002	0.976
	16	0.058	0.000	0.005	0.930	0.000	0.001	0.944
	17	0.049	0.000	0.004	0.911	0.000	0.000	0.871
	18	0.037	0.000	0.003	0.870	0.000	0.000	0.734
	19	0.051	0.000	0.003	0.851	0.000	0.002	0.594
	20	0.052	0.000	0.003	0.742	0.000	0.004	0.378

Remarks: (a) $\Pr(\pi_1$ is selected); (b) $\Pr(\pi_2$ is selected); (c) $\Pr(\pi_3$ is selected) $= \Pr[CS]$.

Table 9.14: (Continued)

n	r	α	Gamma(3)			LN(0.1)		
			(a)	(b)	(c)	(a)	(b)	(c)
10	5	0.082	0.000	0.089	0.863	0.001	0.057	0.631
	6	0.029	0.000	0.040	0.809	0.000	0.027	0.503
	7	0.052	0.000	0.017	0.819	0.000	0.024	0.650
	8	0.060	0.000	0.008	0.761	0.000	0.019	0.677
	9	0.052	0.000	0.005	0.625	0.000	0.013	0.622
	10	0.029	0.000	0.004	0.376	0.000	0.007	0.443
15	6	0.044	0.000	0.098	0.892	0.000	0.045	0.672
	7	0.017	0.000	0.048	0.916	0.000	0.025	0.579
	8	0.038	0.000	0.024	0.948	0.000	0.023	0.766
	9	0.055	0.000	0.010	0.952	0.000	0.018	0.835
	10	0.065	0.000	0.005	0.943	0.000	0.014	0.865
	11	0.069	0.000	0.003	0.917	0.000	0.011	0.868
	12	0.065	0.000	0.003	0.871	0.000	0.008	0.851
	13	0.055	0.000	0.002	0.777	0.000	0.007	0.801
	14	0.038	0.000	0.002	0.616	0.000	0.005	0.696
	15	0.044	0.000	0.003	0.447	0.000	0.005	0.578
20	6	0.052	0.000	0.147	0.853	0.000	0.059	0.746
	7	0.022	0.000	0.094	0.903	0.000	0.035	0.687
	8	0.051	0.000	0.053	0.945	0.000	0.032	0.863
	9	0.023	0.000	0.027	0.967	0.000	0.019	0.811
	10	0.037	0.000	0.016	0.978	0.000	0.016	0.888
	11	0.049	0.000	0.006	0.987	0.000	0.013	0.922
	12	0.058	0.000	0.003	0.989	0.000	0.010	0.939
	13	0.064	0.000	0.001	0.984	0.000	0.007	0.946
	14	0.066	0.000	0.001	0.972	0.000	0.006	0.946
	15	0.064	0.000	0.001	0.952	0.000	0.004	0.939
	16	0.058	0.000	0.001	0.918	0.000	0.004	0.924
	17	0.049	0.000	0.001	0.852	0.000	0.003	0.892
	18	0.037	0.000	0.001	0.745	0.000	0.003	0.834
	19	0.051	0.000	0.002	0.658	0.000	0.003	0.794
	20	0.052	0.000	0.005	0.467	0.000	0.003	0.647

Remarks: (a) $\Pr(\pi_1$ is selected); (b) $\Pr(\pi_2$ is selected); (c) $\Pr(\pi_3$ is selected) $= \Pr[CS]$.

Table 9.15: Estimated probabilities of selection for n_1, n_2, and n_3, with location-shift $\theta_1 = 0.5$, $\theta_2 = 1.0$

				Normal			Exp(1)		
n_1, n_2, n_3	q	r_1, r_2, r_3	α	(a)	(b)	(c)	(a)	(b)	(c)
10,10,15	0.4	4,4,6	0.048	0.000	0.082	0.264	0.000	0.341	0.558
	0.5	5,5,7	0.029	0.000	0.056	0.225	0.000	0.251	0.620
	0.6	6,6,9	0.038	0.000	0.059	0.276	0.000	0.085	0.610
	0.7	7,7,10	0.044	0.000	0.078	0.372	0.000	0.044	0.603
	0.8	8,8,12	0.046	0.000	0.053	0.297	0.000	0.018	0.363
	0.9	9,9,13	0.045	0.000	0.055	0.323	0.000	0.022	0.287
	1.0	10,10,15	0.061	0.000	0.049	0.245	0.000	0.026	0.127
15,15,20	0.3	4,4,6	0.057	0.000	0.080	0.275	0.000	0.448	0.523
	0.4	6,6,8	0.046	0.000	0.093	0.407	0.000	0.331	0.611
	0.5	7,7,10	0.056	0.000	0.080	0.413	0.000	0.161	0.685
	0.6	9,9,12	0.049	0.000	0.070	0.447	0.000	0.051	0.726
	0.7	10,10,14	0.054	0.000	0.071	0.468	0.000	0.016	0.594
	0.8	12,12,16	0.059	0.000	0.073	0.510	0.000	0.022	0.483
	0.9	13,13,18	0.046	0.000	0.056	0.392	0.000	0.027	0.258
	1.0	15,15,20	0.055	0.000	0.034	0.262	0.001	0.019	0.114
15,20,20	0.2	3,4,4	0.103	0.001	0.175	0.330	0.000	0.496	0.500
	0.3	4,6,6	0.048	0.000	0.087	0.391	0.000	0.450	0.521
	0.4	6,8,8	0.041	0.000	0.090	0.469	0.000	0.329	0.613
	0.5	7,10,10	0.035	0.000	0.081	0.428	0.000	0.160	0.688
	0.6	9,12,12	0.062	0.000	0.091	0.527	0.000	0.051	0.740
	0.7	10,14,14	0.052	0.000	0.070	0.489	0.000	0.015	0.603
	0.8	12,16,16	0.057	0.000	0.068	0.503	0.000	0.020	0.478
	0.9	13,18,18	0.051	0.000	0.054	0.405	0.000	0.027	0.272
	1.0	15,20,20	0.039	0.000	0.033	0.270	0.001	0.018	0.110
10,20,30	0.2	2,4,6	0.129	0.001	0.228	0.273	0.000	0.499	0.500
	0.3	3,6,9	0.057	0.000	0.149	0.353	0.000	0.473	0.512
	0.4	4,8,12	0.045	0.000	0.156	0.395	0.000	0.354	0.592
	0.5	5,10,15	0.040	0.000	0.102	0.399	0.000	0.167	0.703
	0.6	6,12,18	0.048	0.000	0.101	0.452	0.000	0.047	0.718
	0.7	7,14,21	0.052	0.000	0.090	0.457	0.000	0.021	0.602
	0.8	8,16,24	0.048	0.000	0.101	0.457	0.000	0.046	0.429
	0.9	9,18,27	0.044	0.000	0.076	0.362	0.000	0.044	0.231
	1.0	10,20,30	0.047	0.000	0.044	0.203	0.000	0.023	0.103
10,15,20	0.2	2,3,4	0.171	0.003	0.215	0.308	0.000	0.495	0.498
	0.3	3,4,6	0.081	0.000	0.143	0.255	0.000	0.458	0.513
	0.4	4,6,8	0.039	0.000	0.083	0.353	0.000	0.345	0.576
	0.5	5,7,10	0.053	0.000	0.100	0.383	0.000	0.185	0.658
	0.6	6,9,12	0.047	0.000	0.102	0.387	0.000	0.069	0.666
	0.7	7,10,14	0.048	0.000	0.093	0.439	0.000	0.027	0.579
	0.8	8,12,16	0.049	0.000	0.078	0.376	0.000	0.027	0.394
	0.9	9,13,18	0.037	0.000	0.054	0.326	0.000	0.024	0.223
	1.0	10,15,20	0.053	0.000	0.035	0.211	0.001	0.019	0.101

Remarks: (a) $\Pr(\pi_1$ is selected); (b) $\Pr(\pi_2$ is selected); (c) $\Pr(\pi_3$ is selected) = $\Pr[CS]$.

Table 9.15: (Continued)

n_1, n_2, n_3	q	r_1, r_2, r_3	α	Gamma(3)			LN(0.1)		
				(a)	(b)	(c)	(a)	(b)	(c)
10,10,15	0.4	4,4,6	0.048	0.000	0.125	0.531	0.000	0.087	0.306
	0.5	5,5,7	0.029	0.000	0.083	0.484	0.000	0.060	0.260
	0.6	6,6,9	0.038	0.000	0.048	0.423	0.000	0.056	0.298
	0.7	7,7,10	0.044	0.000	0.054	0.454	0.000	0.073	0.383
	0.8	8,8,12	0.046	0.000	0.037	0.291	0.000	0.047	0.290
	0.9	9,9,13	0.045	0.000	0.041	0.270	0.000	0.049	0.301
	1.0	10,10,15	0.061	0.000	0.033	0.147	0.000	0.042	0.207
15,15,20	0.3	4,4,6	0.057	0.000	0.164	0.617	0.000	0.090	0.335
	0.4	6,6,8	0.046	0.000	0.092	0.696	0.000	0.093	0.463
	0.5	7,7,10	0.056	0.000	0.051	0.619	0.000	0.075	0.453
	0.6	9,9,12	0.049	0.000	0.040	0.589	0.000	0.066	0.471
	0.7	10,10,14	0.054	0.000	0.041	0.496	0.000	0.065	0.463
	0.8	12,12,16	0.059	0.000	0.050	0.451	0.000	0.066	0.488
	0.9	13,13,18	0.046	0.000	0.038	0.277	0.000	0.050	0.350
	1.0	15,15,20	0.055	0.000	0.023	0.144	0.000	0.029	0.213
15,20,20	0.2	3,4,4	0.103	0.000	0.309	0.580	0.001	0.189	0.399
	0.3	4,6,6	0.048	0.000	0.167	0.641	0.000	0.096	0.453
	0.4	6,8,8	0.041	0.000	0.083	0.726	0.000	0.089	0.530
	0.5	7,10,10	0.035	0.000	0.045	0.642	0.000	0.075	0.472
	0.6	9,12,12	0.062	0.000	0.043	0.641	0.000	0.082	0.547
	0.7	10,14,14	0.052	0.000	0.038	0.513	0.000	0.063	0.489
	0.8	12,16,16	0.057	0.000	0.042	0.448	0.000	0.062	0.478
	0.9	13,18,18	0.051	0.000	0.037	0.293	0.000	0.049	0.363
	1.0	15,20,20	0.039	0.001	0.023	0.144	0.000	0.029	0.218
10,20,30	0.2	2,4,6	0.129	0.000	0.343	0.523	0.001	0.240	0.329
	0.3	3,6,9	0.057	0.000	0.213	0.619	0.000	0.160	0.411
	0.4	4,8,12	0.045	0.000	0.119	0.666	0.000	0.152	0.446
	0.5	5,10,15	0.040	0.000	0.073	0.630	0.000	0.100	0.438
	0.6	6,12,18	0.048	0.000	0.063	0.581	0.000	0.094	0.470
	0.7	7,14,21	0.052	0.000	0.060	0.491	0.000	0.083	0.454
	0.8	8,16,24	0.048	0.000	0.076	0.392	0.000	0.093	0.434
	0.9	9,18,27	0.044	0.000	0.055	0.249	0.000	0.069	0.321
	1.0	10,20,30	0.047	0.000	0.029	0.120	0.000	0.038	0.168
10,15,20	0.2	2,3,4	0.171	0.000	0.340	0.511	0.002	0.234	0.357
	0.3	3,4,6	0.081	0.000	0.225	0.570	0.000	0.155	0.310
	0.4	4,6,8	0.039	0.000	0.120	0.621	0.000	0.090	0.406
	0.5	5,7,10	0.053	0.000	0.083	0.597	0.000	0.099	0.421
	0.6	6,9,12	0.047	0.000	0.066	0.523	0.000	0.095	0.411
	0.7	7,10,14	0.048	0.000	0.058	0.478	0.000	0.083	0.442
	0.8	8,12,16	0.049	0.000	0.050	0.344	0.000	0.071	0.360
	0.9	9,13,18	0.037	0.000	0.033	0.236	0.000	0.048	0.290
	1.0	10,15,20	0.053	0.000	0.023	0.118	0.000	0.031	0.174

Remarks: (a) $\Pr(\pi_1$ is selected); (b) $\Pr(\pi_2$ is selected); (c) $\Pr(\pi_3$ is selected) $= \Pr[CS]$.

Table 9.16: Estimated probabilities of selection for n_1, n_2, and n_3, with location-shift $\theta_1 = 0.0$, $\theta_2 = 1.0$

				Normal			Exp(1)		
n_1, n_2, n_3	q	r_1, r_2, r_3	α	(a)	(b)	(c)	(a)	(b)	(c)
10,10,15	0.4	4, 4, 6	0.048	0.005	0.008	0.258	0.005	0.006	0.491
	0.5	5, 5, 7	0.029	0.004	0.005	0.196	0.001	0.003	0.496
	0.6	6, 6, 9	0.038	0.003	0.004	0.272	0.000	0.001	0.487
	0.7	7, 7,10	0.044	0.002	0.006	0.320	0.000	0.000	0.466
	0.8	8, 8,12	0.046	0.002	0.003	0.310	0.000	0.000	0.384
	0.9	9, 9,13	0.045	0.002	0.003	0.300	0.000	0.001	0.297
	1.0	10,10,15	0.061	0.002	0.004	0.303	0.001	0.006	0.190
15,15,20	0.3	4, 4, 6	0.057	0.010	0.006	0.299	0.008	0.005	0.510
	0.4	6, 6, 8	0.046	0.004	0.006	0.344	0.001	0.001	0.499
	0.5	7, 7,10	0.056	0.003	0.003	0.398	0.000	0.000	0.518
	0.6	9, 9,12	0.049	0.002	0.002	0.370	0.000	0.000	0.495
	0.7	10,10,14	0.054	0.002	0.002	0.406	0.000	0.000	0.500
	0.8	12,12,16	0.059	0.001	0.002	0.402	0.000	0.001	0.422
	0.9	13,13,18	0.046	0.001	0.002	0.387	0.000	0.002	0.345
	1.0	15,15,20	0.055	0.001	0.001	0.329	0.002	0.003	0.176
15,20,20	0.2	3, 4, 4	0.103	0.021	0.027	0.311	0.018	0.022	0.461
	0.3	4, 6, 6	0.048	0.010	0.007	0.305	0.008	0.005	0.511
	0.4	6, 8, 8	0.041	0.003	0.005	0.339	0.001	0.001	0.499
	0.5	7,10,10	0.035	0.003	0.003	0.362	0.000	0.000	0.518
	0.6	9,12,12	0.062	0.002	0.003	0.423	0.000	0.000	0.497
	0.7	10,14,14	0.052	0.002	0.002	0.436	0.000	0.000	0.504
	0.8	12,16,16	0.057	0.001	0.002	0.423	0.000	0.000	0.432
	0.9	13,18,18	0.051	0.001	0.001	0.416	0.000	0.002	0.367
	1.0	15,20,20	0.039	0.001	0.001	0.314	0.002	0.003	0.163
10,20,30	0.2	2, 4, 6	0.129	0.018	0.044	0.320	0.017	0.033	0.461
	0.3	3, 6, 9	0.057	0.007	0.017	0.294	0.004	0.011	0.495
	0.4	4, 8,12	0.045	0.003	0.014	0.343	0.001	0.003	0.506
	0.5	5,10,15	0.040	0.002	0.006	0.352	0.000	0.001	0.509
	0.6	6,12,18	0.048	0.001	0.005	0.408	0.000	0.000	0.509
	0.7	7,14,21	0.052	0.001	0.004	0.425	0.000	0.000	0.494
	0.8	8,16,24	0.048	0.000	0.005	0.400	0.000	0.003	0.437
	0.9	9,18,27	0.044	0.000	0.004	0.402	0.000	0.005	0.343
	1.0	10,20,30	0.047	0.000	0.003	0.369	0.000	0.006	0.219
10,15,20	0.2	2, 3, 4	0.171	0.031	0.048	0.329	0.028	0.039	0.438
	0.3	3, 4, 6	0.081	0.010	0.020	0.273	0.008	0.015	0.482
	0.4	4, 6, 8	0.039	0.007	0.008	0.260	0.002	0.005	0.498
	0.5	5, 7,10	0.053	0.004	0.008	0.314	0.001	0.002	0.501
	0.6	6, 9,12	0.047	0.002	0.007	0.330	0.000	0.000	0.496
	0.7	7,10,14	0.048	0.002	0.005	0.394	0.000	0.000	0.478
	0.8	8,12,16	0.049	0.001	0.004	0.360	0.000	0.001	0.409
	0.9	9,13,18	0.037	0.001	0.003	0.355	0.000	0.002	0.311
	1.0	10,15,20	0.053	0.001	0.002	0.304	0.002	0.004	0.184

Remarks: (a) $\Pr(\pi_1$ is selected); (b) $\Pr(\pi_2$ is selected); (c) $\Pr(\pi_3$ is selected) $= \Pr[CS]$.

Table 9.16: (Continued)

n_1, n_2, n_3	q	r_1, r_2, r_3	α	Gamma(3) (a)	(b)	(c)	LN(0.1) (a)	(b)	(c)
10,10,15	0.4	4, 4, 6	0.048	0.004	0.007	0.454	0.005	0.009	0.291
	0.5	5, 5, 7	0.029	0.001	0.004	0.401	0.004	0.005	0.225
	0.6	6, 6, 9	0.038	0.000	0.001	0.395	0.003	0.003	0.293
	0.7	7, 7,10	0.044	0.000	0.002	0.393	0.002	0.005	0.333
	0.8	8, 8,12	0.046	0.001	0.001	0.330	0.002	0.002	0.311
	0.9	9, 9,13	0.045	0.001	0.003	0.275	0.002	0.003	0.291
	1.0	10,10,15	0.061	0.001	0.006	0.217	0.002	0.005	0.271
15,15,20	0.3	4, 4, 6	0.057	0.008	0.005	0.499	0.008	0.006	0.347
	0.4	6, 6, 8	0.046	0.001	0.002	0.485	0.003	0.005	0.379
	0.5	7, 7,10	0.056	0.000	0.001	0.494	0.002	0.003	0.424
	0.6	9, 9,12	0.049	0.000	0.000	0.451	0.002	0.002	0.387
	0.7	10,10,14	0.054	0.000	0.001	0.452	0.001	0.002	0.415
	0.8	12,12,16	0.059	0.001	0.002	0.396	0.001	0.002	0.399
	0.9	13,13,18	0.046	0.001	0.002	0.339	0.001	0.002	0.371
	1.0	15,15,20	0.055	0.001	0.002	0.211	0.001	0.002	0.288
15,20,20	0.2	3, 4, 4	0.103	0.020	0.023	0.458	0.020	0.026	0.355
	0.3	4, 6, 6	0.048	0.008	0.006	0.509	0.009	0.006	0.353
	0.4	6, 8, 8	0.041	0.001	0.002	0.485	0.002	0.004	0.376
	0.5	7,10,10	0.035	0.001	0.000	0.494	0.002	0.002	0.394
	0.6	9,12,12	0.062	0.000	0.001	0.474	0.001	0.002	0.435
	0.7	10,14,14	0.052	0.000	0.000	0.472	0.001	0.001	0.442
	0.8	12,16,16	0.057	0.000	0.001	0.413	0.001	0.002	0.417
	0.9	13,18,18	0.051	0.001	0.001	0.368	0.001	0.002	0.399
	1.0	15,20,20	0.039	0.002	0.003	0.198	0.001	0.002	0.274
10,20,30	0.2	2, 4, 6	0.129	0.017	0.036	0.456	0.018	0.041	0.363
	0.3	3, 6, 9	0.057	0.003	0.012	0.487	0.006	0.016	0.342
	0.4	4, 8,12	0.045	0.001	0.004	0.490	0.002	0.012	0.384
	0.5	5,10,15	0.040	0.000	0.002	0.478	0.001	0.005	0.385
	0.6	6,12,18	0.048	0.000	0.002	0.475	0.000	0.004	0.424
	0.7	7,14,21	0.052	0.000	0.002	0.456	0.000	0.004	0.430
	0.8	8,16,24	0.048	0.000	0.004	0.405	0.000	0.005	0.397
	0.9	9,18,27	0.044	0.000	0.005	0.344	0.000	0.004	0.384
	1.0	10,20,30	0.047	0.000	0.006	0.252	0.000	0.005	0.328
10,15,20	0.2	2, 3, 4	0.171	0.030	0.043	0.431	0.030	0.046	0.363
	0.3	3, 4, 6	0.081	0.008	0.016	0.471	0.009	0.019	0.317
	0.4	4, 6, 8	0.039	0.002	0.006	0.464	0.005	0.007	0.300
	0.5	5, 7,10	0.053	0.000	0.002	0.461	0.003	0.007	0.346
	0.6	6, 9,12	0.047	0.000	0.002	0.435	0.001	0.006	0.353
	0.7	7,10,14	0.048	0.000	0.002	0.432	0.001	0.005	0.401
	0.8	8,12,16	0.049	0.000	0.003	0.365	0.001	0.004	0.358
	0.9	9,13,18	0.037	0.000	0.003	0.305	0.001	0.002	0.339
	1.0	10,15,20	0.053	0.001	0.004	0.205	0.001	0.003	0.271

Remarks: (a) $\Pr(\pi_1$ is selected); (b) $\Pr(\pi_2$ is selected); (c) $\Pr(\pi_3$ is selected) $= \Pr[CS]$.

Table 9.17: Estimated probabilities of selection for n_1, n_2, and n_3, with location-shift $\theta_1 = 1.0$, $\theta_2 = 2.0$

n_1, n_2, n_3	q	r_1, r_2, r_3	α	Normal (a)	(b)	(c)	Exp(1) (a)	(b)	(c)
10,10,15	0.4	4, 4, 6	0.048	0.000	0.204	0.658	0.000	0.496	0.502
	0.5	5, 5, 7	0.029	0.000	0.160	0.674	0.000	0.482	0.514
	0.6	6, 6, 9	0.038	0.000	0.116	0.778	0.000	0.386	0.583
	0.7	7, 7,10	0.044	0.000	0.100	0.839	0.000	0.301	0.653
	0.8	8, 8,12	0.046	0.000	0.061	0.830	0.000	0.098	0.732
	0.9	9, 9,13	0.045	0.000	0.049	0.851	0.000	0.039	0.724
	1.0	10,10,15	0.061	0.000	0.030	0.668	0.000	0.015	0.356
15,15,20	0.3	4, 4, 6	0.057	0.000	0.220	0.668	0.000	0.501	0.499
	0.4	6, 6, 8	0.046	0.000	0.163	0.801	0.000	0.498	0.502
	0.5	7, 7,10	0.056	0.000	0.110	0.850	0.000	0.471	0.525
	0.6	9, 9,12	0.049	0.000	0.073	0.901	0.000	0.384	0.604
	0.7	10,10,14	0.054	0.000	0.047	0.911	0.000	0.212	0.730
	0.8	12,12,16	0.059	0.000	0.030	0.932	0.000	0.064	0.827
	0.9	13,13,18	0.046	0.000	0.019	0.872	0.000	0.008	0.666
	1.0	15,15,20	0.055	0.000	0.012	0.731	0.000	0.009	0.344
15,20,20	0.2	3, 4, 4	0.103	0.000	0.318	0.609	0.000	0.500	0.500
	0.3	4, 6, 6	0.048	0.000	0.224	0.714	0.000	0.499	0.500
	0.4	6, 8, 8	0.041	0.000	0.150	0.824	0.000	0.500	0.500
	0.5	7,10,10	0.035	0.000	0.093	0.869	0.000	0.476	0.519
	0.6	9,12,12	0.062	0.000	0.057	0.920	0.000	0.385	0.601
	0.7	10,14,14	0.052	0.000	0.036	0.924	0.000	0.212	0.732
	0.8	12,16,16	0.057	0.000	0.023	0.942	0.000	0.064	0.826
	0.9	13,18,18	0.051	0.000	0.013	0.881	0.000	0.008	0.669
	1.0	15,20,20	0.039	0.000	0.008	0.743	0.000	0.009	0.353
10,20,30	0.2	2, 4, 6	0.129	0.000	0.358	0.533	0.000	0.500	0.500
	0.3	3, 6, 9	0.057	0.000	0.284	0.649	0.000	0.500	0.500
	0.4	4, 8,12	0.045	0.000	0.213	0.738	0.000	0.499	0.501
	0.5	5,10,15	0.040	0.000	0.150	0.806	0.000	0.488	0.511
	0.6	6,12,18	0.048	0.000	0.100	0.856	0.000	0.405	0.586
	0.7	7,14,21	0.052	0.000	0.066	0.886	0.000	0.208	0.749
	0.8	8,16,24	0.048	0.000	0.041	0.892	0.000	0.045	0.821
	0.9	9,18,27	0.044	0.000	0.026	0.853	0.000	0.008	0.663
	1.0	10,20,30	0.047	0.000	0.017	0.577	0.000	0.022	0.272
10,15,20	0.2	2, 3, 4	0.171	0.000	0.357	0.529	0.000	0.500	0.500
	0.3	3, 4, 6	0.081	0.000	0.284	0.602	0.000	0.500	0.500
	0.4	4, 6, 8	0.039	0.000	0.207	0.724	0.000	0.497	0.502
	0.5	5, 7,10	0.053	0.000	0.150	0.796	0.000	0.478	0.517
	0.6	6, 9,12	0.047	0.000	0.105	0.841	0.000	0.396	0.584
	0.7	7,10,14	0.048	0.000	0.070	0.872	0.000	0.231	0.707
	0.8	8,12,16	0.049	0.000	0.047	0.873	0.000	0.073	0.773
	0.9	9,13,18	0.037	0.000	0.031	0.839	0.000	0.012	0.646
	1.0	10,15,20	0.053	0.000	0.019	0.635	0.000	0.012	0.305

Remarks: (a) $\Pr(\pi_1$ is selected); (b) $\Pr(\pi_2$ is selected); (c) $\Pr(\pi_3$ is selected) $= \Pr[CS]$.

Table 9.17: (Continued)

n_1, n_2, n_3	q	r_1, r_2, r_3	α	Gamma(3)			LN(0.1)		
				(a)	(b)	(c)	(a)	(b)	(c)
10,10,15	0.4	4, 4, 6	0.048	0.000	0.397	0.586	0.000	0.229	0.683
	0.5	5, 5, 7	0.029	0.000	0.323	0.659	0.000	0.179	0.712
	0.6	6, 6, 9	0.038	0.000	0.168	0.777	0.000	0.113	0.799
	0.7	7, 7,10	0.044	0.000	0.106	0.836	0.000	0.089	0.850
	0.8	8, 8,12	0.046	0.000	0.028	0.810	0.000	0.050	0.823
	0.9	9, 9,13	0.045	0.000	0.016	0.775	0.000	0.038	0.830
	1.0	10,10,15	0.061	0.000	0.018	0.433	0.000	0.026	0.586
15,15,20	0.3	4, 4, 6	0.057	0.000	0.451	0.546	0.000	0.255	0.685
	0.4	6, 6, 8	0.046	0.000	0.363	0.633	0.000	0.175	0.803
	0.5	7, 7,10	0.056	0.000	0.229	0.757	0.000	0.108	0.859
	0.6	9, 9,12	0.049	0.000	0.112	0.868	0.000	0.064	0.911
	0.7	10,10,14	0.054	0.000	0.033	0.904	0.000	0.037	0.914
	0.8	12,12,16	0.059	0.000	0.011	0.900	0.000	0.021	0.926
	0.9	13,13,18	0.046	0.000	0.006	0.732	0.000	0.014	0.834
	1.0	15,15,20	0.055	0.000	0.010	0.445	0.000	0.011	0.631
15,20,20	0.2	3, 4, 4	0.103	0.000	0.490	0.509	0.000	0.356	0.607
	0.3	4, 6, 6	0.048	0.000	0.459	0.538	0.000	0.257	0.703
	0.4	6, 8, 8	0.041	0.000	0.361	0.636	0.000	0.166	0.815
	0.5	7,10,10	0.035	0.000	0.227	0.754	0.000	0.096	0.873
	0.6	9,12,12	0.062	0.000	0.105	0.876	0.000	0.052	0.926
	0.7	10,14,14	0.052	0.000	0.034	0.905	0.000	0.028	0.926
	0.8	12,16,16	0.057	0.000	0.007	0.905	0.000	0.016	0.931
	0.9	13,18,18	0.051	0.000	0.004	0.737	0.000	0.010	0.838
	1.0	15,20,20	0.039	0.000	0.009	0.449	0.000	0.008	0.646
10,20,30	0.2	2, 4, 6	0.129	0.000	0.493	0.507	0.000	0.389	0.549
	0.3	3, 6, 9	0.057	0.000	0.471	0.527	0.000	0.308	0.650
	0.4	4, 8,12	0.045	0.000	0.400	0.595	0.000	0.226	0.740
	0.5	5,10,15	0.040	0.000	0.263	0.726	0.000	0.150	0.815
	0.6	6,12,18	0.048	0.000	0.139	0.837	0.000	0.091	0.870
	0.7	7,14,21	0.052	0.000	0.046	0.899	0.000	0.052	0.896
	0.8	8,16,24	0.048	0.000	0.013	0.864	0.000	0.031	0.888
	0.9	9,18,27	0.044	0.000	0.010	0.730	0.000	0.020	0.821
	1.0	10,20,30	0.047	0.000	0.022	0.337	0.000	0.018	0.489
10,15,20	0.2	2, 3, 4	0.171	0.000	0.492	0.506	0.000	0.387	0.543
	0.3	3, 4, 6	0.081	0.000	0.459	0.537	0.000	0.312	0.625
	0.4	4, 6, 8	0.039	0.000	0.396	0.597	0.000	0.226	0.726
	0.5	5, 7,10	0.053	0.000	0.288	0.695	0.000	0.155	0.799
	0.6	6, 9,12	0.047	0.000	0.153	0.812	0.000	0.099	0.850
	0.7	7,10,14	0.048	0.000	0.062	0.867	0.000	0.061	0.876
	0.8	8,12,16	0.049	0.000	0.020	0.839	0.000	0.038	0.867
	0.9	9,13,18	0.037	0.000	0.009	0.714	0.000	0.025	0.802
	1.0	10,15,20	0.053	0.000	0.015	0.381	0.000	0.017	0.547

Remarks: (a) $\Pr(\pi_1$ is selected); (b) $\Pr(\pi_2$ is selected); (c) $\Pr(\pi_3$ is selected) $= \Pr[CS]$.

Table 9.18: Estimated probabilities of selection for n_1, n_2, and n_3, with location-shift $\theta_1 = 0.5$, $\theta_2 = 1.5$

n_1, n_2, n_3	q	r_1, r_2, r_3	α	Normal			Exp(1)		
				(a)	(b)	(c)	(a)	(b)	(c)
10,10,15	0.4	4, 4, 6	0.048	0.000	0.069	0.454	0.000	0.340	0.561
	0.5	5, 5, 7	0.029	0.000	0.047	0.441	0.000	0.251	0.632
	0.6	6, 6, 9	0.038	0.000	0.035	0.513	0.000	0.084	0.669
	0.7	7, 7,10	0.044	0.000	0.036	0.619	0.000	0.039	0.706
	0.8	8, 8,12	0.046	0.000	0.021	0.541	0.000	0.005	0.558
	0.9	9, 9,13	0.045	0.000	0.019	0.578	0.000	0.002	0.504
	1.0	10,10,15	0.061	0.000	0.015	0.428	0.000	0.011	0.248
15,15,20	0.3	4, 4, 6	0.057	0.000	0.068	0.465	0.000	0.448	0.521
	0.4	6, 6, 8	0.046	0.000	0.053	0.638	0.000	0.329	0.613
	0.5	7, 7,10	0.056	0.000	0.033	0.632	0.000	0.158	0.691
	0.6	9, 9,12	0.049	0.000	0.024	0.699	0.000	0.049	0.764
	0.7	10,10,14	0.054	0.000	0.016	0.660	0.000	0.008	0.667
	0.8	12,12,16	0.059	0.000	0.012	0.703	0.000	0.001	0.633
	0.9	13,13,18	0.046	0.000	0.009	0.569	0.000	0.003	0.442
	1.0	15,15,20	0.055	0.000	0.006	0.467	0.000	0.007	0.232
15,20,20	0.2	3, 4, 4	0.103	0.001	0.147	0.486	0.000	0.494	0.501
	0.3	4, 6, 6	0.048	0.000	0.069	0.553	0.000	0.450	0.522
	0.4	6, 8, 8	0.041	0.000	0.044	0.685	0.000	0.327	0.616
	0.5	7,10,10	0.035	0.000	0.028	0.649	0.000	0.160	0.695
	0.6	9,12,12	0.062	0.000	0.021	0.734	0.000	0.049	0.767
	0.7	10,14,14	0.052	0.000	0.013	0.668	0.000	0.008	0.673
	0.8	12,16,16	0.057	0.000	0.009	0.709	0.000	0.001	0.636
	0.9	13,18,18	0.051	0.000	0.006	0.581	0.000	0.002	0.451
	1.0	15,20,20	0.039	0.000	0.005	0.476	0.000	0.006	0.231
10,20,30	0.2	2, 4, 6	0.129	0.001	0.188	0.406	0.000	0.497	0.502
	0.3	3, 6, 9	0.057	0.000	0.118	0.523	0.000	0.471	0.515
	0.4	4, 8,12	0.045	0.000	0.082	0.593	0.000	0.354	0.592
	0.5	5,10,15	0.040	0.000	0.050	0.626	0.000	0.165	0.708
	0.6	6,12,18	0.048	0.000	0.035	0.654	0.000	0.044	0.736
	0.7	7,14,21	0.052	0.000	0.024	0.651	0.000	0.006	0.676
	0.8	8,16,24	0.048	0.000	0.018	0.630	0.000	0.003	0.579
	0.9	9,18,27	0.044	0.000	0.013	0.561	0.000	0.008	0.431
	1.0	10,20,30	0.047	0.000	0.011	0.390	0.000	0.014	0.229
10,15,20	0.2	2, 3, 4	0.171	0.003	0.189	0.415	0.000	0.493	0.500
	0.3	3, 4, 6	0.081	0.000	0.120	0.435	0.000	0.458	0.513
	0.4	4, 6, 8	0.039	0.000	0.068	0.555	0.000	0.347	0.576
	0.5	5, 7,10	0.053	0.000	0.052	0.605	0.000	0.187	0.663
	0.6	6, 9,12	0.047	0.000	0.038	0.621	0.000	0.064	0.705
	0.7	7,10,14	0.048	0.000	0.028	0.639	0.000	0.014	0.667
	0.8	8,12,16	0.049	0.000	0.020	0.602	0.000	0.003	0.572
	0.9	9,13,18	0.037	0.000	0.013	0.551	0.000	0.003	0.424
	1.0	10,15,20	0.053	0.000	0.009	0.398	0.000	0.008	0.219

Remarks: (a) $\Pr(\pi_1$ is selected); (b) $\Pr(\pi_2$ is selected); (c) $\Pr(\pi_3$ is selected) $= \Pr[CS]$.

Table 9.18: (Continued)

n_1, n_2, n_3	q	r_1, r_2, r_3	α	Gamma(3)			LN(0.1)		
				(a)	(b)	(c)	(a)	(b)	(c)
10,10,15	0.4	4, 4, 6	0.048	0.000	0.118	0.657	0.000	0.074	0.514
	0.5	5, 5, 7	0.029	0.000	0.074	0.713	0.000	0.049	0.507
	0.6	6, 6, 9	0.038	0.000	0.021	0.663	0.000	0.031	0.549
	0.7	7, 7,10	0.044	0.000	0.011	0.688	0.000	0.030	0.636
	0.8	8, 8,12	0.046	0.000	0.006	0.528	0.000	0.017	0.532
	0.9	9, 9,13	0.045	0.000	0.008	0.501	0.000	0.015	0.553
	1.0	10,10,15	0.061	0.000	0.014	0.288	0.000	0.015	0.377
15,15,20	0.3	4, 4, 6	0.057	0.000	0.166	0.667	0.000	0.075	0.536
	0.4	6, 6, 8	0.046	0.000	0.069	0.784	0.000	0.049	0.690
	0.5	7, 7,10	0.056	0.000	0.023	0.738	0.000	0.027	0.664
	0.6	9, 9,12	0.049	0.000	0.007	0.758	0.000	0.017	0.714
	0.7	10,10,14	0.054	0.000	0.003	0.648	0.000	0.012	0.653
	0.8	12,12,16	0.059	0.000	0.002	0.633	0.000	0.010	0.681
	0.9	13,13,18	0.046	0.000	0.005	0.463	0.000	0.008	0.533
	1.0	15,15,20	0.055	0.000	0.007	0.289	0.000	0.006	0.402
15,20,20	0.2	3, 4, 4	0.103	0.000	0.309	0.595	0.000	0.163	0.549
	0.3	4, 6, 6	0.048	0.000	0.166	0.658	0.000	0.077	0.600
	0.4	6, 8, 8	0.041	0.000	0.074	0.780	0.000	0.042	0.723
	0.5	7,10,10	0.035	0.000	0.023	0.741	0.000	0.022	0.675
	0.6	9,12,12	0.062	0.000	0.007	0.772	0.000	0.015	0.743
	0.7	10,14,14	0.052	0.000	0.003	0.666	0.000	0.010	0.664
	0.8	12,16,16	0.057	0.000	0.003	0.647	0.000	0.007	0.685
	0.9	13,18,18	0.051	0.000	0.005	0.481	0.000	0.006	0.546
	1.0	15,20,20	0.039	0.000	0.006	0.293	0.000	0.005	0.410
10,20,30	0.2	2, 4, 6	0.129	0.000	0.340	0.537	0.001	0.203	0.455
	0.3	3, 6, 9	0.057	0.000	0.206	0.631	0.000	0.125	0.568
	0.4	4, 8,12	0.045	0.000	0.100	0.704	0.000	0.077	0.635
	0.5	5,10,15	0.040	0.000	0.038	0.737	0.000	0.044	0.657
	0.6	6,12,18	0.048	0.000	0.012	0.714	0.000	0.028	0.667
	0.7	7,14,21	0.052	0.000	0.005	0.661	0.000	0.019	0.651
	0.8	8,16,24	0.048	0.000	0.007	0.577	0.000	0.015	0.611
	0.9	9,18,27	0.044	0.000	0.011	0.451	0.000	0.012	0.527
	1.0	10,20,30	0.047	0.000	0.014	0.266	0.000	0.013	0.348
10,15,20	0.2	2, 3, 4	0.171	0.000	0.339	0.519	0.002	0.206	0.454
	0.3	3, 4, 6	0.081	0.000	0.208	0.619	0.000	0.128	0.501
	0.4	4, 6, 8	0.039	0.000	0.105	0.690	0.000	0.071	0.598
	0.5	5, 7,10	0.053	0.000	0.048	0.709	0.000	0.046	0.634
	0.6	6, 9,12	0.047	0.000	0.017	0.694	0.000	0.030	0.638
	0.7	7,10,14	0.048	0.000	0.008	0.650	0.000	0.021	0.639
	0.8	8,12,16	0.049	0.000	0.006	0.561	0.000	0.016	0.586
	0.9	9,13,18	0.037	0.000	0.008	0.446	0.000	0.011	0.511
	1.0	10,15,20	0.053	0.000	0.009	0.258	0.000	0.009	0.346

Remarks: (a) $\Pr(\pi_1$ is selected); (b) $\Pr(\pi_2$ is selected); (c) $\Pr(\pi_3$ is selected) $= \Pr[CS]$.

9.5   EVALUATION AND COMPARATIVE REMARKS

From the simulation results, we observe that the probability of CS of the proposed procedure increases with increasing sample sizes as well as with increasing location-shift. These simulation studies show that the procedure can pick up the best population effectively in most cases early in the life-testing experiment.

One may be interested in maximizing $\Pr[CS]$ with respect to q; that is, we decide to determine the best choice of q when we design the experiment. When we compare the estimated $\Pr[CS]$ for different values of q, we find that the optimal choice of q is in the range 60%–80% if the underlying distributions are close to symmetric, such as the normal distribution and lognormal distributions with shape parameter 0.1. However, under some right-skewed distributions such as the exponential distribution and gamma distribution with shape parameter 3.0, we find that the optimal choice of q is in the range 40%–60%. Therefore, as a compromise between these two types of distributions, we recommend the use of this procedure with $q \approx 50\%$–60%.

While this procedure will work very well for the balanced sample situation or near balanced situations, the selection procedure may not be effective when the sample sizes differ too much.

9.6   ILLUSTRATIVE EXAMPLES

Example 9.1. Let us consider X_1-, X_2-, and X_3-samples to be the data from appliance cord life in Tests 1, 2, and 3 (Nelson 1982, p. 510), respectively. These three tests were done using two types of cord. For the purpose of our example, we will consider these three tests to have been done using three different types of cord (A, B, and C, respectively). We are interested in testing the homogeneity of the life distributions of the three types of cord and select the best type if the null hypothesis is rejected.

Suppose we want to test the equality of the lifetime distributions of these 3 cords and select the best type of cord among these three (best meaning the cord producing the longest lifetimes) if their lifetime distributions are not equal. For these data, we have $k = 3$, $n_1 = n_2 = n_3 = 12$. Had we fixed $q = 0.5$, that is, $r = 6$, the experiment would have stopped as soon as the 6th failure from any of the three samples had been observed. Thus, the

Table 9.19: Appliance cord life data from Nelson (1982, p. 510)

Test 1	96.9	100.3	100.8	103.3	103.4	*	*	*	*	*	*	*
Test 2	57.5	77.8	88.0	98.4	102.1	105.3	*	*	*	*	*	*
Test 3	72.4	78.6	81.2	94.0	*	*	*	*	*	*	*	*

Table 9.20: One-way ANOVA table based on log-lifetimes of appliance cord in Tests 1, 2, and 3

Source	Degrees of freedom	Sum of squares	MS	F-test	p-value
Test	2	0.1174	0.0587	2.35	0.137
Error	12	0.2995	0.0250		
Total	14	0.4170			

experiment would have terminated at 105.3 hours (terminated with Test 2). The data thus obtained are presented in Table 9.19.

Parametric Procedure. In analyzing these data, a parametric test for equality of the means based on analysis of variance (ANOVA) can be carried out by assuming independent normal distributions (with homogeneous variance) for the log-lifetime of appliance cord in Tests 1, 2, and 3 (i.e., assuming the lifetime distribution of appliance cord in Tests 1, 2, and 3 to be lognormal). Let the mean log-lifetime of appliance cord in Tests 1, 2, and 3 be θ_1, θ_2, and θ_3, respectively. We are then interested in testing the hypotheses

$$H_0 : \theta_1 = \theta_2 = \theta_3 \quad \text{vs.} \quad H_1 : \text{At least one } \theta_i \neq \theta_j. \tag{9.30}$$

If H_0 in (9.30) is rejected, different pairwise comparison procedures (for example, Tukey's pairwise comparison) can be used to identify which population(s) is (are) different from the others.

Based on the data presented in Table 9.19, the one-way ANOVA table based on log-lifetimes of appliance cord in Tests 1, 2, and 3 is given in Table 9.20. Since the p-value of the F-test is 0.137, we do not reject H_0 in (9.30) and conclude that no one type of cord is superior to the others.

Nonparametric Procedure. If the precedence-type procedure proposed in this chapter is used, we then have $P_{0.5,1} = 5$ and $P_{0.5,3} = 4$, in which case the test statistic is

$$P^*_{(6)} = \min\{5, 4\} = 4.$$

Table 9.21: Survival times of 40 patients receiving two different treatments

X-sample (Treatment 1)	17	28	49	98	119	133	145	146	158	160
	174	211	220	231	252	256	267	322	323	327
Y-sample (Treatment 2)	26	34	47	59	101	112	114	136	154	154
	161	186	197	226	226	243	253	269	308	465

The critical value for $n = 12, r = 6$ is $s = 0$ (with the exact level of significance being 0.03642), and so we will not reject the null hypothesis. This means that no one type of cord is superior to the others. In fact, the p-value of this test is 0.91078. Note that this finding agrees with the parametric result obtained above.

Example 9.2. The data from Lee and Wang (2002, p. 253) of the survival times of 20 patients receiving two different treatments are presented in Table 9.21.

Suppose we want to choose the better one among Treatments 1 and 2 and we set $r = 7$, then the experiment would terminate at 114 (with the Y-sample).

Parametric Procedure. Lee and Wang (2002) assumed that the two populations follow gamma distributions with a common shape parameter 2. Testing the hypothesis that the two populations are equal is equivalent to testing the equality of the two scale parameters, say λ_1 and λ_2, i.e.,

$$H_0 : \lambda_1 = \lambda_2 \quad \text{vs.} \quad H_1 : \lambda_1 \neq \lambda_2. \tag{9.31}$$

Note that the hypothesis in (9.31) is equivalent to testing the hypothesis in (9.1). Based on the complete data presented in Table 9.21, we compute the average survival times $\bar{X} = 181.8$ and $\bar{Y} = 173.55$. Under the null hypothesis in (9.31), $\bar{X}/\bar{Y}$ has a F-distribution with $(80, 80)$ degrees of freedom. We have $\bar{X}/\bar{Y} = 1.048$, which yields the p-value as 0.4172, and so we do not reject H_0.

Nonparametric Procedure. In this case, we have $n_1 = n_2 = 20$, $r = 7$, and $m_1 = m_2 = m_4 = m_5 = 1$ and $m_3 = m_6 = m_7 = 0$. The value of the precedence statistic is, therefore, $P_{(7)} = 4$ and the corresponding p-value is 0.480. Once again, from the large p-value, we conclude that the data do not provide enough evidence to reject H_0 in (9.1). It is important to note that the decision is reached here without making any assumption on the underlying lifetime distributions and having observed less than 35% of the survival times. This conclusion also agrees with the parametric result obtained above by assuming gamma distributions (with the same shape parameter) for the two complete samples.

Chapter 10

Selecting the Best Population Using a Test for Equality Based on Minimal Wilcoxon Rank-sum Precedence Statistic

10.1 INTRODUCTION

In Chapter 9, we discussed a nonparametric procedure based on the precedence statistic to select the best population based on a test for equality. We also discussed the k-sample situation and showed that the procedure works well for different underlying distributions via Monte Carlo simulations. Since the minimal Wilcoxon rank-sum precedence statistic is more powerful than the precedence statistic (as shown in Chapter 6), Ng, Balakrishnan, and Panchapakesan (2006) used the minimal Wilcoxon rank-sum precedence statistic instead of the precedence statistic to select the best population based on a test for equality. They also compared the probability of correct selection of the best population for the minimal Wilcoxon rank-sum precedence statistic with that of the precedence statistic.

In Section 10.2, using the same notation and formulation in Chapter 9, we first describe the nonparametric procedure based on the Wilcoxon rank-sum precedence statistic and present the exact null distribution of the statistic. Both equal and unequal sample size situations are considered and some critical values close to 5% level of significance are tabulated. Exact probability of correct selection for this procedure under the Lehmann alternative is also discussed. In Section 10.3, we discuss the extension of this procedure to the k-sample situation, as given by Ng, Balakrishnan, and Panchapakesan (2006). We discuss the test statistic and present its exact null distribution. We examine the properties of this selection procedure and compare it with the procedure based on the precedence statistic under a location-shift between the populations by means of Monte Carlo simulations for $k = 2$ and 3. Next, we discuss the simulation results and the choice of the value of q in Section 10.4. Finally, we present in Section 10.5 two examples to illustrate the minimal Wilcoxon rank-sum precedence selection procedure.

10.2   TWO-SAMPLE SELECTION PROBLEM

Following the same notation as in Chapter 9, we are interested in testing the hypothesis in (9.1). We will first describe the test procedure based on the Wilcoxon rank-sum precedence statistic for the equal sample size situation (i.e., $n_1 = n_2 = n$) and then present its generalization to unequal sample sizes.

10.2.1   Equal Sample Size Situation

In this subsection, we assume that $n_1 = n_2 = n$. Here, for notational convenience, let us denote $\mathbf{M} = (M_1, M_2, \cdots, M_r)$, $S_r(\mathbf{M}) = \sum_{i=1}^{r} M_i$ and $S_r^*(\mathbf{M}) = \sum_{i=1}^{r} i M_i$, and their realizations by $\mathbf{m} = (m_1, m_2, \cdots, m_r)$, s_r and s_r^*, respectively. The *minimal Wilcoxon rank-sum statistic* is defined by assuming all the remaining $(n_1 - s_r)$ X-failures (Y-failures) occur between the rth and $(r + 1)$th Y-failures (X-failures). The *minimal Wilcoxon rank-sum statistic* is given by (see Chapter 6)

$$W_r(\mathbf{M}) = \frac{1}{2} n_1(n_1 + 2r + 1) - (r + 1)S_r(\mathbf{M}) + S_r^*(\mathbf{M}).$$

As in Eq. (9.2), we denote

$$T_r(n_1, n_2) = T_r = \min(X_{r:n_1}, Y_{r:n_2}).$$

We can then write

$$W_r(\mathbf{M}) = \begin{cases} \frac{1}{2}n_1(n_1 + 2r + 1) - (r+1)S_r(\mathbf{M}) + S_r^*(\mathbf{M}) & \text{if } T = X_{r:n_1} \\ \frac{1}{2}n_2(n_2 + 2r + 1) - (r+1)S_r(\mathbf{M}) + S_r^*(\mathbf{M}) & \text{if } T = Y_{r:n_2}. \end{cases}$$

$$(10.1)$$

The test procedure proposed by Ng, Balakrishnan, and Panchapakesan (2006) is based on the minimal Wilcoxon rank-sum statistic $W_r(\mathbf{M})$ in (10.1). Since we are interested in selecting the best population among π_1 and π_2 (best in the sense of longer life), it is clear that larger values of $W_r(\mathbf{M})$ will lead to the rejection of the null hypothesis H_0 in (9.1). Then, the critical region for the test based on $W_r(\mathbf{M})$ will be $\{s, s+1, \cdots, \frac{1}{2}n_1(n_1 + 2r + 1)\}$. For this reason, a decision rule can be given as follows:

Reject H_0 and choose H_{A1} (select π_1 as the best)

if and only if $T = Y_{r:n_2}$ and $W_r(\mathbf{M}) \geq s$;

Reject H_0 and choose H_{A2} (select π_2 as the best)

if and only if $T = X_{r:n_1}$ and $W_r(\mathbf{M}) \geq s$.   (10.2)

In Chapter 9, we described a nonparametric procedure based on the precedence statistic $P_{(r)} = S_r(\mathbf{M})$. The critical region for the test based on $P_{(r)}$ is $\{0, 1, \cdots, s^*\}$ $(s^* < r)$ and the decision rule was described in (9.4).

Under the null hypothesis that the two populations are the same, i.e., $H_0 : F_X = F_Y$, the probabilities of the life-testing experiment terminating with the X- and Y-samples are as given in (9.5) and (9.6), respectively, and the probabilities under the special case of equal sample sizes are given in (9.7).

Under the null hypothesis, the distribution of $\mathbf{M} = (M_1, M_2, \cdots, M_r)$ is given by

$$\Pr(\mathbf{M} = \mathbf{m}, T = X_{r:n_1} \mid H_0 : F_X = F_Y)$$
$$= \Pr(\mathbf{M} = \mathbf{m}, T = Y_{r:n_2} \mid H_0 : F_X = F_Y)$$
$$= \frac{\dbinom{N - s_r(\mathbf{m}) - r}{n_2 - r}}{\dbinom{N}{n_1}},$$

$$(10.3)$$

and the distribution of $W_r(\mathbf{M})$ can then be expressed as

$$\Pr(W_r(\mathbf{M}) = w, T = X_{r:n_1} \mid H_0 : F_X = F_Y)$$
$$= \Pr(W_r(\mathbf{M}) = w, T = Y_{r:n_2} \mid H_0 : F_X = F_Y)$$
$$= \sum_{\substack{\mathbf{m} \\ W_r(\mathbf{m})=w}} \frac{\binom{N - s_r(\mathbf{m}) - r}{n_2 - r}}{\binom{N}{n_1}}. \tag{10.4}$$

For a fixed level of significance α, the critical region for this test procedure will be $\{s, s+1, \cdots, \frac{1}{2}n_1(n_1 + 2r + 1)\}$, where

$$\begin{aligned} \alpha &= \Pr(W_r(\mathbf{M}) \geq s, T = X_{r:n_1} \mid H_0 : F_X = F_Y) \\ &+ \Pr(W_r(\mathbf{M}) \geq s, T = Y_{r:n_2} \mid H_0 : F_X = F_Y). \end{aligned} \tag{10.5}$$

For specified values of n, r, and α, the critical values can be computed from (10.3)–(10.5). The critical value s and the exact level of significance α (as close to 5% as possible) for different choices of the common sample size n and $r = 5(1)n$ are presented in Table 10.1. The critical values and the exact levels of significance for $r \geq 22$ were obtained based on 10^7 Monte Carlo simulations.

10.2.2   Unequal Sample Size Situation

Now let us consider the situation in which the sizes of the two samples are unequal as in Section 9.2.2. We propose that the life-testing experiment be terminated as soon as the r_1th X-failure or the r_2th Y-failure occurs, where $r_1 = \lfloor n_1 q \rfloor$, $r_2 = \lfloor n_2 q \rfloor$, and $\lfloor h \rfloor$ is the integer part of h. When the experiment is terminated, we count the number of failures from the other sample and calculate the statistic $W_q(\mathbf{M})$ as

$$\begin{cases} \frac{1}{2}n_2(n_2 + 2r_1 + 1) - (r_1 + 1)S_{r_1}(\mathbf{M}) + S_{r_1}^*(\mathbf{M}) & \text{if } T = X_{r_1:n_1} \\ \frac{1}{2}n_1(n_1 + 2r_2 + 1) - (r_2 + 1)S_{r_2}(\mathbf{M}) + S_{r_2}^*(\mathbf{M}) & \text{if } T = Y_{r_2:n_2}, \end{cases} \tag{10.6}$$

where $T = T_{r_1, r_2}(n_1, n_2) = \min(X_{r_1:n_1}, Y_{r_2:n_2})$, as before.

The test procedure is now based on the statistic $W_q(\mathbf{M})$. The critical region for the test will be $\{s_1, s_1 + 1, \cdots, \frac{1}{2}n_2(n_2 + 2r_1 + 1)\}$ if $T = X_{r_1:n_1}$ or $\{s_2, s_2 + 1, \cdots, \frac{1}{2}n_1(n_1 + 2r_2 + 1)\}$ if $T = Y_{r_2:n_2}$. Specifically, the decision rule

Table 10.1: Near 5% critical values and exact levels of significance (in parentheses) for the procedure in (10.2)

r	$n = 10$	$n = 15$	$n = 20$	$n = 30$
5	104(0.0542)	195(0.0421)	310(0.0471)	615(0.0522)
6	112(0.0483)	208(0.0443)	328(0.0508)	643(0.0573)
7	119(0.0478)	220(0.0517)	346(0.0449)	671(0.0524)
8	125(0.0445)	232(0.0480)	363(0.0454)	698(0.0549)
9	129(0.0468)	242(0.0549)	379(0.0479)	725(0.0494)
10	131(0.0474)	252(0.0508)	394(0.0505)	751(0.0472)
11		261(0.0467)	408(0.0527)	776(0.0472)
12		268(0.0479)	422(0.0480)	800(0.0483)
13		273(0.0515)	434(0.0494)	823(0.0497)
14		277(0.0502)	445(0.0498)	845(0.0513)
15		278(0.0512)	455(0.0491)	867(0.0485)
16			463(0.0511)	887(0.0501)
17			470(0.0510)	906(0.0515)
18			476(0.0486)	925(0.0492)
19			479(0.0494)	942(0.0500)
20			479(0.0503)	958(0.0503)
21				973(0.0502)
22				987(0.0495)
23				999(0.0505)
24				1010(0.0506)
25				1020(0.0498)
26				1028(0.0500)
27				1035(0.0488)
28				1039(0.0495)
29				1041(0.0495)
30				1039(0.0502)

is as follows:

Reject H_0 and choose H_{A1} (select π_1 as the best)

if and only if $T = Y_{r_2:n_2}$ and $W_q(\mathbf{M}) \geq s_2$;

Reject H_0 and choose H_{A2} (select π_2 as the best)

if and only if $T = X_{r_1:n_1}$ and $W_q(\mathbf{M}) \geq s_1$. (10.7)

Note that in Chapter 9, we discussed a selection procedure based on the precedence statistic defined as

$$P_q = \begin{cases} S_{r_1}(\mathbf{M}) & \text{if } T = X_{r_1:n_1} \\ S_{r_2}(\mathbf{M}) & \text{if } T = Y_{r_2:n_2}, \end{cases}$$

where $T = T_{r_1,r_2}(n_1, n_2) = \min(X_{r_1:n_1}, Y_{r_2:n_2})$. The critical region for the test is $\{0, 1, \cdots, s_1\}$ if $T = X_{r_1:n_1}$ or $\{0, 1, \cdots, s_2\}$ if $T = Y_{r_2:n_2}$, where $s_1 < r_2$ and $s_2 < r_1$. The decision rule was described in (9.11).

Under the null hypothesis $H_0 : F_X = F_Y$, the probabilities of the life-testing experiment terminating with the X- and Y-samples are as given in (9.5) and (9.6), respectively, and the null distribution of $\mathbf{M}$ is given by

$$\Pr(\mathbf{M} = \mathbf{m}, T = X_{r_1:n_1} \mid H_0 : F_X = F_Y)$$

$$= \frac{\binom{N - s_{r_1}(\mathbf{m}) - r_1}{n_2 - r_1}}{\binom{N}{n_1}}; \quad (10.8)$$

$$\Pr(\mathbf{M} = \mathbf{m}, T = Y_{r_2:n_2} \mid H_0 : F_X = F_Y)$$

$$= \frac{\binom{N - s_{r_2}(\mathbf{m}) - r_2}{n_2 - r_2}}{\binom{N}{n_1}}. \quad (10.9)$$

For a fixed level of significance α, the critical region for this test will be $\{s_1, s_1 + 1, \cdots, \frac{1}{2}n_2(n_2 + 2r_1 + 1)\}$ if the experiment terminated with the X-sample, and $\{s_2, s_2 + 1, \cdots, \frac{1}{2}n_1(n_1 + 2r_2 + 1)\}$ if the experiment terminated with the Y-sample, where

$$\begin{aligned} \alpha = \ & \Pr(W_q(\mathbf{M}) \geq s_1, T = X_{r_1:n_1} \mid H_0 : F_X = F_Y) \\ & + \Pr(W_q(\mathbf{M}) \geq s_2, T = Y_{r_2:n_2} \mid H_0 : F_X = F_Y). \end{aligned} \quad (10.10)$$

In order to obtain the critical values s_1 and s_2 for a specified level of significance α, we may solve (10.10) by using (9.5), (9.6), (10.8), and (10.9). However, if the probabilities of the experiment terminating with the X- and Y-samples are close to 0.5, in order to obtain the critical values s_1 and s_2 at α (or close to α) level of significance, we can find s_1 and s_2 such that

$$
\begin{aligned}
\alpha \;=\; & \alpha_1 \Pr(T = X_{r_1:n_1} \mid H_0 : F_X = F_Y) \\
& + \alpha_2 \Pr(T = Y_{r_2:n_2} \mid H_0 : F_X = F_Y),
\end{aligned}
$$

where

$$
\begin{aligned}
\alpha_1 \;=\; & \Pr(W_q(\mathbf{M}) \geq s_1 \mid T = X_{r_1:n_1}, H_0 : F_X = F_Y) \\
\;=\; & \frac{\displaystyle\sum_{\substack{\mathbf{m} \\ W_q(\mathbf{m}) \geq s_1}} \binom{N - s_{r_1}(\mathbf{m}) + r_1}{n_1 - r_1}}{\displaystyle\sum_{\substack{\mathbf{m} \\ s_{r_1}(\mathbf{m}) < r_2}} \binom{N - s_{r_1}(\mathbf{m}) + r_1}{n_1 - r_1}}
\end{aligned}
$$

and

$$
\begin{aligned}
\alpha_2 \;=\; & \Pr(W_q(\mathbf{M}) \geq s_2 \mid T = Y_{r_2:n_2}, H_0 : F_X = F_Y) \\
\;=\; & \frac{\displaystyle\sum_{\substack{\mathbf{m} \\ W_q(\mathbf{m}) \geq s_2}} \binom{N - s_{r_2}(\mathbf{m}) + r_2}{n_2 - r_2}}{\displaystyle\sum_{\substack{\mathbf{m} \\ s_{r_2}(\mathbf{m}) < r_1}} \binom{N - s_{r_2}(\mathbf{m}) + r_2}{n_2 - r_2}},
\end{aligned}
$$

with α_1 and α_2 close or equal to α. This indeed is the case when the difference between n_1 and n_2 is not too large (see Table 10.2).

Note that when $n_1 = n_2 = n$, the test procedure and all the computational formulas reduce readily to those presented earlier in Section 10.2.1. The probabilities of terminating with the X- and Y-samples, critical values s_1 and s_2, α_1 and α_2, and the exact level of significance α as close to 5% as possible, are all presented in Table 10.2 for different choices of the sample sizes (n_1, n_2) and q.

10.2.3 Performance Under Lehmann Alternative

The probability of correct selection can be expressed as

$$
\Pr[CS \mid H_1 : F_X \neq F_Y]
$$

Table 10.2: Near 5% critical values and exact levels of significance (in parentheses) for the procedure in (10.7) when $n_1 \neq n_2$

q	r_1	r_2	$\Pr(T = X_{r_1:n_1})$	$\Pr(T = Y_{r_2:n_2})$	$s_1(\alpha_1)$	$s_2(\alpha_2)$	α
			$n_1 = 10, n_2 = 15$				
0.5	5	7	0.4660	0.5340	192(0.0484)	123(0.0603)	0.0547
0.6	6	9	0.5340	0.4660	203(0.0494)	139(0.0536)	0.0514
0.7	7	10	0.4697	0.5303	213(0.0539)	146(0.0445)	0.0489
0.8	8	12	0.5447	0.4553	221(0.0471)	157(0.0527)	0.0496
0.9	9	13	0.4676	0.5324	227(0.0532)	160(0.0531)	0.0532
1.0	10	15	0.6000	0.4000	229(0.0478)	166(0.0478)	0.0478
			$n_1 = 15, n_2 = 20$				
0.30	4	6	0.6058	0.3942	290(0.0430)	210(0.0606)	0.0499
0.35	5	7	0.5597	0.4403	308(0.0458)	224(0.0402)	0.0433
0.40	6	8	0.5183	0.4817	325(0.0538)	236(0.0534)	0.0536
0.45	6	9	0.6347	0.3653	325(0.0439)	250(0.0440)	0.0439
0.50	7	10	0.5959	0.4041	341(0.0495)	261(0.0517)	0.0504
0.55	8	11	0.5582	0.4418	356(0.0523)	272(0.0493)	0.0510
0.60	9	12	0.5206	0.4794	370(0.0533)	282(0.0484)	0.0510
0.65	9	13	0.6370	0.3630	369(0.0508)	293(0.0514)	0.0510
0.70	10	14	0.6030	0.3970	382(0.0481)	301(0.0524)	0.0498
0.75	11	15	0.5671	0.4329	393(0.0496)	308(0.0527)	0.0510
0.80	12	16	0.5274	0.4726	403(0.0485)	314(0.0519)	0.0501
0.85	12	17	0.6679	0.3321	400(0.0518)	323(0.0511)	0.0515
0.90	13	18	0.6404	0.3596	408(0.0492)	327(0.0489)	0.0491
0.95	14	19	0.6096	0.3904	413(0.0505)	328(0.0518)	0.0510
1.00	15	20	0.5714	0.4286	415(0.0518)	328(0.0484)	0.0504

$$= \Pr[\text{Do not reject } H_{A1} \mid H_{A1} : F_X < F_Y]$$
$$+ \Pr[\text{Do not reject } H_{A2} \mid H_{A2} : F_X > F_Y]$$
$$= \Pr[W_q(\mathbf{M}) < s_2, T = Y_{r_2:n_2} \mid H_{A1} : F_X < F_Y]$$
$$+ \Pr[W_q(\mathbf{M}) < s_1, T = X_{r_1:n_1} \mid H_{A2} : F_X > F_Y]. \quad (10.11)$$

For specific distributions F_X and F_Y, one can compute $\Pr[CS]$ from (10.11). We can, however, derive an explicit expression for $\Pr[CS]$ under the Lehmann alternative, $H_1 : [F_X]^\gamma = F_Y$ for some $\gamma > 1$.

From Chapter 6, under the Lehmann alternative $H_1 : F_X^\gamma = F_Y$, $\gamma > 1$, we have

$$\Pr[\mathbf{M} = \mathbf{m}, T = X_{r_1:n_1} \mid H_1 : F_X^\gamma = F_Y]$$

$$= \frac{n_1! n_2!}{m_1!(n_1 - r_1)! \gamma^{r_1}} \left\{ \prod_{j=1}^{r_1-1} \frac{\Gamma\left(\sum\limits_{i=1}^{j} m_i + j/\gamma\right)}{\Gamma\left(\sum\limits_{i=1}^{j+1} m_i + j/\gamma + 1\right)} \right\}$$

$$\times \left\{ \sum_{k=0}^{n_1-r_1} \binom{n_1 - r_1}{k} (-1)^k \frac{\Gamma\left(\sum\limits_{i=1}^{r} m_i + (r_1 + k)/\gamma\right)}{\Gamma\left(n_2 + (r_1 + k)/\gamma + 1\right)} \right\},$$

$$(10.12)$$

where $\Gamma(\cdot)$ denotes the complete gamma function.

Therefore, the probability of correct selection under the Lehmann alternative $H_1 : F_X^\gamma = F_Y$, $\gamma > 1$, is

$$\Pr[CS \mid H_1 : F_X^\gamma = F_Y]$$
$$= \Pr[W_q(\mathbf{M}) < s_1, T = X_{r_1:n_1} \mid H_1 : F_X^\gamma = F_Y]$$
$$= \sum_{\substack{\mathbf{m} \\ W_q(\mathbf{m}) < s_1}} \Pr[\mathbf{M} = \mathbf{m}, T = X_{r_1:n_1} \mid H_1 : F_X^\gamma = F_Y]. \quad (10.13)$$

Now, we compare the testing and the selection procedure for the best population based on the precedence statistic and the minimal Wilcoxon rank-sum precedence statistic in terms of $\Pr[CS]$ under the Lehmann alternatives. For this purpose, we computed $\Pr[CS]$ by means of (10.12) and (10.13) for different values of γ. For $n_1 = n_2 = n = 10, 20$, $r = 5(1)n$ and $\gamma = 2(1)7$, the exact value of $\Pr[CS]$ computed from the above expression for the Wilcoxon rank-sum precedence statistic and the exact value of $\Pr[CS]$ based on the precedence statistic taken from Table 9.3 are presented in Table 10.3. We observe that the values of $\Pr[CS]$ for the procedure based on the Wilcoxon rank-sum precedence statistic are consistently higher than the corresponding values for the procedure based on the precedence statistic.

10.3 k-SAMPLE SELECTION PROBLEM

Following the formulation in Section 9.3, we are interested in selecting the best population among $\pi_1, \pi_2, \cdots, \pi_k$ if H_0 in (9.15) is rejected, and for this purpose we consider the alternatives in (9.16). If H_{Ai} is chosen, then the population corresponding to F_i is chosen as the best population.

For the reason described earlier in Section 9.2, we wish to have a balanced setting (that is, $n_1 = n_2 = \cdots = n_k = n$) for our procedure. However, this may not be the case in practice, and so we describe the test procedure for unequal sample sizes.

10.3.1 Selection Procedure and Null Distribution

Suppose k samples are placed simultaneously on a life-testing experiment and a value q, $0 < q < 1$, is fixed prior to the experiment, where q is the upper limit for the proportion of units or subjects allowed to fail from any one sample. Then, the experiment is terminated as soon as the r_ith failure from the X_i-sample is observed, where $r_i = \lfloor n_i q \rfloor$ for $i = 1, 2, \cdots, k$. When the life-testing experiment is terminated, we count the number of failures from the jth sample, $j = 1, 2, \cdots, k$, $j \neq i$, that occurred before the first, between the first and the second, $\cdots$, between the $(r_i - 1)$th and r_ith failures from the X_i-sample. Specifically, let $M_{j,1}^{(i)}, M_{j,2}^{(i)}, \cdots, M_{j,r_i}^{(i)}$ denote the number of failures from the X_j-sample that occurred before the first, between the first and the second, $\cdots$, between the $(r_i - 1)$th and r_ith failures from the X_i-sample, respectively. For notational convenience, let us denote

$$\mathbf{M}_{i,j} = (M_{j,1}^{(i)}, M_{j,2}^{(i)}, \cdots, M_{j,r_i}^{(i)}), \ S_{r_i}(\mathbf{M}_{i,j}) = \sum_{l=1}^{r_i} M_{j,l}^{(i)} \text{ and } S_{r_i}^*(\mathbf{M}_{i,j}) = \sum_{l=1}^{r_i} l M_{j,l}^{(i)},$$

and their realizations by $\mathbf{m}_{i,j} = (m_{j,1}^{(i)}, m_{j,2}^{(i)}, \cdots, m_{j,r_i}^{(i)})$, s_{r_i} and $s_{r_i}^*$.

Suppose the experiment is terminated with the sample from population π_i. Then, let $T = T_{r_1,\cdots,r_k}(n_1, \cdots, n_k) = \min_{1 \le i \le k} X_{r_i:n_i}$ (once again, we use the simpler notation for convenience). Then, the minimal Wilcoxon rank-sum

Table 10.3: $\Pr[CS]$ under the Lehmann alternative for $n_1 = n_2 = n$

n	r	$\gamma = 2$		$\gamma = 3$		$\gamma = 4$	
		P_r	W_r	P_r	W_r	P_r	W_r
10	5	0.1772	0.2527	0.4087	0.5238	0.5997	0.7139
	6	0.2397	0.2500	0.4905	0.5307	0.6730	0.7253
	7	0.2458	0.2540	0.4809	0.5377	0.6521	0.7308
	8	0.2168	0.2395	0.4200	0.5131	0.5771	0.7044
	9	0.1611	0.2372	0.3153	0.4980	0.4460	0.6804
	10	0.0839	0.2206	0.1682	0.4520	0.2481	0.6146
20	5	0.3676	0.3676	0.7231	0.7231	0.8917	0.8917
	6	0.2523	0.4175	0.6056	0.7827	0.8196	0.9283
	7	0.3868	0.4245	0.7591	0.8028	0.9180	0.9424
	8	0.4580	0.4496	0.8140	0.8303	0.9435	0.9562
	9	0.3321	0.4722	0.7101	0.8491	0.8922	0.9639
	10	0.3646	0.4908	0.7344	0.8625	0.9028	0.9687
	11	0.3769	0.5025	0.7367	0.8693	0.9006	0.9708
	12	0.3741	0.4861	0.7236	0.8594	0.8891	0.9671
	13	0.3593	0.4891	0.6976	0.8591	0.8683	0.9662
	14	0.3342	0.4858	0.6585	0.8539	0.8365	0.9631
	15	0.3000	0.4760	0.6053	0.8434	0.7900	0.9573

n	r	$\gamma = 5$		$\gamma = 6$		$\gamma = 7$	
		P_r	W_r	P_r	W_r	P_r	W_r
10	5	0.7325	0.8288	0.8202	0.8959	0.8775	0.9352
	6	0.7896	0.8399	0.8622	0.9049	0.9076	0.9420
	7	0.7646	0.8429	0.8373	0.9060	0.8849	0.9419
	8	0.6883	0.8191	0.7661	0.8861	0.8210	0.9258
	9	0.5483	0.7921	0.6272	0.8596	0.6882	0.9014
	10	0.3182	0.7173	0.3783	0.7825	0.4298	0.8254
20	5	0.9578	0.9578	0.9831	0.9831	0.9929	0.9929
	6	0.9201	0.9762	0.9642	0.9917	0.9835	0.9969
	7	0.9723	0.9832	0.9903	0.9949	0.9964	0.9983
	8	0.9826	0.9887	0.9943	0.9969	0.9980	0.9991
	9	0.9606	0.9913	0.9851	0.9978	0.9941	0.9994
	10	0.9645	0.9928	0.9865	0.9982	0.9946	0.9995
	11	0.9622	0.9933	0.9849	0.9983	0.9936	0.9995
	12	0.9549	0.9921	0.9808	0.9979	0.9913	0.9994
	13	0.9418	0.9915	0.9731	0.9976	0.9869	0.9993
	14	0.9205	0.9900	0.9597	0.9970	0.9787	0.9989
	15	0.8869	0.9872	0.9369	0.9956	0.9634	0.9982

statistics are

$$
\begin{aligned}
W_{q,j}^{(i)} &= W_{q,j}^{(i)}(\mathbf{M}_{i,j}) \\
&= \frac{1}{2}n_j(n_j + 2r_i + 1) - (r_i + 1)S_{r_i}(\mathbf{M}_{i,j}) + S_{r_i}^*(\mathbf{M}_{i,j}), \text{ if } T = X_{r_i:n_i},
\end{aligned}
$$

for $j = 1, \cdots, k, \ j \neq i$.

The selection procedure proposed by Ng, Balakrishnan, and Panchapakesan (2006) is based on the test statistic

$$
W_q^{*(i)} = \max_{\substack{1 \leq j \leq k \\ j \neq i}} \left(\frac{W_{q,j}^{(i)} - E(W_{q,j}^{(i)}|H_0)}{\sqrt{Var(W_{q,j}^{(i)}|H_0)}} \right),
$$

where

$$
E(W_{q,j}^{(i)}|H_0) = \sum_{\mathbf{m}_{i,j}} W_{q,j}^{(i)}(\mathbf{m}_{i,j}) \frac{\binom{n_i + n_j - s_{r_i}(\mathbf{m}_{i,j}) - r_i}{n_j - r_i}}{\binom{n_i + n_j}{n_j}},
$$

$$
E(W_{q,j}^{(i)^2}|H_0) = \sum_{\mathbf{m}_{i,j}} W_{q,j}^{(i)^2}(\mathbf{m}_{i,j}) \frac{\binom{n_i + n_j - s_{r_i}(\mathbf{m}_{i,j}) - r_i}{n_j - r_i}}{\binom{n_i + n_j}{n_j}},
$$

$$
Var(W_{q,j}^{(i)^2}|H_0) = E(W_{q,j}^{(i)^2}|H_0) - \left[E(W_{q,j}^{(i)}|H_0) \right]^2.
$$

Large values of $W_q^{*(i)}$ will lead to the rejection of H_0 in (9.20) and will choose at least one alternative hypothesis in (9.21), that is, we will select at least one of the populations as the best. The critical region for this procedure will be the interval $[s_i, \infty]$ if $T = X_{r_i:n_i}$. Therefore, the decision rule can be written as follows:

Reject H_0 and choose H_{Aj} (select π_j as the best) if and only if

$$
T = X_{r_i:n_i}, i \neq j, \text{ and } \frac{W_{q,j}^{(i)} - E(W_{q,j}^{(i)}|H_0)}{\sqrt{Var(W_{q,j}^{(i)}|H_0)}} = W_q^{*(i)} \geq s_i,
$$

$$
\tag{10.14}
$$

for $j = 1, 2, \cdots, k$.

Under the null hypothesis $H_0 : F_1 = F_2 = \cdots = F_k$, the probability that the experiment terminates with the X_i-sample as given in (9.24). Under the null hypothesis, we have

$$\Pr(\mathbf{M}_{i,j} = \mathbf{m}_{i,j}, j = 1, 2, \cdots, k, j \neq i, T = X_{r_i:n_i} | H_0 : F_1 = F_2 = \cdots = F_k)$$

$$= \frac{1}{\binom{N}{n_1, n_2, \cdots, n_k}} \left\{ \prod_{l=1}^{r_i} \binom{\sum\limits_{\substack{j=1 \\ j \neq i}}^{k} m_{jl}^{(i)}}{m_{1,l}^{(i)}, m_{2,l}^{(i)}, \cdots, m_{k,l}^{(i)}} \right\}$$

$$\times \binom{N - \sum\limits_{\substack{j=1 \\ j \neq i}}^{k} \sum\limits_{l=1}^{r_i} m_{j,l}^{(i)} - r_i}{n_1^*, \cdots, n_{i-1}^*, n_{i+1}^*, \cdots, n_k^*, n_i - r_i},$$

where $n_j^* = n_j - \sum\limits_{l=1}^{r_i} m_{j,l}^{(i)}$.

The null distribution of $W_q^{*(i)}$ is then given by

$$\Pr(W_q^{*(i)} = w, T = X_{r_i:n_i} \mid H_0 : F_1 = F_2 = \cdots = F_k)$$

$$= \sum_{\substack{\mathbf{M}_{i,j}(j=1,\cdots,k)=0, j \neq i \\ W_q^{*(i)}=w}} \Pr(\mathbf{M}_{i,j} = \mathbf{m}_{i,j}, j = 1, 2, \cdots, k, j \neq i,$$

$$T = X_{r_i:n_i} | H_0 : F_1 = F_2 = \cdots = F_k). \qquad (10.15)$$

Let

$$\alpha_i = \sum_{w \geq s_i} \Pr(W_q^{*(i)} = w \mid T = X_{r_i:n_i}, H_0 : F_1 = F_2 = \cdots = F_k); \qquad (10.16)$$

then, for specified values of n, r, and s_i, $i = 1, \cdots, k$, an expression for the level of significance α is given by

$$\alpha = \sum_{i=1}^{k} \alpha_i \Pr(T = X_{r_i:n_i} \mid H_0 : F_1 = F_2 = \cdots = F_k). \qquad (10.17)$$

For a fixed level of significance α, the critical region for this procedure, $[s_i, \infty]$, $i = 1, 2, \cdots, k$, can be determined from (10.15)–(10.17). For equal sample sizes $n_1 = \cdots = n_k = n$, we have $r_1 = r_2 = \cdots = r_k = r$, $\alpha_1 = \alpha_2 = \cdots = \alpha_k$, and

$$\Pr(T = X_{r_i:n_i} \mid H_0 : F_1 = F_2 = \cdots = F_k) = \frac{1}{k}, \quad i = 1, 2, \cdots, k. \qquad (10.18)$$

As a result, we can determine the critical value $s_1 = s_2 = \cdots = s_k = s$ by solving

$$\sum_{w \geq s} \Pr(W_q^{*(i)} = w \mid T = X_{r_i:n_i}, H_0 : F_1 = F_2 = \cdots = F_k) = \alpha. \quad (10.19)$$

Note that the test statistic in this special case reduces to

$$W_r^{*(i)} = \max_{\substack{1 \leq j \leq k \\ j \neq i}} \left(W_{q,j}^{(i)} \right),$$

and the null hypothesis is rejected when $W_r^{*(i)} \geq s$. For $k = 3$, the critical value s and the level of significance α as close to 5% as possible for the sample sizes $n_1 = n_2 = n_3 = n = 10, 15, 20, 30$ and $r = (1)n$ are presented in Table 10.4. The critical values and the levels of significance for $r \geq 12$ were obtained based on 10^7 Monte Carlo simulations.

For unequal sample sizes, one can find the critical values $s_1, s_2, \cdots, s_k$ for fixed level of significance α by solving (10.16) and (10.17). If $\Pr(T = X_{r_i:n_i} \mid H_0 : F_1 = F_2 = \cdots = F_k)$, $i = 1, 2, \cdots, k$, are close to $\frac{1}{k}$, the formula in (10.17) can still be used to obtain the critical values. This is the case when the differences among $n_1, n_2, \cdots, n_k$ are not too large (see Table 10.5).

For $k = 3$, the critical values s_1, s_2, s_3, values of $\alpha_1, \alpha_2, \alpha_3$, and the exact level of significance α as close to 5% as possible for different choices of the sample sizes n_1, n_2, n_3, and q are presented in Table 10.5.

10.3.2 Handling Ties

In practice, it is possible that two or more X_j-samples have the same value of $\frac{W_{q,j}^{(i)} - E(W_{q,j}^{(i)} \mid H_0)}{\sqrt{Var(W_{q,j}^{(i)} \mid H_0)}}$ equaling $W_q^{*(i)}$ and that H_0 is rejected. When such ties occur, randomization may be used in which case we will randomly pick one of the corresponding populations as the best. For example, when $k = 3$, suppose π_1 is the best population among the three populations; then the $\Pr[CS]$ can be computed in this case as

$$\Pr[CS \mid H_1 : F_1 < F_2, F_1 < F_3]$$
$$= \Pr\left[T = X_{r_2:n_2}, \frac{W_{q,1}^{(2)} - E(W_{q,1}^{(2)} \mid H_0)}{\sqrt{Var(W_{q,1}^{(2)} \mid H_0)}} \geq s_2, \right.$$

Table 10.4: Near 5% critical values and exact levels of significance (in parentheses) for the procedure in (10.14) with $n_1 = n_2 = n_3 = n$

r	$n = 10$	$n = 15$	$n = 20$	$n = 30$
4	95 (0.1998)	180 (0.2252)	290 (0.2376)	585 (0.2496)
5	105 (0.0821)	195 (0.1040)	310 (0.1150)	615 (0.1260)
6	114 (0.0483)	210 (0.0442)	330 (0.0523)	645 (0.0607)
7	121 (0.0590)	223 (0.0445)	348 (0.0546)	674 (0.0428)
8	128 (0.0471)	235 (0.0495)	366 (0.0472)	702 (0.0421)
9	133 (0.0461)	247 (0.0439)	383 (0.0464)	729 (0.0472)
10	135 (0.0517)	257 (0.0492)	399 (0.0488)	755 (0.0552)
11		266 (0.0516)	414 (0.0507)	781 (0.0528)
12		274 (0.0508)	428 (0.0522)	806 (0.0528)
13		280 (0.0529)	442 (0.0466)	831 (0.0477)
14		285 (0.0490)	454 (0.0469)	854 (0.0499)
15		286 (0.0519)	464 (0.0509)	876 (0.0520)
16			474 (0.0479)	898 (0.0496)
17			481 (0.0511)	918 (0.0516)
18			487 (0.0509)	938 (0.0494)
19			491 (0.0498)	956 (0.0506)
20			491 (0.0513)	973 (0.0512)
21				989 (0.0512)
22				1004 (0.0504)
23				1018 (0.0490)
24				1030 (0.0490)
25				1040 (0.0501)
26				1049 (0.0498)
27				1056 (0.0498)
28				1061 (0.0494)
29				1063 (0.0497)
30				1061 (0.0498)

Table 10.5: Near 5% critical values and exact levels of significance (in parentheses) for the procedure in (10.14) with $k = 3$ and unequal sample sizes

q	(r_1, r_2, r_3)	$\Pr(T = X_{r_1:n_1})$ $s_1 \ (\alpha_1)$	$\Pr(T = X_{r_2:n_2})$ $s_2 \ (\alpha_2)$	$\Pr(T = X_{r_3:n_3})$ $s_3 \ (\alpha_3)$	α
		$n_1 = n_2 = 10, n_3 = 15$			
0.4	(4,4,6)	0.3569	0.3569	0.2863	
		1.9192 (0.0426)	1.9192 (0.0426)	1.7521 (0.1451)	0.0720
0.5	(5,5,7)	0.3245	0.3245	0.3510	
		1.9410 (0.0418)	1.9410 (0.0418)	1.9917 (0.0630)	0.0492
0.6	(6,6,9)	0.3585	0.3585	0.2831	
		2.0539 (0.0462)	2.0539 (0.0462)	2.0348 (0.0522)	0.0479
0.7	(7,7,10)	0.3265	0.3265	0.3469	
		1.9931 (0.0543)	1.9931 (0.0543)	2.0371 (0.0498)	0.0528
0.8	(8,8,12)	0.3632	0.3632	0.2737	
		2.0236 (0.0495)	2.0236 (0.0495)	2.0635 (0.0503)	0.0497
		$n_1 = n_2 = 15, n_3 = 20$			
0.30	(4,4, 6)	0.3896	0.3896	0.2208	
		1.7917 (0.0596)	1.7917 (0.0596)	1.7540 (0.1502)	0.0796
0.40	(6,6, 8)	0.3486	0.3486	0.3027	
		2.0387 (0.0438)	2.0387 (0.0438)	2.0623 (0.0452)	0.0442
0.50	(7,7,10)	0.3847	0.3847	0.2306	
		1.9608 (0.0494)	1.9608 (0.0494)	2.0698 (0.0476)	0.0490
		$n_1 = 10, n_2 = 15, n_3 = 20$			
0.40	(4,6, 8)	0.3980	0.3225	0.2794	
		1.9192 (0.0555)	1.7773 (0.0519)	1.9301 (0.0356)	0.0488
0.50	(5,7,10)	0.3624	0.3956	0.2421	
		1.9053 (0.0557)	1.9917 (0.0504)	2.0347 (0.0507)	0.0524

$$\left. \frac{W_{q,1}^{(2)} - E(W_{q,1}^{(2)}|H_0)}{\sqrt{Var(W_{q,1}^{(2)}|H_0)}} > \frac{W_{q,3}^{(2)} - E(W_{q,3}^{(2)}|H_0)}{\sqrt{Var(W_{q,3}^{(2)}|H_0)}} \right| H_1 : F_1 < F_2, F_1 < F_3 \right]$$

$$+ \Pr\left[T = X_{r_3:n_3}, \frac{W_{q,1}^{(3)} - E(W_{q,1}^{(3)}|H_0)}{\sqrt{Var(W_{q,1}^{(3)}|H_0)}} \geq s_2, \right.$$

$$\left. \frac{W_{q,1}^{(3)} - E(W_{q,1}^{(3)}|H_0)}{\sqrt{Var(W_{q,1}^{(3)}|H_0)}} < \frac{W_{q,2}^{(3)} - E(W_{q,2}^{(3)}|H_0)}{\sqrt{Var(W_{q,2}^{(3)}|H_0)}} \right| F_1 < F_2, F_1 < F_3 \right]$$

$$+ \frac{1}{2}\Pr\left[T = X_{r_2:n_2}, \right.$$

$$\left. \frac{W_{q,1}^{(2)} - E(W_{q,1}^{(2)}|H_0)}{\sqrt{Var(W_{q,1}^{(2)}|H_0)}} = \frac{W_{q,3}^{(2)} - E(W_{q,3}^{(2)}|H_0)}{\sqrt{Var(W_{q,3}^{(2)}|H_0)}} \geq s_2 \right| H_1 : F_1 < F_2, F_1 < F_3 \right]$$

$$+ \frac{1}{2}\Pr\left[T = X_{r_3:n_3}, \right.$$

$$\left. \frac{W_{q,1}^{(3)} - E(W_{q,1}^{(3)}|H_0)}{\sqrt{Var(W_{q,1}^{(3)}|H_0)}} = \frac{W_{q,2}^{(3)} - E(W_{q,2}^{(3)}|H_0)}{\sqrt{Var(W_{q,2}^{(3)}|H_0)}} \geq s_3 \right| H_1 : F_1 < F_2, F_1 < F_3 \right].$$

10.4 MONTE CARLO SIMULATION UNDER LOCATION-SHIFT

To further examine the performance of the proposed test procedure and compare it with the procedure proposed earlier in Chapter 9, we consider the location-shift alternative. For the two-sample problem, we consider $H_1 :$ $F_X(x) = F_Y(x+\theta)$ for some $\theta > 0$, where θ is the shift in location. For the k-sample problem with $k = 3$, we consider $H_1 : F_1(x) = F_2(x+\theta_1) = F_3(x+\theta_2)$ for some $\theta_1, \theta_2 > 0$, where θ_1 and θ_2 are shifts in the location parameters of F_2 and F_3, respectively.

The $\Pr[CS]$ for the proposed procedure was estimated through Monte Carlo simulations when $\theta = 0.5$ and 1.0 for the two-sample problem and $(\theta_1, \theta_2) = (0.5, 1.0)$ and $(1.0, 2.0)$ for the k-sample problem with $k = 3$. The following lifetime distributions were used for $F_X(x)$ and $F_1(x)$ in this study:

1. Standard normal distribution

2. Standard exponential distribution

3. Gamma distribution with shape parameter 3 and standardized by mean 3 and standard deviation $\sqrt{3}$

4. Lognormal distribution with shape parameter 0.1 and standardized by mean $e^{0.005}$ and standard deviation $\sqrt{e^{0.01}(e^{0.01} - 1)}$

A brief description of these distributions and their properties has been provided in Section 2.5. For different choices of sample sizes, we generated 100,000 sets of data in order to obtain the estimated probabilities of selection of each population.

In Tables 10.6–10.13, we have presented the estimated probabilities of selecting π_1 and π_2 for the two-sample case for different choices of n_1, n_2, and q for the distributions listed above with location-shift θ being equal to 0.5 and 1.0.

In Tables 10.14–10.17, we have presented the estimated probabilities of selection of π_1, π_2, and π_3 for the three-sample case for different choices of $n_1 = n_2 = n_3 = n$ and q for the distributions listed above with the location-shift (θ_1, θ_2) being equal to $(0.5, 1.0)$ and $(1.0, 2.0)$. For comparison purposes, the corresponding exact levels of significance are also included in these tables.

10.5   EVALUATION AND COMPARATIVE REMARKS

From the simulation results, we observe that $\Pr[CS]$ of the proposed procedure increases with increasing sample sizes as well as with increasing location-shift. These simulation studies show that the procedure can pick up the best population very effectively in most cases early in the life-testing experiment. Compared to the procedure based on the precedence statistic discussed earlier in Chapter 9, the procedure based on the minimal Wilcoxon rank-sum precedence statistic gives higher $\Pr[CS]$, especially for large values of r.

One may be interested in maximizing $\Pr[CS]$ with respect to q; that is, we want to determine the best choice of q in designing the experiment. By comparing $\Pr[CS]$ for different choices of q, we find that the optimal choice of q is in the range 60%–70% if the underlying distributions are close to symmetry, such as the normal distribution and lognormal distribution with shape parameter 0.1. However, under some right-skewed distributions

Table 10.6: Estimated probabilities of selection under location-shift with $n_1 = n_2 = 10$

		Normal		Exp(1)		Gamma(3)		LN(0.1)	
r	α	(a)	(b)	(a)	(b)	(a)	(b)	(a)	(b)
		Precedence statistic, location-shift = 0.5							
5	0.033	0.001	0.092	0.000	0.490	0.000	0.174	0.001	0.102
6	0.057	0.003	0.150	0.001	0.392	0.001	0.216	0.002	0.159
7	0.070	0.003	0.178	0.002	0.307	0.002	0.213	0.003	0.183
8	0.070	0.003	0.177	0.003	0.225	0.003	0.185	0.003	0.176
9	0.057	0.003	0.147	0.003	0.145	0.003	0.135	0.003	0.141
10	0.033	0.002	0.088	0.003	0.069	0.003	0.070	0.002	0.081
		Minimal Wilcoxon rank-sum statistic, location-shift = 0.5							
5	0.054	0.003	0.135	0.000	0.582	0.001	0.250	0.002	0.149
6	0.048	0.002	0.137	0.000	0.503	0.001	0.242	0.002	0.151
7	0.048	0.002	0.148	0.000	0.445	0.001	0.237	0.001	0.160
8	0.044	0.001	0.150	0.000	0.373	0.001	0.216	0.001	0.159
9	0.047	0.001	0.163	0.000	0.330	0.001	0.211	0.001	0.168
10	0.047	0.001	0.169	0.000	0.284	0.001	0.196	0.001	0.169
		Precedence statistic, location-shift = 1.0							
5	0.033	0.000	0.283	0.000	0.917	0.000	0.602	0.000	0.327
6	0.057	0.000	0.424	0.000	0.838	0.000	0.621	0.000	0.457
7	0.070	0.000	0.481	0.000	0.724	0.000	0.578	0.000	0.496
8	0.070	0.000	0.480	0.000	0.577	0.000	0.495	0.000	0.476
9	0.057	0.000	0.424	0.000	0.396	0.000	0.374	0.000	0.403
10	0.033	0.000	0.285	0.000	0.190	0.000	0.205	0.000	0.254
		Minimal Wilcoxon rank-sum statistic, location-shift = 1.0							
5	0.054	0.000	0.375	0.000	0.941	0.000	0.718	0.000	0.428
6	0.048	0.000	0.400	0.000	0.901	0.000	0.705	0.000	0.452
7	0.048	0.000	0.447	0.000	0.860	0.000	0.689	0.000	0.491
8	0.044	0.000	0.470	0.000	0.789	0.000	0.648	0.000	0.502
9	0.047	0.000	0.509	0.000	0.719	0.000	0.619	0.000	0.526
10	0.047	0.000	0.524	0.000	0.616	0.000	0.558	0.000	0.521

Remarks: (a) $\Pr(\pi_1$ is selected); (b) $\Pr(\pi_2$ is selected) $= \Pr[CS]$.

Table 10.7: Estimated probabilities of selection under location-shift $= 0.5$ with $n_1 = n_2 = 15$

		Normal		Exp(1)		Gamma(3)		LN(0.1)	
r	α	(a)	(b)	(a)	(b)	(a)	(b)	(a)	(b)
				Precedence statistic					
5	0.042	0.002	0.125	0.000	0.839	0.000	0.309	0.001	0.146
6	0.080	0.003	0.221	0.000	0.768	0.001	0.394	0.002	0.246
7	0.035	0.001	0.134	0.000	0.596	0.000	0.254	0.001	0.151
8	0.050	0.001	0.183	0.000	0.517	0.000	0.281	0.001	0.197
9	0.060	0.001	0.211	0.000	0.440	0.001	0.281	0.001	0.222
10	0.066	0.002	0.226	0.000	0.369	0.001	0.266	0.001	0.232
11	0.066	0.002	0.227	0.001	0.301	0.001	0.242	0.001	0.227
12	0.060	0.002	0.213	0.001	0.233	0.002	0.205	0.002	0.207
13	0.050	0.001	0.181	0.002	0.167	0.002	0.158	0.001	0.172
14	0.035	0.001	0.136	0.002	0.104	0.001	0.106	0.001	0.125
15	0.042	0.002	0.125	0.004	0.079	0.003	0.087	0.002	0.110
				Minimal Wilcoxon rank-sum statistic					
5	0.042	0.002	0.125	0.000	0.839	0.000	0.309	0.001	0.146
6	0.044	0.001	0.144	0.000	0.804	0.000	0.343	0.001	0.167
7	0.052	0.001	0.177	0.000	0.775	0.000	0.382	0.001	0.203
8	0.048	0.001	0.184	0.000	0.714	0.000	0.362	0.001	0.207
9	0.055	0.001	0.214	0.000	0.681	0.000	0.380	0.001	0.238
10	0.051	0.001	0.216	0.000	0.620	0.000	0.357	0.001	0.236
11	0.047	0.001	0.215	0.000	0.560	0.000	0.334	0.000	0.231
12	0.048	0.001	0.227	0.000	0.517	0.000	0.326	0.000	0.240
13	0.052	0.001	0.245	0.000	0.481	0.000	0.323	0.001	0.254
14	0.050	0.001	0.247	0.000	0.432	0.000	0.303	0.000	0.251
15	0.051	0.001	0.249	0.000	0.385	0.000	0.282	0.000	0.247

Remarks: (a) $\Pr(\pi_1$ is selected); (b) $\Pr(\pi_2$ is selected) $= \Pr[CS]$.

Table 10.8: Estimated probabilities of selection under location-shift $= 1.0$ with $n_1 = n_2 = 15$

r	α	Normal (a)	Normal (b)	Exp(1) (a)	Exp(1) (b)	Gamma(3) (a)	Gamma(3) (b)	LN(0.1) (a)	LN(0.1) (b)
		\multicolumn{8}{c}{Precedence statistic}							
5	0.042	0.000	0.378	0.000	0.997	0.000	0.857	0.000	0.454
6	0.080	0.000	0.572	0.000	0.993	0.000	0.884	0.000	0.637
7	0.035	0.000	0.447	0.000	0.973	0.000	0.775	0.000	0.506
8	0.050	0.000	0.547	0.000	0.944	0.000	0.774	0.000	0.592
9	0.060	0.000	0.601	0.000	0.900	0.000	0.751	0.000	0.630
10	0.066	0.000	0.625	0.000	0.832	0.000	0.710	0.000	0.638
11	0.066	0.000	0.625	0.000	0.739	0.000	0.651	0.000	0.624
12	0.060	0.000	0.601	0.000	0.621	0.000	0.572	0.000	0.585
13	0.050	0.000	0.546	0.000	0.476	0.000	0.464	0.000	0.514
14	0.035	0.000	0.445	0.000	0.310	0.000	0.327	0.000	0.401
15	0.042	0.000	0.376	0.001	0.198	0.000	0.233	0.000	0.321
		\multicolumn{8}{c}{Minimal Wilcoxon rank-sum statistic}							
5	0.042	0.000	0.378	0.000	0.997	0.000	0.857	0.000	0.454
6	0.044	0.000	0.434	0.000	0.995	0.000	0.885	0.000	0.517
7	0.052	0.000	0.522	0.000	0.992	0.000	0.900	0.000	0.603
8	0.048	0.000	0.557	0.000	0.985	0.000	0.886	0.000	0.630
9	0.055	0.000	0.621	0.000	0.976	0.000	0.889	0.000	0.682
10	0.051	0.000	0.639	0.000	0.960	0.000	0.870	0.000	0.692
11	0.047	0.000	0.651	0.000	0.934	0.000	0.845	0.000	0.695
12	0.048	0.000	0.679	0.000	0.903	0.000	0.827	0.000	0.711
13	0.052	0.000	0.709	0.000	0.861	0.000	0.807	0.000	0.729
14	0.050	0.000	0.717	0.000	0.795	0.000	0.765	0.000	0.722
15	0.051	0.000	0.710	0.000	0.696	0.000	0.690	0.000	0.694

Remarks: (a) $\Pr(\pi_1$ is selected); (b) $\Pr(\pi_2$ is selected) $= \Pr[CS]$.

Table 10.9: Estimated probabilities of selection under location-shift $= 0.5$ with $n_1 = n_2 = 20$

r	α	Normal (a)	Normal (b)	Exp(1) (a)	Exp(1) (b)	Gamma(3) (a)	Gamma(3) (b)	LN(0.1) (a)	LN(0.1) (b)
				Precedence statistic					
5	0.047	0.002	0.152	0.000	0.963	0.000	0.434	0.001	0.181
6	0.020	0.000	0.092	0.000	0.904	0.000	0.294	0.000	0.112
7	0.044	0.001	0.176	0.000	0.857	0.000	0.396	0.001	0.205
8	0.065	0.001	0.243	0.000	0.807	0.000	0.437	0.001	0.273
9	0.031	0.001	0.162	0.000	0.672	0.000	0.308	0.000	0.183
10	0.041	0.001	0.200	0.000	0.604	0.000	0.327	0.000	0.220
11	0.048	0.001	0.230	0.000	0.538	0.000	0.329	0.001	0.246
12	0.054	0.001	0.249	0.000	0.474	0.000	0.320	0.001	0.260
13	0.056	0.001	0.259	0.000	0.412	0.001	0.304	0.001	0.265
14	0.056	0.001	0.258	0.000	0.350	0.001	0.279	0.001	0.259
15	0.054	0.001	0.249	0.001	0.291	0.001	0.248	0.001	0.244
16	0.048	0.000	0.231	0.001	0.232	0.001	0.212	0.001	0.221
17	0.041	0.000	0.201	0.001	0.174	0.001	0.170	0.000	0.188
18	0.065	0.001	0.243	0.003	0.178	0.002	0.189	0.002	0.223
19	0.044	0.001	0.176	0.002	0.111	0.002	0.124	0.001	0.156
20	0.047	0.002	0.149	0.005	0.084	0.004	0.096	0.002	0.128
				Minimal Wilcoxon rank-sum statistic					
5	0.047	0.002	0.152	0.000	0.963	0.000	0.434	0.001	0.181
6	0.051	0.001	0.175	0.000	0.950	0.000	0.482	0.001	0.209
7	0.045	0.001	0.178	0.000	0.926	0.000	0.476	0.001	0.212
8	0.045	0.001	0.197	0.000	0.900	0.000	0.479	0.001	0.233
9	0.048	0.001	0.219	0.000	0.873	0.000	0.483	0.001	0.254
10	0.050	0.001	0.241	0.000	0.846	0.000	0.488	0.000	0.275
11	0.053	0.001	0.260	0.000	0.816	0.000	0.488	0.000	0.294
12	0.048	0.000	0.258	0.000	0.772	0.000	0.464	0.000	0.288
13	0.049	0.000	0.274	0.000	0.738	0.000	0.458	0.000	0.302
14	0.050	0.000	0.285	0.000	0.700	0.000	0.449	0.000	0.309
15	0.049	0.000	0.293	0.000	0.661	0.000	0.434	0.000	0.313
16	0.051	0.000	0.307	0.000	0.627	0.000	0.429	0.000	0.324
17	0.051	0.000	0.314	0.000	0.586	0.000	0.413	0.000	0.326
18	0.049	0.000	0.312	0.000	0.539	0.000	0.390	0.000	0.320
19	0.049	0.000	0.317	0.000	0.496	0.000	0.371	0.000	0.319
20	0.050	0.000	0.315	0.000	0.446	0.000	0.344	0.000	0.309

Remarks: (a) $\Pr(\pi_1$ is selected); (b) $\Pr(\pi_2$ is selected) $= \Pr[CS]$.

Table 10.10: Estimated probabilities of selection under location-shift = 1.0 with $n_1 = n_2 = 20$

r	α	Normal (a)	(b)	Exp(1) (a)	(b)	Gamma(3) (a)	(b)	LN(0.1) (a)	(b)
				Precedence statistic					
5	0.047	0.000	0.439	0.000	1.000	0.000	0.951	0.000	0.537
6	0.020	0.000	0.335	0.000	1.000	0.000	0.894	0.000	0.423
7	0.044	0.000	0.543	0.000	0.999	0.000	0.926	0.000	0.627
8	0.065	0.000	0.663	0.000	0.997	0.000	0.932	0.000	0.728
9	0.031	0.000	0.556	0.000	0.990	0.000	0.867	0.000	0.621
10	0.041	0.000	0.633	0.000	0.980	0.000	0.863	0.000	0.683
11	0.048	0.000	0.681	0.000	0.962	0.000	0.850	0.000	0.717
12	0.054	0.000	0.711	0.000	0.934	0.000	0.828	0.000	0.733
13	0.056	0.000	0.724	0.000	0.893	0.000	0.797	0.000	0.735
14	0.056	0.000	0.724	0.000	0.831	0.000	0.754	0.000	0.725
15	0.054	0.000	0.712	0.000	0.751	0.000	0.697	0.000	0.700
16	0.048	0.000	0.683	0.000	0.652	0.000	0.624	0.000	0.659
17	0.041	0.000	0.632	0.000	0.529	0.000	0.531	0.000	0.594
18	0.065	0.000	0.664	0.000	0.472	0.000	0.510	0.000	0.610
19	0.044	0.000	0.546	0.000	0.311	0.000	0.363	0.000	0.478
20	0.047	0.000	0.439	0.001	0.198	0.000	0.250	0.000	0.366
				Minimal Wilcoxon rank-sum statistic					
5	0.047	0.000	0.439	0.000	1.000	0.000	0.951	0.000	0.537
6	0.051	0.000	0.502	0.000	1.000	0.000	0.967	0.000	0.603
7	0.045	0.000	0.531	0.000	1.000	0.000	0.966	0.000	0.633
8	0.045	0.000	0.592	0.000	0.999	0.000	0.965	0.000	0.688
9	0.048	0.000	0.647	0.000	0.999	0.000	0.964	0.000	0.732
10	0.050	0.000	0.693	0.000	0.997	0.000	0.964	0.000	0.768
11	0.053	0.000	0.731	0.000	0.995	0.000	0.961	0.000	0.796
12	0.048	0.000	0.741	0.000	0.992	0.000	0.953	0.000	0.801
13	0.049	0.000	0.769	0.000	0.987	0.000	0.948	0.000	0.819
14	0.050	0.000	0.790	0.000	0.978	0.000	0.941	0.000	0.831
15	0.049	0.000	0.804	0.000	0.964	0.000	0.930	0.000	0.838
16	0.051	0.000	0.823	0.000	0.946	0.000	0.920	0.000	0.848
17	0.051	0.000	0.833	0.000	0.918	0.000	0.901	0.000	0.850
18	0.049	0.000	0.836	0.000	0.877	0.000	0.874	0.000	0.842
19	0.049	0.000	0.839	0.000	0.818	0.000	0.832	0.000	0.833
20	0.050	0.000	0.815	0.000	0.722	0.000	0.750	0.000	0.788

Remarks: (a) $\Pr(\pi_1$ is selected); (b) $\Pr(\pi_2$ is selected) $= \Pr[CS]$.

Table 10.11: Estimated probabilities of selection under location-shift with $n_1 = 10$ and $n_2 = 15$

r_1	r_2	α	Normal (a)	(b)	Exp(1) (a)	(b)	Gamma(3) (a)	(b)	LN(0.1) (a)	(b)
			\multicolumn{8}{c}{Precedence statistic, location-shift = 0.5}						
5	7	0.036	0.001	0.138	0.000	0.541	0.000	0.251	0.001	0.154
6	9	0.038	0.001	0.145	0.000	0.399	0.001	0.220	0.001	0.156
7	10	0.047	0.002	0.131	0.001	0.272	0.001	0.173	0.002	0.137
8	12	0.047	0.001	0.177	0.001	0.230	0.001	0.187	0.001	0.177
9	13	0.038	0.001	0.116	0.002	0.123	0.002	0.110	0.002	0.113
10	15	0.045	0.002	0.145	0.003	0.107	0.003	0.114	0.002	0.133
			\multicolumn{8}{c}{Minimal Wilcoxon rank-sum statistic, location-shift = 0.5}						
5	7	0.055	0.003	0.141	0.000	0.645	0.001	0.304	0.002	0.161
6	9	0.051	0.002	0.175	0.000	0.594	0.001	0.320	0.001	0.195
7	10	0.049	0.001	0.185	0.000	0.519	0.001	0.302	0.001	0.202
8	12	0.050	0.001	0.197	0.000	0.462	0.001	0.289	0.001	0.210
9	13	0.053	0.001	0.200	0.000	0.392	0.001	0.263	0.001	0.208
10	15	0.048	0.001	0.224	0.000	0.372	0.001	0.266	0.001	0.226
			\multicolumn{8}{c}{Precedence statistic, location-shift = 1.0}						
5	7	0.036	0.000	0.421	0.000	0.930	0.000	0.717	0.000	0.475
6	9	0.038	0.000	0.454	0.000	0.845	0.000	0.650	0.000	0.490
7	10	0.047	0.000	0.437	0.000	0.709	0.000	0.550	0.000	0.456
8	12	0.047	0.000	0.514	0.000	0.587	0.000	0.521	0.000	0.508
9	13	0.038	0.000	0.391	0.000	0.369	0.000	0.346	0.000	0.370
10	15	0.045	0.000	0.396	0.001	0.252	0.000	0.279	0.000	0.352
			\multicolumn{8}{c}{Minimal Wilcoxon rank-sum statistic, location-shift = 1.0}						
5	7	0.055	0.000	0.430	0.000	0.957	0.000	0.803	0.000	0.501
6	9	0.051	0.000	0.520	0.000	0.936	0.000	0.806	0.000	0.579
7	10	0.049	0.000	0.555	0.000	0.894	0.000	0.778	0.000	0.603
8	12	0.050	0.000	0.592	0.000	0.848	0.000	0.750	0.000	0.624
9	13	0.053	0.000	0.610	0.000	0.762	0.000	0.698	0.000	0.624
10	15	0.048	0.000	0.645	0.000	0.708	0.000	0.669	0.000	0.640

Remarks: (a) $\Pr(\pi_1 \text{ is selected})$; (b) $\Pr(\pi_2 \text{ is selected}) = \Pr[CS]$.

Table 10.12: Estimated probabilities of selection under location-shift $= 0.5$ with $n_1 = 15$ and $n_2 = 20$

r_1	r_2	α	Normal (a)	Normal (b)	Exp(1) (a)	Exp(1) (b)	Gamma(3) (a)	Gamma(3) (b)	LN(0.1) (a)	LN(0.1) (b)
						Precedence statistic				
4	6	0.050	0.002	0.145	0.000	0.931	0.000	0.409	0.001	0.173
5	7	0.052	0.000	0.227	0.000	0.873	0.000	0.464	0.000	0.259
6	8	0.045	0.001	0.131	0.000	0.734	0.000	0.299	0.001	0.153
6	9	0.061	0.001	0.270	0.000	0.798	0.000	0.462	0.000	0.299
7	10	0.046	0.001	0.163	0.000	0.631	0.000	0.301	0.001	0.183
8	11	0.033	0.000	0.177	0.000	0.524	0.000	0.283	0.000	0.193
9	12	0.039	0.001	0.175	0.000	0.420	0.000	0.251	0.001	0.186
9	13	0.050	0.000	0.278	0.000	0.506	0.000	0.353	0.000	0.290
10	14	0.048	0.000	0.255	0.000	0.404	0.000	0.300	0.000	0.261
11	15	0.044	0.001	0.224	0.000	0.305	0.001	0.242	0.001	0.225
12	16	0.039	0.001	0.184	0.001	0.213	0.001	0.180	0.001	0.179
12	17	0.059	0.001	0.269	0.001	0.281	0.001	0.254	0.001	0.261
13	18	0.046	0.001	0.202	0.002	0.183	0.001	0.175	0.001	0.191
14	19	0.045	0.001	0.196	0.002	0.146	0.001	0.151	0.001	0.179
15	20	0.050	0.002	0.145	0.006	0.089	0.005	0.101	0.003	0.128
					Minimal Wilcoxon rank-sum statistic					
4	6	0.050	0.002	0.145	0.000	0.931	0.000	0.409	0.001	0.173
5	7	0.043	0.001	0.163	0.000	0.898	0.000	0.440	0.001	0.195
6	8	0.054	0.001	0.197	0.000	0.867	0.000	0.463	0.001	0.231
6	9	0.044	0.000	0.197	0.000	0.867	0.000	0.463	0.000	0.231
7	10	0.050	0.001	0.223	0.000	0.832	0.000	0.467	0.000	0.256
8	11	0.051	0.001	0.239	0.000	0.785	0.000	0.456	0.000	0.270
9	12	0.051	0.001	0.245	0.000	0.733	0.000	0.437	0.000	0.273
9	13	0.051	0.000	0.268	0.000	0.754	0.000	0.465	0.000	0.299
10	14	0.050	0.001	0.264	0.000	0.695	0.000	0.433	0.000	0.290
11	15	0.051	0.000	0.271	0.000	0.643	0.000	0.413	0.000	0.292
12	16	0.050	0.001	0.265	0.000	0.579	0.000	0.381	0.000	0.281
12	17	0.052	0.000	0.314	0.000	0.636	0.000	0.434	0.000	0.331
13	18	0.049	0.000	0.306	0.000	0.578	0.000	0.402	0.000	0.319
14	19	0.051	0.000	0.308	0.000	0.523	0.000	0.379	0.000	0.314
15	20	0.050	0.000	0.301	0.000	0.458	0.000	0.341	0.000	0.300

Remarks: (a) $\Pr(\pi_1$ is selected); (b) $\Pr(\pi_2$ is selected) $= \Pr[CS]$.

Table 10.13: Estimated probabilities of selection under location-shift $= 1.0$ with $n_1 = 15$ and $n_2 = 20$

			Normal		Exp(1)		Gamma(3)		LN(0.1)	
r_1	r_2	α	(a)	(b)	(a)	(b)	(a)	(b)	(a)	(b)
				Precedence statistic						
4	6	0.050	0.000	0.419	0.000	0.999	0.000	0.925	0.000	0.511
5	7	0.052	0.000	0.591	0.000	0.998	0.000	0.931	0.000	0.669
6	8	0.045	0.000	0.453	0.000	0.991	0.000	0.846	0.000	0.529
6	9	0.061	0.000	0.670	0.000	0.994	0.000	0.920	0.000	0.731
7	10	0.046	0.000	0.535	0.000	0.976	0.000	0.830	0.000	0.596
8	11	0.033	0.000	0.573	0.000	0.947	0.000	0.795	0.000	0.619
9	12	0.039	0.000	0.584	0.000	0.895	0.000	0.745	0.000	0.616
9	13	0.050	0.000	0.715	0.000	0.920	0.000	0.821	0.000	0.736
10	14	0.048	0.000	0.694	0.000	0.850	0.000	0.757	0.000	0.703
11	15	0.044	0.000	0.654	0.000	0.747	0.000	0.672	0.000	0.650
12	16	0.039	0.000	0.590	0.000	0.610	0.000	0.559	0.000	0.573
12	17	0.059	0.000	0.693	0.000	0.671	0.000	0.643	0.000	0.671
13	18	0.046	0.000	0.600	0.000	0.505	0.000	0.505	0.000	0.564
14	19	0.045	0.000	0.560	0.000	0.377	0.000	0.414	0.000	0.508
15	20	0.050	0.000	0.424	0.001	0.219	0.001	0.262	0.000	0.363
				Minimal Wilcoxon rank-sum statistic						
4	6	0.050	0.000	0.419	0.000	0.999	0.000	0.925	0.000	0.511
5	7	0.043	0.000	0.480	0.000	0.999	0.000	0.939	0.000	0.577
6	8	0.054	0.000	0.567	0.000	0.997	0.000	0.943	0.000	0.661
6	9	0.044	0.000	0.567	0.000	0.997	0.000	0.943	0.000	0.661
7	10	0.050	0.000	0.633	0.000	0.995	0.000	0.942	0.000	0.712
8	11	0.051	0.000	0.674	0.000	0.991	0.000	0.935	0.000	0.743
9	12	0.051	0.000	0.700	0.000	0.983	0.000	0.923	0.000	0.759
9	13	0.051	0.000	0.727	0.000	0.986	0.000	0.933	0.000	0.782
10	14	0.050	0.000	0.737	0.000	0.974	0.000	0.916	0.000	0.783
11	15	0.051	0.000	0.756	0.000	0.956	0.000	0.901	0.000	0.791
12	16	0.050	0.000	0.760	0.000	0.925	0.000	0.876	0.000	0.786
12	17	0.052	0.000	0.802	0.000	0.948	0.000	0.903	0.000	0.826
13	18	0.049	0.000	0.803	0.000	0.914	0.000	0.877	0.000	0.817
14	19	0.051	0.000	0.808	0.000	0.858	0.000	0.840	0.000	0.810
15	20	0.050	0.000	0.790	0.000	0.757	0.000	0.760	0.000	0.772

Remarks: (a) $\Pr(\pi_1$ is selected); (b) $\Pr(\pi_2$ is selected) $= \Pr[CS]$.

Table 10.14: Estimated probabilities of selection for $k = 3$ with $n_1 = n_2 = n_3 = 10, 15$, and location-shift $\theta_1 = 0.5$, $\theta_2 = 1.0$

n	r	α	Normal			Exp(1)		
			(a)	(b)	(c)	(a)	(b)	(c)
				Precedence statistic				
10	6	0.0291	0.000	0.033	0.178	0.000	0.133	0.697
	7	0.0518	0.001	0.046	0.275	0.000	0.058	0.640
	8	0.0598	0.000	0.047	0.309	0.000	0.024	0.515
	9	0.0518	0.000	0.039	0.281	0.000	0.015	0.337
	10	0.0291	0.000	0.023	0.185	0.000	0.011	0.151
15	6	0.0442	0.000	0.053	0.267	0.000	0.342	0.653
	7	0.0173	0.000	0.029	0.186	0.000	0.246	0.730
	8	0.0377	0.000	0.047	0.322	0.000	0.151	0.798
	9	0.0546	0.000	0.055	0.410	0.000	0.078	0.823
	10	0.0654	0.000	0.057	0.459	0.000	0.034	0.791
	11	0.0690	0.000	0.055	0.476	0.000	0.016	0.708
	12	0.0654	0.000	0.050	0.465	0.000	0.012	0.583
	13	0.0546	0.000	0.042	0.424	0.000	0.013	0.431
	14	0.0377	0.000	0.031	0.341	0.000	0.013	0.268
	15	0.0442	0.000	0.028	0.294	0.001	0.015	0.164
				Minimal Wilcoxon rank-sum statistic				
10	6	0.0483	0.000	0.047	0.242	0.000	0.133	0.745
	7	0.0590	0.000	0.054	0.310	0.000	0.060	0.780
	8	0.0471	0.000	0.048	0.309	0.000	0.026	0.711
	9	0.0461	0.000	0.046	0.333	0.000	0.016	0.624
	10	0.0517	0.000	0.046	0.363	0.000	0.016	0.532
15	6	0.0442	0.000	0.053	0.267	0.000	0.343	0.653
	7	0.0445	0.000	0.053	0.306	0.000	0.245	0.746
	8	0.0495	0.000	0.058	0.368	0.000	0.150	0.832
	9	0.0439	0.000	0.054	0.391	0.000	0.077	0.890
	10	0.0492	0.000	0.057	0.446	0.000	0.034	0.912
	11	0.0516	0.000	0.056	0.485	0.000	0.014	0.905
	12	0.0508	0.000	0.054	0.508	0.000	0.008	0.870
	13	0.0529	0.000	0.052	0.535	0.000	0.007	0.820
	14	0.0490	0.000	0.048	0.534	0.000	0.007	0.745
	15	0.0519	0.000	0.045	0.534	0.000	0.008	0.658

Remarks: (a) $\Pr(\pi_1$ is selected); (b) $\Pr(\pi_2$ is selected); (c) $\Pr(\pi_3$ is selected) $= \Pr[CS]$.

Table 10.14: (Continued)

n	r	α	Gamma(3) (a)	(b)	(c)	LN(0.1) (a)	(b)	(c)
			\multicolumn{6}{c}{Precedence statistic}					
10	6	0.0291	0.000	0.046	0.388	0.000	0.034	0.204
	7	0.0518	0.000	0.040	0.419	0.000	0.046	0.297
	8	0.0598	0.000	0.035	0.371	0.000	0.045	0.316
	9	0.0518	0.000	0.028	0.279	0.000	0.036	0.274
	10	0.0291	0.000	0.016	0.145	0.000	0.021	0.170
15	6	0.0442	0.000	0.097	0.661	0.000	0.060	0.325
	7	0.0173	0.000	0.053	0.543	0.000	0.033	0.230
	8	0.0377	0.000	0.041	0.622	0.000	0.049	0.368
	9	0.0546	0.000	0.037	0.631	0.000	0.055	0.443
	10	0.0654	0.000	0.035	0.606	0.000	0.054	0.480
	11	0.0690	0.000	0.034	0.557	0.000	0.051	0.484
	12	0.0654	0.000	0.033	0.485	0.000	0.045	0.460
	13	0.0546	0.000	0.029	0.387	0.000	0.038	0.403
	14	0.0377	0.000	0.021	0.266	0.000	0.027	0.310
	15	0.0442	0.000	0.020	0.187	0.000	0.025	0.251
			\multicolumn{6}{c}{Minimal Wilcoxon rank-sum statistic}					
10	6	0.0483	0.000	0.055	0.498	0.000	0.049	0.277
	7	0.0590	0.000	0.044	0.558	0.001	0.054	0.348
	8	0.0471	0.000	0.034	0.499	0.000	0.046	0.341
	9	0.0461	0.000	0.030	0.462	0.000	0.043	0.351
	10	0.0517	0.000	0.028	0.427	0.000	0.041	0.367
15	6	0.0442	0.000	0.096	0.662	0.000	0.060	0.325
	7	0.0445	0.000	0.062	0.727	0.000	0.058	0.370
	8	0.0495	0.000	0.043	0.766	0.000	0.059	0.434
	9	0.0439	0.000	0.031	0.752	0.000	0.054	0.452
	10	0.0492	0.000	0.027	0.754	0.000	0.055	0.499
	11	0.0516	0.000	0.024	0.741	0.000	0.052	0.530
	12	0.0508	0.000	0.024	0.713	0.000	0.049	0.543
	13	0.0529	0.000	0.023	0.685	0.000	0.047	0.558
	14	0.0490	0.000	0.022	0.634	0.000	0.042	0.543
	15	0.0519	0.000	0.021	0.573	0.000	0.039	0.527

Remarks: (a) $\Pr(\pi_1$ is selected); (b) $\Pr(\pi_2$ is selected); (c) $\Pr(\pi_3$ is selected) $= \Pr[CS]$.

Table 10.15: Estimated probabilities of selection for $k = 3$ with $n_1 = n_2 = n_3 = 20$, and location-shift $\theta_1 = 0.5$, $\theta_2 = 1.0$

n	r	α	Normal (a)	(b)	(c)	Exp(1) (a)	(b)	(c)
				Precedence statistic				
20	6	0.0523	0.000	0.064	0.320	0.000	0.451	0.549
	7	0.0224	0.000	0.039	0.241	0.000	0.404	0.596
	8	0.0512	0.000	0.063	0.415	0.000	0.329	0.670
	9	0.0227	0.000	0.040	0.324	0.000	0.241	0.752
	10	0.0368	0.000	0.050	0.433	0.000	0.159	0.825
	11	0.0491	0.000	0.055	0.505	0.000	0.092	0.875
	12	0.0584	0.000	0.056	0.555	0.000	0.045	0.892
	13	0.0642	0.000	0.055	0.581	0.000	0.021	0.870
	14	0.0661	0.000	0.052	0.595	0.000	0.010	0.816
	15	0.0642	0.000	0.049	0.587	0.000	0.008	0.733
	16	0.0584	0.000	0.044	0.565	0.000	0.009	0.623
	17	0.0491	0.000	0.038	0.523	0.000	0.012	0.494
	18	0.0368	0.000	0.031	0.452	0.000	0.013	0.350
	19	0.0512	0.000	0.032	0.452	0.000	0.018	0.273
	20	0.0523	0.000	0.027	0.360	0.001	0.018	0.167
				Minimal Wilcoxon rank-sum statistic				
20	6	0.0523	0.000	0.064	0.320	0.000	0.453	0.547
	7	0.0546	0.000	0.066	0.366	0.000	0.402	0.598
	8	0.0472	0.000	0.062	0.389	0.000	0.327	0.673
	9	0.0464	0.000	0.061	0.436	0.000	0.243	0.756
	10	0.0488	0.000	0.062	0.487	0.000	0.159	0.839
	11	0.0507	0.000	0.063	0.531	0.000	0.093	0.903
	12	0.0522	0.000	0.061	0.569	0.000	0.047	0.945
	13	0.0466	0.000	0.056	0.579	0.000	0.020	0.964
	14	0.0469	0.000	0.053	0.606	0.000	0.008	0.965
	15	0.0509	0.000	0.052	0.639	0.000	0.003	0.956
	16	0.0479	0.000	0.048	0.649	0.000	0.002	0.933
	17	0.0511	0.000	0.047	0.672	0.000	0.003	0.904
	18	0.0509	0.000	0.045	0.680	0.000	0.003	0.863
	19	0.0498	0.000	0.042	0.677	0.000	0.004	0.804
	20	0.0513	0.000	0.038	0.656	0.000	0.004	0.721

Remarks: (a) $\Pr(\pi_1 \text{ is selected})$; (b) $\Pr(\pi_2 \text{ is selected})$; (c) $\Pr(\pi_3 \text{ is selected}) = \Pr[CS]$.

Table 10.15: (Continued)

n	r	α	Gamma(3) (a)	(b)	(c)	LN(0.1) (a)	(b)	(c)
			Precedence statistic					
20	6	0.0523	0.000	0.146	0.764	0.000	0.075	0.400
	7	0.0224	0.000	0.094	0.727	0.000	0.045	0.313
	8	0.0512	0.000	0.064	0.810	0.000	0.065	0.493
	9	0.0227	0.000	0.040	0.734	0.000	0.042	0.395
	10	0.0368	0.000	0.031	0.763	0.000	0.050	0.495
	11	0.0491	0.000	0.028	0.764	0.000	0.053	0.554
	12	0.0584	0.000	0.026	0.751	0.000	0.052	0.589
	13	0.0642	0.000	0.027	0.724	0.000	0.052	0.603
	14	0.0661	0.000	0.026	0.682	0.000	0.049	0.606
	15	0.0642	0.000	0.027	0.625	0.000	0.044	0.590
	16	0.0584	0.000	0.026	0.556	0.000	0.040	0.554
	17	0.0491	0.000	0.024	0.465	0.000	0.034	0.500
	18	0.0368	0.000	0.020	0.354	0.000	0.028	0.416
	19	0.0512	0.000	0.024	0.307	0.000	0.030	0.399
	20	0.0523	0.000	0.021	0.208	0.000	0.024	0.301
			Minimal Wilcoxon rank-sum statistic					
20	6	0.0523	0.000	0.147	0.763	0.000	0.074	0.401
	7	0.0546	0.000	0.097	0.835	0.000	0.071	0.456
	8	0.0472	0.000	0.060	0.867	0.000	0.065	0.481
	9	0.0464	0.000	0.040	0.884	0.000	0.062	0.529
	10	0.0488	0.000	0.027	0.895	0.000	0.060	0.576
	11	0.0507	0.000	0.021	0.897	0.000	0.059	0.612
	12	0.0522	0.000	0.017	0.893	0.000	0.057	0.642
	13	0.0466	0.000	0.016	0.876	0.000	0.052	0.645
	14	0.0469	0.000	0.015	0.863	0.000	0.049	0.663
	15	0.0509	0.000	0.014	0.853	0.000	0.047	0.686
	16	0.0479	0.000	0.014	0.828	0.000	0.043	0.686
	17	0.0511	0.000	0.015	0.809	0.000	0.041	0.698
	18	0.0509	0.000	0.014	0.778	0.000	0.038	0.696
	19	0.0498	0.000	0.014	0.732	0.000	0.035	0.679
	20	0.0513	0.000	0.012	0.661	0.000	0.031	0.643

Remarks: (a) $\Pr(\pi_1$ is selected); (b) $\Pr(\pi_2$ is selected); (c) $\Pr(\pi_3$ is selected) $= \Pr[CS]$.

Table 10.16: Estimated probabilities of selection for $k = 3$ with $n_1 = n_2 = n_3 = 10, 15$, and location-shift $\theta_1 = 1.0$, $\theta_2 = 2.0$

n	r	α	Normal			Exp(1)		
			(a)	(b)	(c)	(a)	(b)	(c)
			Precedence statistic					
10	5	0.0325	0.000	0.156	0.679	0.000	0.458	0.542
	6	0.0573	0.000	0.100	0.627	0.000	0.389	0.610
	7	0.0698	0.000	0.086	0.778	0.000	0.280	0.712
	8	0.0698	0.000	0.064	0.826	0.000	0.150	0.809
	9	0.0573	0.000	0.043	0.823	0.000	0.054	0.788
	10	0.0325	0.000	0.023	0.705	0.000	0.011	0.524
15	5	0.0421	0.000	0.202	0.710	0.000	0.500	0.500
	6	0.0801	0.000	0.148	0.710	0.000	0.493	0.507
	7	0.0352	0.000	0.102	0.684	0.000	0.483	0.517
	8	0.0502	0.000	0.095	0.835	0.000	0.444	0.556
	9	0.0604	0.000	0.077	0.891	0.000	0.381	0.619
	10	0.0656	0.000	0.058	0.921	0.000	0.289	0.710
	11	0.0656	0.000	0.043	0.939	0.000	0.183	0.813
	12	0.0604	0.000	0.032	0.947	0.000	0.093	0.888
	13	0.0502	0.000	0.023	0.944	0.000	0.035	0.893
	14	0.0352	0.000	0.016	0.915	0.000	0.009	0.773
	15	0.0421	0.000	0.011	0.849	0.000	0.004	0.530
			Minimal Wilcoxon rank-sum statistic					
10	5	0.0821	0.000	0.156	0.679	0.000	0.459	0.541
	6	0.0483	0.000	0.111	0.702	0.000	0.392	0.608
	7	0.0590	0.000	0.088	0.806	0.000	0.278	0.720
	8	0.0471	0.000	0.064	0.848	0.000	0.149	0.841
	9	0.0461	0.000	0.046	0.884	0.000	0.055	0.913
	10	0.0517	0.000	0.032	0.900	0.000	0.015	0.888
15	5	0.1040	0.000	0.201	0.711	0.000	0.499	0.501
	6	0.0442	0.000	0.149	0.709	0.000	0.495	0.505
	7	0.0445	0.000	0.119	0.778	0.000	0.478	0.522
	8	0.0495	0.000	0.098	0.847	0.000	0.447	0.553
	9	0.0439	0.000	0.076	0.888	0.000	0.382	0.618
	10	0.0492	0.000	0.060	0.921	0.000	0.290	0.710
	11	0.0516	0.000	0.046	0.943	0.000	0.185	0.815
	12	0.0508	0.000	0.034	0.957	0.000	0.093	0.906
	13	0.0529	0.000	0.024	0.968	0.000	0.035	0.960
	14	0.0490	0.000	0.017	0.972	0.000	0.009	0.973
	15	0.0519	0.000	0.012	0.966	0.000	0.002	0.933

Remarks: (a) $\Pr(\pi_1$ is selected); (b) $\Pr(\pi_2$ is selected); (c) $\Pr(\pi_3$ is selected) $= \Pr[CS]$.

Table 10.16: (Continued)

n	r	α	Gamma(3)			LN(0.1)		
			(a)	(b)	(c)	(a)	(b)	(c)
			Precedence statistic					
10	5	0.0325	0.000	0.299	0.696	0.000	0.172	0.720
	6	0.0573	0.000	0.200	0.773	0.000	0.108	0.687
	7	0.0698	0.000	0.110	0.852	0.000	0.082	0.808
	8	0.0698	0.000	0.048	0.875	0.000	0.054	0.840
	9	0.0573	0.000	0.019	0.803	0.000	0.035	0.810
	10	0.0325	0.000	0.008	0.557	0.000	0.019	0.651
15	5	0.0421	0.000	0.427	0.573	0.000	0.234	0.726
	6	0.0801	0.000	0.367	0.633	0.000	0.174	0.751
	7	0.0352	0.000	0.284	0.715	0.000	0.121	0.752
	8	0.0502	0.000	0.201	0.797	0.000	0.097	0.863
	9	0.0604	0.000	0.127	0.870	0.000	0.071	0.908
	10	0.0656	0.000	0.070	0.925	0.000	0.049	0.935
	11	0.0656	0.000	0.034	0.954	0.000	0.035	0.948
	12	0.0604	0.000	0.014	0.958	0.000	0.025	0.951
	13	0.0502	0.000	0.006	0.925	0.000	0.017	0.939
	14	0.0352	0.000	0.004	0.813	0.000	0.012	0.886
	15	0.0421	0.000	0.006	0.619	0.000	0.009	0.778
			Minimal Wilcoxon rank-sum statistic					
10	5	0.0821	0.000	0.302	0.693	0.000	0.171	0.720
	6	0.0483	0.000	0.199	0.790	0.000	0.115	0.756
	7	0.0590	0.000	0.110	0.880	0.000	0.082	0.850
	8	0.0471	0.000	0.048	0.928	0.000	0.053	0.882
	9	0.0461	0.000	0.017	0.936	0.000	0.034	0.903
	10	0.0517	0.000	0.007	0.896	0.000	0.022	0.897
15	5	0.1040	0.000	0.427	0.573	0.000	0.233	0.728
	6	0.0442	0.000	0.364	0.635	0.000	0.174	0.751
	7	0.0445	0.000	0.285	0.715	0.000	0.131	0.820
	8	0.0495	0.000	0.202	0.797	0.000	0.096	0.881
	9	0.0439	0.000	0.128	0.872	0.000	0.068	0.917
	10	0.0492	0.000	0.071	0.929	0.000	0.049	0.943
	11	0.0516	0.000	0.034	0.965	0.000	0.033	0.960
	12	0.0508	0.000	0.013	0.984	0.000	0.022	0.971
	13	0.0529	0.000	0.004	0.989	0.000	0.015	0.977
	14	0.0490	0.000	0.001	0.981	0.000	0.010	0.976
	15	0.0519	0.000	0.001	0.945	0.000	0.006	0.960

Remarks: (a) $\Pr(\pi_1$ is selected); (b) $\Pr(\pi_2$ is selected); (c) $\Pr(\pi_3$ is selected) $= \Pr[CS]$.

Table 10.17: Estimated probabilities of selection for $k = 3$ with $n_1 = n_2 = n_3 = 20$, and location-shift $\theta_1 = 1.0$, $\theta_2 = 2.0$

n	r	α	Normal (a)	(b)	(c)	Exp(1) (a)	(b)	(c)
					Precedence statistic			
20	5	0.0471	0.000	0.231	0.712	0.000	0.500	0.500
	6	0.0202	0.000	0.180	0.726	0.000	0.498	0.502
	7	0.0436	0.000	0.137	0.725	0.000	0.499	0.501
	8	0.0648	0.000	0.122	0.849	0.000	0.495	0.505
	9	0.0310	0.000	0.093	0.856	0.000	0.492	0.508
	10	0.0407	0.000	0.076	0.906	0.000	0.473	0.527
	11	0.0484	0.000	0.059	0.933	0.000	0.436	0.564
	12	0.0536	0.000	0.047	0.949	0.000	0.377	0.623
	13	0.0562	0.000	0.035	0.962	0.000	0.300	0.700
	14	0.0562	0.000	0.027	0.970	0.000	0.208	0.791
	15	0.0536	0.000	0.021	0.976	0.000	0.124	0.874
	16	0.0484	0.000	0.015	0.981	0.000	0.061	0.929
	17	0.0407	0.000	0.011	0.981	0.000	0.024	0.941
	18	0.0648	0.000	0.009	0.974	0.000	0.007	0.888
	19	0.0436	0.000	0.007	0.965	0.000	0.002	0.774
	20	0.0471	0.000	0.005	0.904	0.000	0.005	0.532
				Minimal Wilcoxon rank-sum statistic				
20	5	0.1150	0.000	0.231	0.712	0.000	0.502	0.498
	6	0.0523	0.000	0.180	0.726	0.000	0.502	0.498
	7	0.0546	0.000	0.149	0.786	0.000	0.502	0.498
	8	0.0472	0.000	0.120	0.833	0.000	0.497	0.503
	9	0.0464	0.000	0.097	0.878	0.000	0.489	0.511
	10	0.0488	0.000	0.078	0.910	0.000	0.474	0.526
	11	0.0507	0.000	0.062	0.931	0.000	0.437	0.563
	12	0.0522	0.000	0.050	0.946	0.000	0.381	0.619
	13	0.0466	0.000	0.039	0.959	0.000	0.301	0.699
	14	0.0469	0.000	0.030	0.968	0.000	0.209	0.791
	15	0.0509	0.000	0.023	0.976	0.000	0.126	0.874
	16	0.0479	0.000	0.017	0.982	0.000	0.061	0.939
	17	0.0511	0.000	0.012	0.986	0.000	0.023	0.976
	18	0.0509	0.000	0.009	0.990	0.000	0.007	0.990
	19	0.0498	0.000	0.006	0.991	0.000	0.001	0.985
	20	0.0513	0.000	0.004	0.986	0.000	0.000	0.945

Remarks: (a) $\Pr(\pi_1 \text{ is selected})$; (b) $\Pr(\pi_2 \text{ is selected})$; (c) $\Pr(\pi_3 \text{ is selected}) = \Pr[CS]$.

Table 10.17: (Continued)

n	r	α	Gamma(3)			LN(0.1)		
			(a)	(b)	(c)	(a)	(b)	(c)
			Precedence statistic					
20	5	0.0471	0.000	0.478	0.522	0.000	0.273	0.707
	6	0.0202	0.000	0.447	0.553	0.000	0.217	0.747
	7	0.0436	0.000	0.400	0.600	0.000	0.165	0.772
	8	0.0648	0.000	0.342	0.658	0.000	0.133	0.856
	9	0.0310	0.000	0.271	0.729	0.000	0.101	0.878
	10	0.0407	0.000	0.200	0.800	0.000	0.076	0.916
	11	0.0484	0.000	0.134	0.866	0.000	0.054	0.941
	12	0.0536	0.000	0.083	0.917	0.000	0.039	0.959
	13	0.0562	0.000	0.046	0.953	0.000	0.027	0.970
	14	0.0562	0.000	0.022	0.976	0.000	0.019	0.978
	15	0.0536	0.000	0.010	0.986	0.000	0.014	0.982
	16	0.0484	0.000	0.004	0.985	0.000	0.010	0.983
	17	0.0407	0.000	0.002	0.969	0.000	0.008	0.979
	18	0.0648	0.000	0.001	0.920	0.000	0.006	0.963
	19	0.0436	0.000	0.003	0.844	0.000	0.005	0.940
	20	0.0471	0.000	0.005	0.645	0.000	0.005	0.831
			Minimal Wilcoxon rank-sum statistic					
20	5	0.1150	0.000	0.476	0.524	0.000	0.274	0.706
	6	0.0523	0.000	0.442	0.558	0.000	0.217	0.747
	7	0.0546	0.000	0.401	0.599	0.000	0.173	0.805
	8	0.0472	0.000	0.340	0.660	0.000	0.133	0.853
	9	0.0464	0.000	0.270	0.730	0.000	0.100	0.894
	10	0.0488	0.000	0.199	0.801	0.000	0.074	0.924
	11	0.0507	0.000	0.136	0.864	0.000	0.053	0.945
	12	0.0522	0.000	0.082	0.918	0.000	0.039	0.960
	13	0.0466	0.000	0.046	0.954	0.000	0.027	0.972
	14	0.0469	0.000	0.023	0.977	0.000	0.020	0.980
	15	0.0509	0.000	0.009	0.990	0.000	0.013	0.986
	16	0.0479	0.000	0.003	0.996	0.000	0.009	0.990
	17	0.0511	0.000	0.001	0.998	0.000	0.006	0.992
	18	0.0509	0.000	0.000	0.997	0.000	0.004	0.993
	19	0.0498	0.000	0.000	0.990	0.000	0.003	0.992
	20	0.0513	0.000	0.000	0.963	0.000	0.002	0.979

Remarks: (a) $\Pr(\pi_1$ is selected); (b) $\Pr(\pi_2$ is selected); (c) $\Pr(\pi_3$ is selected) $= \Pr[CS]$.

such as the exponential distribution and the gamma distribution with shape parameter 3.0, we find that the optimal choice of q is in the range 30%–40%. Therefore, as a compromise between these two types of distributions, we recommend the use of this procedure with $q \approx 50\%$.

While this procedure will work very well for the balanced sample situation or near balanced situations, the selection procedure may not be effective when the sample sizes differ too much.

10.6 ILLUSTRATIVE EXAMPLES

Example 10.1. For the purpose of illustration, let us reconsider the data in Example 9.1 wherein we stopped the experiment as soon as the 6th failure from any of the three samples is observed. In this case, $k = 3$ and $n_1 = n_2 = n_3 = 12$, the termination of the experiment occurred with the X_2-sample, and we then find

$$
\begin{aligned}
\mathbf{M}_{2,1} &= (M_{1,1}^{(2)}, M_{1,2}^{(2)}, \cdots, M_{1,6}^{(2)}) = (0, 0, 0, 1, 2, 2), \\
\mathbf{M}_{2,3} &= (M_{3,1}^{(2)}, M_{3,2}^{(2)}, \cdots, M_{3,6}^{(2)}) = (0, 1, 2, 1, 0, 0), \\
S_{r_2}(\mathbf{M}_{2,1}) &= 5, S_{r_2}^*(\mathbf{M}_{2,1}) = 26, \\
S_{r_2}(\mathbf{M}_{2,3}) &= 4, S_{r_2}^*(\mathbf{M}_{2,3}) = 12, \\
W_{6,1}^{(2)} &= 150 - 35 + 26 = 141, \\
W_{6,3}^{(2)} &= 150 - 28 + 12 = 134,
\end{aligned}
$$

and consequently the minimal Wilcoxon rank-sum precedence test statistic is

$$
W_6^{*(2)} = \max\{141, 134\} = 141.
$$

The critical value for $n = 12$ and $r = 6$ is $s = 149$ (with the exact level of significance being 0.05868), and so we will not reject the null hypothesis in this case. This means that no one type of cord is superior to the others. In fact, the p-value of this test is 0.5399. This finding agrees with the result of the parametric test as well as with the procedure based on the precedence statistic described in Chapter 9.

Example 10.2. Let us consider X_1-, X_2- and X_3-samples to be the natural logs of times to breakdown of an insulating fluid at 30kV, 35kV, and

Table 10.18: Insulating fluid breakdown times (in natural logs of seconds) from Nelson (1982, p. 278)

30kV	**3.912**	4.898	5.231	6.782	7.279	7.293
	7.736	7.983	8.338	9.668	10.282	11.363
35kV	**3.401**	**3.497**	**3.715**	4.466	4.533	4.585
	4.754	5.553	6.133	7.073	7.208	7.313
40kV	**0.000**	**0.000**	**0.693**	**1.099**	**2.485**	**3.219**
	3.829	**4.025**	4.220	4.691	5.778	6.033

40kV, respectively [Nelson (1982, p. 278)]. We are interested in testing the homogeneity of the life distributions of the insulating fluid under three different voltages and selecting the best working voltage if the null hypothesis is rejected. The data are presented in Table 10.18.

For these data, $k = 3$ and $n_1 = n_2 = n_3 = 12$. Had we fixed $q = 2/3$, that is, $r = 8$, the experiment would have stopped as soon as the 8th failure from any of the three samples occurred. In other words, the experiment would have terminated with the X_3-sample at time $e^{4.025} = 55.98$ seconds (with 40kV). The observed failure times are shown in bold type in Table 10.18. We then have

$$
\begin{aligned}
\mathbf{M}_{3,1} &= (M_{1,1}^{(3)}, M_{1,2}^{(3)}, \cdots, M_{1,8}^{(3)}) = (0, 0, 0, 0, 0, 0, 3, 0), \\
\mathbf{M}_{3,2} &= (M_{2,1}^{(3)}, M_{2,2}^{(3)}, \cdots, M_{2,8}^{(3)}) = (0, 0, 0, 0, 0, 0, 0, 1), \\
S_{r_3}(\mathbf{M}_{3,1}) &= 1, S_{r_3}^*(\mathbf{M}_{3,1}) = 8, \\
S_{r_3}(\mathbf{M}_{3,2}) &= 3, S_{r_3}^*(\mathbf{M}_{3,2}) = 21, \\
W_{8,1}^{(3)} &= 174 - 9 + 8 = 173, \\
W_{8,2}^{(3)} &= 174 - 27 + 21 = 168,
\end{aligned}
$$

and consequently the minimal Wilcoxon rank-sum precedence test statistic is

$$
W_8^{*(3)} = \max\{173, 168\} = 173.
$$

The critical value for $n = 12$ and $r = 8$ is $s = 168$ (with the exact level of significance being 0.05086), and so we will reject the null hypothesis and select the first population as the best population. In other words, 30kV is the best working condition compared to the other two voltages. In fact, the p-value of this test is 0.00662.

In this example, the precedence test statistic is

$$P_{(8)}^{*(3)} = \min\{3, 1\} = 1$$

which gives the p-value as 0.02557. Based on this result, we will once again reject the null hypothesis and select the first population as the best population.

Appendix

Computational Formulas for the Moments of Test Statistics $W_{\min,r}$, $W_{\max,r}$, and $W_{E,r}$

The moments of M_i's are denoted by

$$
\begin{aligned}
\mu_a &= E(M_i^a), \\
\mu_{a,b} &= E(M_i^a M_j^b), \\
\mu_{a,b,c} &= E(M_i^a M_j^b M_k^c), \\
\mu_{a,b,c,d} &= E(M_i^a M_j^b M_k^c M_j^d).
\end{aligned}
$$

To derive the first, second, third and fourth moments of the Wilcoxon-type rank-sum statistics, we require the following:

$$
\begin{aligned}
\mu_1 &= \frac{n_1}{n_2 + 1}, \\
\mu_2 &= \frac{n_1(n_2 + 2n_1)}{(n_2 + 1)(n_2 + 2)}, \\
\mu_3 &= \frac{3n_1(n_1 - 1)(n_1 - 2)}{n_2 + 3} - (3n_1^2 - 6n_1 + 2)\mu_1 + 3(n_1 - 1)\mu_2, \\
\mu_4 &= -\frac{4n_1(n_1 - 1)(n_1 - 2)(n_1 - 3)}{n_2 + 4} + 2(2n_1^3 - 9n_1^2 + 11n_1 - 3)\mu_1 \\
&\quad -(6n_1^2 - 18n_1 + 11)\mu_2 + 2(2n_1 - 3)\mu_3, \\
\mu_{1,1} &= \frac{n_1(n_1 - 1)}{(n_2 + 1)(n_2 + 2)}, \\
\mu_{2,1} &= \frac{1}{n_2}(n_1\mu_2 - \mu_3),
\end{aligned}
$$

$$\mu_{3,1} = \frac{1}{n_2}(n_1\mu_3 - \mu_4),$$

$$\mu_{2,2} = -\frac{2n_1(n_1 - 1)(n_1 - 2)(n_1 - 3)(2n_2 + 3)}{3(n_2 + 3)(n_2 + 4)}$$

$$+\frac{2}{3}(2n_1^3 - 9n_1^2 + 11n_1 - 3)\mu_1$$

$$-\frac{1}{3}(6n_1^2 + 18n_1 + 11)(\mu_2 + \mu_{1,1})$$

$$+\frac{2}{3}(2n_1 - 3)(\mu_3 + 3\mu_{2,1}) - \frac{4}{3}\mu_{3,1} - \frac{1}{3}\mu_4,$$

$$\mu_{1,1,1} = \frac{1}{n_2(n_2^2 - 1)}[n_1^3 - (n_2 + 1)\mu_3 - 3n_2(n_2 + 1)\mu_{2,1}],$$

$$\mu_{2,1,1} = -\frac{n_1(n_1 - 1)(n_1 - 2)(n_1 - 3)(n_2^2 + 2n_2 + 2)}{3(n_2 + 2)(n_2 + 3)(n_2 + 4)}$$

$$+\frac{1}{6}(2n_1^3 - 9n_1^2 + 11n_1 - 3)\mu_1$$

$$-\frac{1}{12}(6n_1^2 - 18n_1 + 11)(\mu_2 + 2\mu_{1,1})$$

$$+\frac{1}{6}(2n_1 - 3)(\mu_3 + 6\mu_{2,1} + 2\mu_{1,1,1})$$

$$-\frac{1}{6}(4\mu_{3,1} + 3\mu_{2,2}) - \frac{1}{12}\mu_4,$$

$$\mu_{1,1,1,1} = \frac{1}{n_2(n_2^2 - 1)(n_2 - 2)}$$

$$\times\left[n_1^4 - (n_2 + 1)\mu_4 - n_2(n_2 + 1)(4\mu_{3,1} + 3\mu_{2,2})\right.$$

$$\left. -6n_2(n_2^2 - 1)\mu_{2,1,1}\right].$$

Since $M_1, M_2, \cdots, M_r$ are exchangeable random variables, we readily have

$$\mu_{2,1} = \mu_{1,2},$$
$$\mu_{3,1} = \mu_{1,3},$$
$$\mu_{2,1,1} = \mu_{1,2,1} = \mu_{1,1,2}.$$

We further define the following quantities:

$$S_r(1) = \sum_{i=1}^{r} i = \frac{r(r + 1)}{2},$$

$$S_r(2) = \sum_{i=1}^{r} i^2 = \frac{r(r+1)(2r+1)}{6},$$

$$S_r(3) = \sum_{i=1}^{r} i^3 = \frac{r^2(r+1)^2}{4},$$

$$S_r(4) = \sum_{i=1}^{r} i^4 = \frac{r(r+1)(2r+1)(3r^2+3r-1)}{30},$$

$$S_r(1,1) = \sum_{i=1}^{r} \sum_{\substack{j=1 \\ i \neq j}}^{r} ij = [S_r(1)]^2 - S_r(2) = \frac{r(r^2-1)(3r+2)}{12},$$

$$S_r(2,1) = \sum_{i=1}^{r} \sum_{\substack{j=1 \\ i \neq j}}^{r} i^2 j = S_r(1)S_r(2) - S_r(3) = \frac{r^2(r+1)^2(r-1)}{6},$$

$$S_r(2,2) = \sum_{i=1}^{r} \sum_{\substack{j=1 \\ i \neq j}}^{r} i^2 j^2$$

$$= [S_r(2)]^2 - S_r(4) = \frac{r(r^2-1)[3r^2(5r+7)-4]}{120},$$

$$S_r(3,1) = \sum_{i=1}^{r} \sum_{\substack{j=1 \\ i \neq j}}^{r} i^3 j$$

$$= S_r(1)S_r(3) - S_r(4) = \frac{r(r^2-1)(4r^2-1)(5r+6)}{180},$$

$$S_r(1,1,1) = \sum_{i=1}^{r} \sum_{\substack{j=1 \\ i \neq j \neq k}}^{r} \sum_{k=1}^{r} ijk$$

$$= [S_r(1)]^3 - 3S_r(2,1) - S_r(3) = \frac{r^2(r+1)^2(r-1)(r-2)}{8},$$

$$S_r(2,1,1) = \sum_{i=1}^{r} \sum_{\substack{j=1 \\ i \neq j \neq k}}^{r} \sum_{k=1}^{r} i^2 jk$$

$$= [S_r(1)]^2 S_r(2) - 2S_r(3,1) - S_r(2,2) - S_r(4),$$

$$S_r(1,1,1,1) = \sum_{i=1}^{r} \sum_{\substack{j=1 \\ i \neq j \neq k \neq l}}^{r} \sum_{k=1}^{r} \sum_{l=1}^{r} ijkl$$

$$= [S_r(1)]^4 - 6S_r(2,1,1) - 3S_r(2,2) - 4S_r(3,1) - S_r(4),$$

and

$$\nu_{a,b} = E\left[\left(\sum_{i=1}^{r} M_i\right)^a \left(\sum_{i=1}^{r} iM_i\right)^b\right].$$

Now we can express all the required $\nu_{a,b}$ in terms of S_r's and μ's as follows:

$$\nu_{1,0} = r\mu_1,$$

$$\nu_{0,1} = S_r(1)\mu_1,$$

$$\nu_{2,0} = r\mu_2 + r(r-1)\mu_{1,1},$$

$$\nu_{0,2} = S_r(2)\mu_2 + S_r(1,1)\mu_{1,1},$$

$$\nu_{1,1} = S_r(1)\mu_2 + (r-1)S_r(1)\mu_{1,1},$$

$$\nu_{3,0} = r\mu_3 + 3r(r-1)\mu_{2,1} + r(r-1)(r-2)\mu_{1,1,1},$$

$$\nu_{0,3} = S_r(3)\mu_3 + 3S_r(2,1)\mu_{2,1} + S_r(1,1,1)\mu_{1,1,1},$$

$$\nu_{2,1} = S_r(1)\mu_3 + 3(r-1)S_r(1)\mu_{2,1} + (r-1)(r-2)S_r(1)\mu_{1,1,1},$$

$$\nu_{1,2} = S_r(2)\mu_3 + 2S_r(1,1)\mu_{2,1} + (r-1)S_r(2)\mu_{2,1}$$
$$\quad\quad + (r-2)S_r(1,1)\mu_{1,1,1},$$

$$\nu_{4,0} = r\mu_4 + 4r(r-1)\mu_{3,1} + 3r(r-1)\mu_{2,2} + 6r(r-1)(r-2)\mu_{2,1,1}$$
$$\quad\quad + r(r-1)(r-2)(r-3)\mu_{1,1,1,1},$$

$$\nu_{0,4} = S_r(4)\mu_4 + 4S_r(3,1)\mu_{3,1} + 3S_r(2,2)\mu_{2,2} + 6S_r(2,1,1)\mu_{2,1,1}$$
$$\quad\quad + S_r(1,1,1,1)\mu_{1,1,1,1},$$

$$\nu_{3,1} = S_r(1)\mu_4 + 4(r-1)S_r(1)\mu_{3,1} + 3(r-1)S_r(1)\mu_{2,2}$$
$$\quad\quad + 6(r-1)(r-2)S_r(1)\mu_{2,1,1} + (r-1)(r-2)(r-3)S_r(1)\mu_{1,1,1,1},$$

$$\nu_{1,3} = S_r(3)\mu_4 + (r-1)S_r(3)\mu_{3,1} + 3S_r(2,1)\mu_{3,1} + 3S_r(2,1)\mu_{2,2}$$
$$\quad\quad + 3(r-2)S_r(2,1)\mu_{2,1,1} + 3S_r(1,1,1)\mu_{2,1,1} + (r-3)S_r(1,1,1)\mu_{1,1,1,1},$$

$$\nu_{2,2} = S_r(2)\mu_4 + 2(r-1)S_r(2)\mu_{3,1} + 2S_r(1,1)\mu_{3,1} + (r-1)S_r(2)\mu_{2,2}$$
$$\quad\quad + 2S_r(1,1)\mu_{2,2} + (r-1)(r-2)S_r(2)\mu_{2,1,1} + (r-2)S_r(1,1)\mu_{2,1,1}$$
$$\quad\quad + 4(r-2)S_r(1,1)\mu_{2,1,1} + (r-2)(r-3)S_r(1,1)\mu_{1,1,1,1}.$$

By taking expectation of the conditional expectation, we get the first, second, third, and fourth moments of the minimal rank-sum statistic as

$$E\left(W_{\min,r}\right) = \frac{n_1(n_1 + 2r + 1)}{2} - (r+1)\nu_{1,0} + \nu_{0,1},$$

$$
E\left(W_{\min,r}^2\right) = \left[\frac{n_1(n_1 + 2r + 1)}{2}\right]^2 + (r+1)^2\nu_{2,0} + \nu_{0,2} - 2(r+1)\nu_{1,1}
$$

$$
+ n_1(n_1 + 2r + 1)\{\nu_{0,1} - (r+1)\nu_{1,0}\},
$$

$$
E\left(W_{\min,r}^3\right) = \left[\frac{n_1(n_1 + 2r + 1)}{2}\right]^3 + (r+1)^3\nu_{3,0} + \nu_{0,3}
$$

$$
+ 3\left\{\left[\frac{n_1(n_1 + 2r + 1)}{2}\right]^2[\nu_{0,1} - (r+1)\nu_{1,0}]\right.
$$

$$
+ \frac{n_1(n_1 + 2r + 1)}{2}\left[\nu_{0,2} + (r+1)^2\nu_{2,0}\right]
$$

$$
\left. + (r+1)^2\nu_{2,1} - (r+1)\nu_{1,2}\right\}
$$

$$
- 3n_1(n_1 + 2r + 1)(r+1)\nu_{1,1},
$$

$$
E\left(W_{\min,r}^4\right) = \left[\frac{n_1(n_1 + 2r + 1)}{2}\right]^4 + (r+1)^4\nu_{4,0} + \nu_{0,4}
$$

$$
+ 4\left\{\left[\frac{n_1(n_1 + 2r + 1)}{2}\right]^3[\nu_{0,1} - (r+1)\nu_{1,0}]\right.
$$

$$
+ \frac{n_1(n_1 + 2r + 1)}{2}\left[\nu_{0,3} - (r+1)^3\nu_{3,0}\right]
$$

$$
\left. - (r+1)^3\nu_{3,1} - (r+1)\nu_{0,3}\right\}
$$

$$
+ 6\left\{\left[\frac{n_1(n_1 + 2r + 1)}{2}\right]^2\left[\nu_{0,2} + (r+1)^2\nu_{2,0}\right] + (r+1)^2\nu_{2,2}\right\}
$$

$$
+ 6n_1(n_1 + 2r + 1)(r+1)\left[(r+1)\nu_{2,1} - \nu_{1,2}\right.
$$

$$
\left. - \frac{n_1(n_1 + 2r + 1)}{2}\nu_{1,1}\right].
$$

Similarly, the first, second, third, and fourth moments of the maximal rank-sum statistic are found from these expressions with r replaced by n_2, and those of the expected rank-sum statistic are with r replaced by $(n_2 + r)/2$.

Bibliography

Aggarwala, R. and Balakrishnan, N. (1998). Some properties of progressive censored order statistics from arbitrary and uniform distributions with applications to inference and simulation, *Journal of Statistical Planning and Inference*, **70**, 35-49.

Arnold, B. C., Balakrishnan, N., and Nagaraja, H. N. (1992). *A First Course in Order Statistics*. New York: John Wiley & Sons.

Bain, L. J. and Engelhardt, M. (1991). *Statistical Analysis of Reliability and Life-testing Models*, second edition. New York: Marcel Dekker.

Balakrishnan, N. and Aggarwala, R. (2000). *Progressive Censoring: Theory, Methods and Applications*. Boston: Birkhäuser.

Balakrishnan, N. and Basu, A. P. (Eds.) (1995). *The Exponential Distribution: Theory, Methods and Applications*. Langhorne, PA: Gordon and Breach.

Balakrishnan, N. and Cohen, A. C. (1991). *Order Statistics and Inference: Estimation Methods*. San Diego: Academic Press.

Balakrishnan, N. and Frattina, R. (2000). Precedence test and maximal precedence test. In *Recent Advances in Reliability Theory: Methodology, Practice, and Inference* (Eds., N. Limnios and M. Nikulin), pp. 355–378. Boston: Birkhäuser.

Balakrishnan, N. and Koutras, M. V. (2002). *Runs and Scans with Applications*. New York: John Wiley & Sons.

Balakrishnan, N. and Ng. H. K. T. (2001). A general maximal precedence test. In *System and Bayesian Reliability* (Eds., Y. Hayakawa, T. Irony and M. Xie), pp. 105–122. Singapore: World Scientific Publishing Co.

Balakrishnan, N. and Sandhu, R. A. (1995). A simple simulational algorithm for generating progressive Type-II censored samples, *The American Statistician*, **49**, 229–230.

Balakrishnan, N., Ng, H. K. T., and Panchapakesan, S. (2006). A nonparametric procedure based on early failures for selecting the best population using a test for equality, *Journal of Statistical Planning and Inference* (to appear).

Bartholomew, D. J. (1963). The sampling distribution of an estimate arising in life testing, *Technometrics*, **5**, 361–374.

Bechhofer, R. E., Santner, T. J., and Goldsman, D. M. (1995). *Design and Analysis of Experiments for Statistical Selection, Screening, and Multiple Comparisons*. New York: John Wiley & Sons.

Bickel, P. J. (1974). Edgeworth expansions in nonparametric statistics, *Annals of Statistics*, **2**, 1–20.

Bowker, A. H. (1944). Note on consistency of a proposed test for the problem of two samples, *Annals of Mathematical Statistics*, **15**, 98–101.

Bowman, K. O. and Shenton, L. R. (1988). *Properties of Estimators for the Gamma Distribution*. New York: Marcel Dekker.

Chakraborti, S. and Mukerjee, R. (1989). A confidence interval for a measure associated with the comparison of a treatment with a control, *South African Statistical Journal*, **23**, 219–230.

Chakraborti, S. and van der Laan, P. (1996). Precedence tests and confidence bounds for complete data: An overview and some results, *The Statistician*, **45**, 351–369.

Chakraborti, S. and van der Laan, P. (1997). An overview of precedence-type tests for censored data, *Biometrical Journal*, **39**, 99–116.

Chakraborti, S. and van der Laan, P. (2000). Precedence probability and prediction intervals, *The Statistician*, **49**, 219–228.

Chakraborti, S., van der Laan, P., and van de Wiel, M. A. (2004). A class of distribution-free control charts, *Applied Statistics*, **53**, 443–462.

Chakravarti, I. M., Leone, F. C., and Alanen, J. D. (1961). Relative efficiency of Mood's two sample tests against some parametric alternatives, *Bulletin of the International Statistical Institute, 33rd Session*.

Cohen, A. C. (1991). *Truncated and Censored Samples: Theory and Applications*. New York: Marcel Dekker.

Cohen, A. C. and Whitten, B. J. (1988). *Parameter Estimation in Reliability and Life Span Models*. New York: Marcel Dekker.

Crow, E. L. and Shimizu, K. (1988). *Lognormal Distributions*. New York: Marcel Dekker.

David, H. A. and Nagaraja, H. N. (2003). *Order Statistics*, third edition. Hoboken, NJ: John Wiley & Sons.

Davies, R. B. (1971). Rank tests for "Lehmann alternative", *Journal of the American Statistical Association*, **66**, 879–883.

Eilbott, J. and Nadler, J. (1965). On precedence life testing, *Technometrics*, **7**, 359–377.

Epstein, B. (1954). Tables for the distribution of the number of exceedances, *Annals of Mathematical Statistics*, **25**, 762–768.

Epstein, B. (1955). Comparison of some nonparametric tests against normal alternatives with an application to life testing, *Journal of the American Statistical Association*, **50**, 894-900.

Epstein, B. and Sobel, M. (1953). Life testing, *Journal of the American Statistical Association*, **48**, 485–502.

Epstein, B. and Sobel, M. (1954). Some theorems relevant to life testing from an exponential distribution, *Annals of Mathematical Statistics*, **25**, 373–381.

Fix, E. and Hodges J. L. (1955). Significance probabilities of the Wilcoxon test, *Annals of Mathematical Statistics*, **26**, 301–312.

Fligner, M. A. and Wolfe, D. A. (1976). Some applications of sample analogues to the probability integral transformation and a coverage property, *The American Statistician*, **30**, 78–85.

Gart, J. J. (1963). A median test with sequential application, *Biometrika*, **50**, 55–62.

Gastwirth, J. L. (1968). The first-median test: A two-sided version of the control median test, *Journal of the American Statistical Association*, **63**, 692–706.

Gibbons, J. D. and Chakraborti, S. (2003). *Nonparametric Statistical Inference*, fourth edition. New York: Marcel Dekker.

Gibbons, J. D., Olkin, I., and Sobel, M. (1999). *Selecting and Ordering Populations: A New Statistical Methodology*, Classics in Applied Mathematics, 26. Philadelphia: Society for Industrial and Applied Mathematics. (Unabridged reproduction of the same title, New York; John Wiley & Sons, 1977.)

Gumbel, E. J. and von Schelling, H. (1950). The distribution of the number of exceedances, *Annals of Mathematical Statistics*, **21**, 247–262.

Gupta, S. S. and Panchapakesan, S. (2002). *Multiple Decision Procedures: Theory and Methodology of Selecting and Ranking Populations*, Classics in Applied Mathematics, 44. Philadelphia: Society for Industrial and Applied Mathematics. (Unabridged reproduction of the same title, New York; John Wiley & Sons, 1979.)

Hackl, W. and Katzenbeisser, W. (1984). A note on the power of two-sample tests for dispersion based on exceedance statistics, *Computational Statistics Quarterly*, **1**, 333–341.

Haga, T. (1959). A two-sample rank test on location, *Annals of the Institute of Statistical Mathematics*, **11**, 211–219.

Hájèk, J. and Sidák, Z. (1967). *Theory of Rank Tests*. New York: Academic Press.

Hall, P. (1992). *The Bootstrap and Edgeworth Expansion*. New York: Springer-Verlag.

Harris, L. B. (1952). On a limiting case for the distribution of exceedances, with an application to life-testing, *Annals of Mathematical Statistics*, **23**, 295–298.

Hettmansperger, T. P. and McKean, J. W. (1998). *Robust Nonparametric Statistical Methods.* London: Arnold.

Hollander, M. and Wolfe, D. A. (1999). *Nonparametric Statistical Methods,* second edition. New York: John Wiley & Sons.

Johnson, N. L., Kotz, S., and Balakrishnan, N. (1994). *Continuous Univariate Distributions, Vol. 1,* second edition. New York: John Wiley & Sons.

Johnson, N. L., Kotz, S., and Balakrishnan, N. (1995). *Continuous Univariate Distributions, Vol. 2,* second edition. New York: John Wiley & Sons.

Kamps, U. (1995a). *A Concept of Generalized Order Statistics.* Stuttgart, Germany: Teubner.

Kamps, U. (1995b). *A concept of generalized order statistics, Journal of Statistical Planning and Inference,* **48**, 1–23.

Katzenbeisser, W. (1985). The distribution of two-sample location exceedance test statistics under Lehmann alternatives, *Statistische Hefte,* **26**, 131–138.

Katzenbeisser, W. (1989). The exact power of two-sample location tests based on exceedance statistics against shift alternatives, *Mathematische Operationsforschung und Statistik, Series Statistics,* **20**, 47–54.

Kimball, A. W., Burnett, W. T., and Doherty, D. G (1957). Chemical protection against ionizing radiation. I. Sampling methods for screening compounds in radiation protection studies with mice, *Radiation Research,* **7**, 1–12.

Lee, E. T. and Wang, J. W. (2002). *Statistical Methods for Survival Data Analysis,* third edition. New York: John Wiley & Sons.

Lehmann, E. L. (1953). The Power of rank tests, *Annals of Mathematical Statistics,* **24**, 23–42.

Lehmann, E. L. (1975). *Nonparametrics: Statistical Methods Based on Ranks.* New York: McGraw-Hill.

Lin, C. H. and Sukhatme, S. (1992). On the choice of precedence tests, *Communications in Statistics—Theory and Methods*, **21**, 2949–2968.

Little, R. E. (1974). Tables for making an early decision in precedence tests. *Journal of Testing and Evaluation*, **2**, 84–86.

Liu, J. (1992). Precedence probabilities and their applications, *Communications in Statistics—Theory and Methods*, **21**, 1667–1682.

Malmquist, S. (1950). On a property of order statistics from a rectangular distribution, *Skandinavisk Aktuarietidskrift*, **33**, 214–222.

Mann, H. B. and Whitney, D. R. (1947). On a test of whether one of two random variables is stochastically larger than the other, *Annals of Mathematical Statistics*, **18**, 50–60.

Massey, F. J. (1951). A note on a two sample test, *Annals of Mathematical Statistics*, **22**, 304–306.

Mathisen, H. C. (1943). A method of testing the hypothesis that two samples are from the same population, *Annals of Mathematical Statistics*, **14**, 188–194.

Mood, A. M. (1954). On the asymptotic efficiency of certain nonparametric two sample tests, *Annals of Mathematical Statistics*, **25**, 514–522.

Murthy, D. N. P., Xie, M., and Jiang, R. (2003). *Weibull Models*. New York: John Wiley & Sons.

Nelson, L. S. (1963). Tables of a precedence life test, *Technometrics*, **5**, 491–499.

Nelson, L. S. (1986). Precedence life test. In *Encyclopedia of Statistical Sciences* **7** (Eds., S. Kotz and N. L. Johnson), pp. 134–136. New York: John Wiley & Sons.

Nelson, L. S. (1993). Tests on early failures: The precedence life test, *Journal of Quality Technology*, **25**, 140–143.

Nelson, W. (1982). *Applied Life Data Analysis*. New York: John Wiley & Sons.

Ng, H. K. T. and Balakrishnan, N. (2002). Wilcoxon-type rank-sum precedence tests: Large-sample approximation and evaluation, *Applied Stochastic Models in Business and Industry*, **18**, 271–286.

Ng, H. K. T. and Balakrishnan, N. (2004). Wilcoxon-type rank-sum precedence tests, *Australia and New Zealand Journal of Statistics*, **46**, 631–648.

Ng, H. K. T. and Balakrishnan, N. (2005). Weighted precedence and maximal precedence tests and an extension to progressive censoring, *Journal of Statistical Planning and Inference*, **135**, 197–221.

Ng, H. K. T. and Balakrishnan, N. (2006). Precedence testing. In *Encyclopedia of Statistical Sciences* **9**, second edition (Eds., S. Kotz, N. Balakrishnan, C. B. Read, and B. Vidakovic), pp. 6317–6323. Hoboken, NJ: John Wiley & Sons.

Ng, H. K. T., Balakrishnan, N., and Panchapakesan, S. (2006). Selecting the best population using a test for equality based on minimal Wilcoxon rank-sum precedence statistics, *Submitted for publication*.

Orban, J. J. and Wolfe, D. A. (1982). A class of distribution-free tests based on placements, *Journal of the American Statistical Association*, **77**, 666–671.

Randles, R. H. and Wolfe, D. A. (1979). *Introduction to the Theory of Nonparametric Statistics*. New York: John Wiley & Sons.

Rosenbaum, S. (1954). Tables for a nonparametric test of location, *Annals of Mathematical Statistics*, **25**, 146–150.

Sarkadi, K. (1957). On the distribution of the number of exceedances, *Annals of Mathematical Statistics*, **28**, 1021–1023.

Schlittgen, R. (1979). Use of a median test for a generalized Behrens-Fisher problem, *Metrika*, **26**, 95–103.

Shorack, R. A. (1967). On the power of precedence life tests, *Technometrics*, **9**, 154–158.

Slud, E. V. (1992). Best precedence tests for censored data, *Journal of Statistical Planning and Inference*, **31**, 283–293.

Sukhatme, P. V. (1937). Tests of significance for samples of the χ^2 population with two degrees of freedom, *Annals of Eugenics*, **8**, 52–56.

Sukhatme, S. (1992). Powers of two-sample rank tests under Lehmann alternatives, *The American Statistician*, **46**, 212–214.

Thomas, D. R. and Wilson, W. M. (1972). Linear order statistic estimation for the two-parameter Weibull and extreme value distributions from Type-II progressively censored samples, *Technometrics*, **14**, 679–691.

van der Laan, P. (1970). Simple distribution-free confidence intervals for a difference in location, Doctoral thesis, Eindhoven University of Technology, Eindhoven.

van der Laan, P. and Chakraborti, S. (1999). Best precedence tests against Lehmann alternatives, *Bulletin of the International Statistical Institute*, **58**, 395–396.

van der Laan, P. and Chakraborti, S. (2001). Precedence tests and Lehmann alternatives, *Statistical Papers*, **42**, 301–312.

Viveros, R. and Balakrishnan, N. (1994). Interval estimation of parameters of life from progressively censored data, *Technometrics*, **36**, 84–91.

Wilcoxon, F. (1945). Individual comparisons by ranking methods, *Biometrics Bulletin*, **1**, 80–83.

Wilcoxon, F., Katti, S. K. and Wilcox, R. A. (1970). *Selected Tables in Mathematical Statistics, Vol. 1.* Providence, RI: American Mathematical Society.

Young, D. H. (1973). A note on some asymptotic properties of the precedence test and applications to a selection problem concerning quantiles, *Sankhyā, Series B*, **35**, 35–44.

Zelen, M. and Dannemiller, M. (1961). The robustness of life testing procedures derived from the exponential distribution, *Technometrics*, **3**, 29–49.

Author Index

283

Subject Index

WILEY SERIES IN PROBABILITY AND STATISTICS

ESTABLISHED BY WALTER A. SHEWHART AND SAMUEL S. WILKS

Editors: *David J. Balding, Noel A. C. Cressie, Nicholas I. Fisher,*
Iain M. Johnstone, J. B. Kadane, Geert Molenberghs. Louise M. Ryan,
David W. Scott, Adrian F. M. Smith, Jozef L. Teugels
Editors Emeriti: *Vic Barnett, J. Stuart Hunter, David G. Kendall*

The **Wiley Series in Probability and Statistics** is well established and authoritative. It covers many topics of current research interest in both pure and applied statistics and probability theory. Written by leading statisticians and institutions, the titles span both state-of-the-art developments in the field and classical methods.

Reflecting the wide range of current research in statistics, the series encompasses applied, methodological and theoretical statistics, ranging from applications and new techniques made possible by advances in computerized practice to rigorous treatment of theoretical approaches.

This series provides essential and invaluable reading for all statisticians, whether in academia, industry, government, or research.

*Now available in a lower priced paperback edition in the Wiley Classics Library.
†Now available in a lower priced paperback edition in the Wiley–Interscience Paperback Series.

BELSLEY · Conditioning Diagnostics: Collinearity and Weak Data in Regression
† BELSLEY, KUH, and WELSCH · Regression Diagnostics: Identifying Influential
  Data and Sources of Collinearity
BENDAT and PIERSOL · Random Data: Analysis and Measurement Procedures,
  Third Edition
BERRY, CHALONER, and GEWEKE · Bayesian Analysis in Statistics and
  Econometrics: Essays in Honor of Arnold Zellner
BERNARDO and SMITH · Bayesian Theory
BHAT and MILLER · Elements of Applied Stochastic Processes, *Third Edition*
BHATTACHARYA and WAYMIRE · Stochastic Processes with Applications
† BIEMER, GROVES, LYBERG, MATHIOWETZ, and SUDMAN · Measurement Errors
  in Surveys
BILLINGSLEY · Convergence of Probability Measures, *Second Edition*
BILLINGSLEY · Probability and Measure, *Third Edition*
BIRKES and DODGE · Alternative Methods of Regression
BLISCHKE AND MURTHY (editors) · Case Studies in Reliability and Maintenance
BLISCHKE AND MURTHY · Reliability: Modeling, Prediction, and Optimization
BLOOMFIELD · Fourier Analysis of Time Series: An Introduction, *Second Edition*
BOLLEN · Structural Equations with Latent Variables
BOLLEN and CURRAN · Latent Curve Models: A Structural Equation Perspective
BOROVKOV · Ergodicity and Stability of Stochastic Processes
BOULEAU · Numerical Methods for Stochastic Processes
BOX · Bayesian Inference in Statistical Analysis
BOX · R. A. Fisher, the Life of a Scientist
BOX and DRAPER · Empirical Model-Building and Response Surfaces
* BOX and DRAPER · Evolutionary Operation: A Statistical Method for Process
  Improvement
BOX, HUNTER, and HUNTER · Statistics for Experimenters: Design, Innovation,
  and Discovery, *Second Editon*
BOX and LUCEÑO · Statistical Control by Monitoring and Feedback Adjustment
BRANDIMARTE · Numerical Methods in Finance: A MATLAB-Based Introduction
BROWN and HOLLANDER · Statistics: A Biomedical Introduction
BRUNNER, DOMHOF, and LANGER · Nonparametric Analysis of Longitudinal Data in
  Factorial Experiments
BUCKLEW · Large Deviation Techniques in Decision, Simulation, and Estimation
CAIROLI and DALANG · Sequential Stochastic Optimization
CASTILLO, HADI, BALAKRISHNAN, and SARABIA · Extreme Value and Related
  Models with Applications in Engineering and Science
CHAN · Time Series: Applications to Finance
CHARALAMBIDES · Combinatorial Methods in Discrete Distributions
CHATTERJEE and HADI · Sensitivity Analysis in Linear Regression
CHATTERJEE and PRICE · Regression Analysis by Example, *Third Edition*
CHERNICK · Bootstrap Methods: A Practitioner's Guide
CHERNICK and FRIIS · Introductory Biostatistics for the Health Sciences
CHILÈS and DELFINER · Geostatistics: Modeling Spatial Uncertainty
CHOW and LIU · Design and Analysis of Clinical Trials: Concepts and Methodologies,
  Second Edition
CLARKE and DISNEY · Probability and Random Processes: A First Course with
  Applications, *Second Edition*
* COCHRAN and COX · Experimental Designs, *Second Edition*
CONGDON · Applied Bayesian Modelling
CONGDON · Bayesian Models for Categorical Data
CONGDON · Bayesian Statistical Modelling
CONOVER · Practical Nonparametric Statistics, *Third Edition*

*Now available in a lower priced paperback edition in the Wiley Classics Library.
†Now available in a lower priced paperback edition in the Wiley–Interscience Paperback Series.

COOK · Regression Graphics

COOK and WEISBERG · Applied Regression Including Computing and Graphics

COOK and WEISBERG · An Introduction to Regression Graphics

CORNELL · Experiments with Mixtures, Designs, Models, and the Analysis of Mixture
Data, *Third Edition*

COVER and THOMAS · Elements of Information Theory

COX · A Handbook of Introductory Statistical Methods

* COX · Planning of Experiments

CRESSIE · Statistics for Spatial Data, *Revised Edition*

CSÖRGŐ and HORVÁTH · Limit Theorems in Change Point Analysis

DANIEL · Applications of Statistics to Industrial Experimentation

DANIEL · Biostatistics: A Foundation for Analysis in the Health Sciences, *Eighth Edition*

* DANIEL · Fitting Equations to Data: Computer Analysis of Multifactor Data,
Second Edition

DASU and JOHNSON · Exploratory Data Mining and Data Cleaning

DAVID and NAGARAJA · Order Statistics, *Third Edition*

* DEGROOT, FIENBERG, and KADANE · Statistics and the Law

DEL CASTILLO · Statistical Process Adjustment for Quality Control

DeMARIS · Regression with Social Data: Modeling Continuous and Limited Response
Variables

DEMIDENKO · Mixed Models: Theory and Applications

DENISON, HOLMES, MALLICK and SMITH · Bayesian Methods for Nonlinear
Classification and Regression

DETTE and STUDDEN · The Theory of Canonical Moments with Applications in
Statistics, Probability, and Analysis

DEY and MUKERJEE · Fractional Factorial Plans

DILLON and GOLDSTEIN · Multivariate Analysis: Methods and Applications

DODGE · Alternative Methods of Regression

* DODGE and ROMIG · Sampling Inspection Tables, *Second Edition*

* DOOB · Stochastic Processes

DOWDY, WEARDEN, and CHILKO · Statistics for Research, *Third Edition*

DRAPER and SMITH · Applied Regression Analysis, *Third Edition*

DRYDEN and MARDIA · Statistical Shape Analysis

DUDEWICZ and MISHRA · Modern Mathematical Statistics

DUNN and CLARK · Basic Statistics: A Primer for the Biomedical Sciences,
Third Edition

DUPUIS and ELLIS · A Weak Convergence Approach to the Theory of Large Deviations

EDLER and KITSOS · Recent Advances in Quantitative Methods in Cancer and Human
Health Risk Assessment

* ELANDT-JOHNSON and JOHNSON · Survival Models and Data Analysis

ENDERS · Applied Econometric Time Series

† ETHIER and KURTZ · Markov Processes: Characterization and Convergence

EVANS, HASTINGS, and PEACOCK · Statistical Distributions, *Third Edition*

FELLER · An Introduction to Probability Theory and Its Applications, Volume I,
Third Edition, Revised; Volume II, *Second Edition*

FISHER and VAN BELLE · Biostatistics: A Methodology for the Health Sciences

FITZMAURICE, LAIRD, and WARE · Applied Longitudinal Analysis

* FLEISS · The Design and Analysis of Clinical Experiments

FLEISS · Statistical Methods for Rates and Proportions, *Third Edition*

† FLEMING and HARRINGTON · Counting Processes and Survival Analysis

FULLER · Introduction to Statistical Time Series, *Second Edition*

FULLER · Measurement Error Models

GALLANT · Nonlinear Statistical Models

*Now available in a lower priced paperback edition in the Wiley Classics Library.

†Now available in a lower priced paperback edition in the Wiley–Interscience Paperback Series.

GEISSER · Modes of Parametric Statistical Inference

GELMAN and MENG · Applied Bayesian Modeling and Causal Inference from Incomplete-Data Perspectives

GEWEKE · Contemporary Bayesian Econometrics and Statistics

GHOSH, MUKHOPADHYAY, and SEN · Sequential Estimation

GIESBRECHT and GUMPERTZ · Planning, Construction, and Statistical Analysis of Comparative Experiments

GIFI · Nonlinear Multivariate Analysis

GIVENS and HOETING · Computational Statistics

GLASSERMAN and YAO · Monotone Structure in Discrete-Event Systems

GNANADESIKAN · Methods for Statistical Data Analysis of Multivariate Observations, *Second Edition*

GOLDSTEIN and LEWIS · Assessment: Problems, Development, and Statistical Issues

GREENWOOD and NIKULIN · A Guide to Chi-Squared Testing

GROSS and HARRIS · Fundamentals of Queueing Theory, *Third Edition*

* HAHN and SHAPIRO · Statistical Models in Engineering

HAHN and MEEKER · Statistical Intervals: A Guide for Practitioners

HALD · A History of Probability and Statistics and their Applications Before 1750

HALD · A History of Mathematical Statistics from 1750 to 1930

† HAMPEL · Robust Statistics: The Approach Based on Influence Functions

HANNAN and DEISTLER · The Statistical Theory of Linear Systems

HEIBERGER · Computation for the Analysis of Designed Experiments

HEDAYAT and SINHA · Design and Inference in Finite Population Sampling

HEDEKER and GIBBONS · Longitudinal Data Analysis

HELLER · MACSYMA for Statisticians

HINKELMANN and KEMPTHORNE · Design and Analysis of Experiments, Volume 1: Introduction to Experimental Design

HINKELMANN and KEMPTHORNE · Design and Analysis of Experiments, Volume 2: Advanced Experimental Design

HOAGLIN, MOSTELLER, and TUKEY · Exploratory Approach to Analysis of Variance

* HOAGLIN, MOSTELLER, and TUKEY · Exploring Data Tables, Trends and Shapes

* HOAGLIN, MOSTELLER, and TUKEY · Understanding Robust and Exploratory Data Analysis

HOCHBERG and TAMHANE · Multiple Comparison Procedures

HOCKING · Methods and Applications of Linear Models: Regression and the Analysis of Variance, *Second Edition*

HOEL · Introduction to Mathematical Statistics, *Fifth Edition*

HOGG and KLUGMAN · Loss Distributions

HOLLANDER and WOLFE · Nonparametric Statistical Methods, *Second Edition*

HOSMER and LEMESHOW · Applied Logistic Regression, *Second Edition*

HOSMER and LEMESHOW · Applied Survival Analysis: Regression Modeling of Time to Event Data

† HUBER · Robust Statistics

HUBERTY · Applied Discriminant Analysis

HUNT and KENNEDY · Financial Derivatives in Theory and Practice, *Revised Edition*

HUSKOVA, BERAN, and DUPAC · Collected Works of Jaroslav Hajek— with Commentary

HUZURBAZAR · Flowgraph Models for Multistate Time-to-Event Data

IMAN and CONOVER · A Modern Approach to Statistics

† JACKSON · A User's Guide to Principle Components

JOHN · Statistical Methods in Engineering and Quality Assurance

JOHNSON · Multivariate Statistical Simulation

*Now available in a lower priced paperback edition in the Wiley Classics Library.

†Now available in a lower priced paperback edition in the Wiley–Interscience Paperback Series.

*Now available in a lower priced paperback edition in the Wiley Classics Library.
†Now available in a lower priced paperback edition in the Wiley–Interscience Paperback Series.

---

*Now available in a lower priced paperback edition in the Wiley Classics Library.

†Now available in a lower priced paperback edition in the Wiley–Interscience Paperback Series.

OCHI · Applied Probability and Stochastic Processes in Engineering and Physical Sciences

OKABE, BOOTS, SUGIHARA, and CHIU · Spatial Tesselations: Concepts and Applications of Voronoi Diagrams, *Second Edition*

OLIVER and SMITH · Influence Diagrams, Belief Nets and Decision Analysis

PALTA · Quantitative Methods in Population Health: Extensions of Ordinary Regressions

PANKRATZ · Forecasting with Dynamic Regression Models

PANKRATZ · Forecasting with Univariate Box-Jenkins Models: Concepts and Cases

* PARZEN · Modern Probability Theory and Its Applications

PEÑA, TIAO, and TSAY · A Course in Time Series Analysis

PIANTADOSI · Clinical Trials: A Methodologic Perspective

PORT · Theoretical Probability for Applications

POURAHMADI · Foundations of Time Series Analysis and Prediction Theory

PRESS · Bayesian Statistics: Principles, Models, and Applications

PRESS · Subjective and Objective Bayesian Statistics, *Second Edition*

PRESS and TANUR · The Subjectivity of Scientists and the Bayesian Approach

PUKELSHEIM · Optimal Experimental Design

PURI, VILAPLANA, and WERTZ · New Perspectives in Theoretical and Applied Statistics

† PUTERMAN · Markov Decision Processes: Discrete Stochastic Dynamic Programming

QIU · Image Processing and Jump Regression Analysis

* RAO · Linear Statistical Inference and Its Applications, *Second Edition*

RAUSAND and HØYLAND · System Reliability Theory: Models, Statistical Methods, and Applications, *Second Edition*

RENCHER · Linear Models in Statistics

RENCHER · Methods of Multivariate Analysis, *Second Edition*

RENCHER · Multivariate Statistical Inference with Applications

* RIPLEY · Spatial Statistics

* RIPLEY · Stochastic Simulation

ROBINSON · Practical Strategies for Experimenting

ROHATGI and SALEH · An Introduction to Probability and Statistics, *Second Edition*

ROLSKI, SCHMIDLI, SCHMIDT, and TEUGELS · Stochastic Processes for Insurance and Finance

ROSENBERGER and LACHIN · Randomization in Clinical Trials: Theory and Practice

ROSS · Introduction to Probability and Statistics for Engineers and Scientists

ROSSI, ALLENBY, and McCULLOCH · Bayesian Statistics and Marketing

† ROUSSEEUW and LEROY · Robust Regression and Outlier Detection

* RUBIN · Multiple Imputation for Nonresponse in Surveys

RUBINSTEIN · Simulation and the Monte Carlo Method

RUBINSTEIN and MELAMED · Modern Simulation and Modeling

RYAN · Modern Regression Methods

RYAN · Statistical Methods for Quality Improvement, *Second Edition*

SALEH · Theory of Preliminary Test and Stein-Type Estimation with Applications

* SCHEFFE · The Analysis of Variance

SCHIMEK · Smoothing and Regression: Approaches, Computation, and Application

SCHOTT · Matrix Analysis for Statistics, *Second Edition*

SCHOUTENS · Levy Processes in Finance: Pricing Financial Derivatives

SCHUSS · Theory and Applications of Stochastic Differential Equations

SCOTT · Multivariate Density Estimation: Theory, Practice, and Visualization

† SEARLE · Linear Models for Unbalanced Data

† SEARLE · Matrix Algebra Useful for Statistics

† SEARLE, CASELLA, and McCULLOCH · Variance Components

SEARLE and WILLETT · Matrix Algebra for Applied Economics

SEBER and LEE · Linear Regression Analysis, *Second Edition*

*Now available in a lower priced paperback edition in the Wiley Classics Library.

†Now available in a lower priced paperback edition in the Wiley–Interscience Paperback Series.

† SEBER · Multivariate Observations
† SEBER and WILD · Nonlinear Regression
SENNOTT · Stochastic Dynamic Programming and the Control of Queueing Systems
* SERFLING · Approximation Theorems of Mathematical Statistics
SHAFER and VOVK · Probability and Finance: It's Only a Game!
SILVAPULLE and SEN · Constrained Statistical Inference: Inequality, Order, and Shape
  Restrictions
SMALL and McLEISH · Hilbert Space Methods in Probability and Statistical Inference
SRIVASTAVA · Methods of Multivariate Statistics
STAPLETON · Linear Statistical Models
STAUDTE and SHEATHER · Robust Estimation and Testing
STOYAN, KENDALL, and MECKE · Stochastic Geometry and Its Applications, *Second
  Edition*
STOYAN and STOYAN · Fractals, Random Shapes and Point Fields: Methods of
  Geometrical Statistics
STYAN · The Collected Papers of T. W. Anderson: 1943–1985
SUTTON, ABRAMS, JONES, SHELDON, and SONG · Methods for Meta-Analysis in
  Medical Research
TAKEZAWA · Introduction to Nonparametric Regression
TANAKA · Time Series Analysis: Nonstationary and Noninvertible Distribution Theory
THOMPSON · Empirical Model Building
THOMPSON · Sampling, *Second Edition*
THOMPSON · Simulation: A Modeler's Approach
THOMPSON and SEBER · Adaptive Sampling
THOMPSON, WILLIAMS, and FINDLAY · Models for Investors in Real World Markets
TIAO, BISGAARD, HILL, PEÑA, and STIGLER (editors) · Box on Quality and
  Discovery: with Design, Control, and Robustness
TIERNEY · LISP-STAT: An Object-Oriented Environment for Statistical Computing
  and Dynamic Graphics
TSAY · Analysis of Financial Time Series, *Second Edition*
UPTON and FINGLETON · Spatial Data Analysis by Example, Volume II:
  Categorical and Directional Data
VAN BELLE · Statistical Rules of Thumb
VAN BELLE, FISHER, HEAGERTY, and LUMLEY · Biostatistics: A Methodology for
  the Health Sciences, *Second Edition*
VESTRUP · The Theory of Measures and Integration
VIDAKOVIC · Statistical Modeling by Wavelets
VINOD and REAGLE · Preparing for the Worst: Incorporating Downside Risk in Stock
  Market Investments
WALLER and GOTWAY · Applied Spatial Statistics for Public Health Data
WEERAHANDI · Generalized Inference in Repeated Measures: Exact Methods in
  MANOVA and Mixed Models
WEISBERG · Applied Linear Regression, *Third Edition*
WELSH · Aspects of Statistical Inference
WESTFALL and YOUNG · Resampling-Based Multiple Testing: Examples and
  Methods for *p*-Value Adjustment
WHITTAKER · Graphical Models in Applied Multivariate Statistics
WINKER · Optimization Heuristics in Economics: Applications of Threshold Accepting
WONNACOTT and WONNACOTT · Econometrics, *Second Edition*
WOODING · Planning Pharmaceutical Clinical Trials: Basic Statistical Principles
WOODWORTH · Biostatistics: A Bayesian Introduction
WOOLSON and CLARKE · Statistical Methods for the Analysis of Biomedical Data,
  Second Edition

*Now available in a lower priced paperback edition in the Wiley Classics Library.
†Now available in a lower priced paperback edition in the Wiley–Interscience Paperback Series.

WU and HAMADA · Experiments: Planning, Analysis, and Parameter Design Optimization

WU and ZHANG · Nonparametric Regression Methods for Longitudinal Data Analysis

YANG · The Construction Theory of Denumerable Markov Processes

ZELTERMAN · Discrete Distributions—Applications in the Health Sciences

* ZELLNER · An Introduction to Bayesian Inference in Econometrics

ZHOU, OBUCHOWSKI, and McCLISH · Statistical Methods in Diagnostic Medicine